Career Competencies
in Marketing Series

Accessories

SECOND
EDITION

Marilyn G. Winn

Educational Consultant
Las Vegas, Nevada

Consulting Editor

Richard L. Lynch

Virginia Polytechnic Institute
and State University
Blacksburg, Virginia

GREGG DIVISION
McGraw-Hill Publishing Company

New York Atlanta Dallas St. Louis San Francisco
Auckland Bogotá Caracas Hamburg Lisbon
London Madrid Mexico Milan Montreal New Delhi
Paris San Juan São Paulo Singapore
Sydney Tokyo Toronto

Sponsoring Editor: Sylvia L. Weber
Editing Supervisor: Melonie Parnes
Design and Art Supervisor: Joseph Piliero
Production Supervisor: Kathleen Donnelly

Text Designer: Robin H. Hoffman
Cover Designer: BC Graphics

Library of Congress Cataloging-in-Publication Data
Winn, Marilyn G.
Apparel and accessories/Marilyn G. Winn—2nd ed.
p. cm.--(Career competencies in marketing series)
ISBN 0-07-040908-0
1. Clothing Trade. I. Title. II Series.
TT497.W56 1990
687'.068'8--dc19
89-30711
CIP

The manuscript for this book was processed electronically.

Apparel and Accessories, Second Edition

2 3 4 5 6 7 8 9 0 SEMSEM 9 6 5 4 3 2 1 0

ISBN 0-07-040908-0

CONTENTS

TO THE INSTRUCTOR

Businesses and individuals who subscribe to the marketing concept— fulfillment of customers' wants and needs at a fair profit—have never found their philosophy more applicable than in today's marketing culture. The *Career Competencies in Marketing (CCM) Series,* Second Edition, is designed to give its readers a competitive advantage as they apply this concept to marketing themselves to potential employers in their chosen occupational areas. *Apparel and Accessories,* Second Edition, is the CCM module for those interested in marketing clothing and accessories.

Competency-Based Instruction

Like the other books in the series, *Apparel and Accessories*, Second Edition, helps develop competencies identified in the National Curriculum Framework and Core Competencies. The focus is on preparing students for positions at the career-sustaining and marketing specialist levels, as described in the Framework, but also considered are competencies needed for entry-level and managerial positions. The *CCM Series* can be coordinated with the competency-based competitive events sponsored by the Distributive Education Clubs of America (DECA) and with the learning activity packages (LAPs) developed by the MarkED Resource Center.

Apparel and Accessories, Second Edition, introduces each chapter with a recent real-life example and refers throughout the text to current industry trends and to the experiences of real apparel and accessories marketing businesses. Learning experiences are interspersed throughout each chapter to give the student an opportunity to develop and demonstrate needed competencies. Each unit concludes with a project or a case that ties together the topics covered in the chapters of that unit.

Scope of the Book

Apparel and Accessories, Second Edition, is organized into four units:

Unit 1: Introduction to Apparel and Accessories Chapter 1 provides an overview of the apparel and accessories industry and a description of various career options. Chapter 2, new to this edition, explains how marketing principles are applied within this industry.

Unit 2: Foundation Skills In Chapter 3, human relations and communications are described as they occur between employees and in employee-supervisor, employee-customer, and employee-supplier encounters. Chapter 4 presents some of the mathematical skills required for apparel and accessories careers.

Unit 3: Product and Service Technology Specific clothing and accessory products are the topics of the chapters in Unit 3. Chapter 5 provides information about the fabrics used for clothing. Women's clothing, menswear, and infants' and children's wear are discussed respectively in Chapters 6, 7, and 8. Chapter 9 discusses shoes, and Chapter 10 covers accessories, including handbags, belts, and jewelry.

Unit 4: Marketing Skills The last unit focuses on the marketing and management skills required for a career in the retailing of apparel and accessories. Chapter 11 discusses buying and pricing. Physical distribution of fashion merchandise is the topic of Chapter 12. Three chapters are devoted to promotional activities. Chapter 13 describes advertising and sales promotion. Visual merchandising and personal selling are treated in Chapters 14 and 15, respectively. Chapter 16 discusses management and administration in a clothing and accessories store or department. The unit ends with a culminating project—preparing a marketing plan to launch a new product. To give students experience in using the computer as a marketing decision-making tool, a software program for this project is available in IBM and Apple versions.

Supporting Materials

The software program, *Career Competencies in Marketing: Marketing Plan,* can be used with the other five industry-based modules in the *CCM Series,* as well as with *Apparel and Accessories,* Second Edition. (A separate software program is available for use with *Entrepreneurship,* Third Edition.)

The *Teacher's Manual and Key for Apparel and Accessories,* Second Edition, contains general teaching suggestions; specific teaching suggestions for each chapter, with the competencies considered in the chapter identified; an answer key to the learning experiences and end-of-unit activities; a hands-on project; and a 100-item objective test and key.

Acknowledgments

The author wishes to express her appreciation to the many apparel and accessories retailers and educators who provided help, suggestions, and cooperation.

Special thanks are extended to Sallie Hook, University of South Carolina, Columbia, South Carolina, for her review of the first edition and her suggestions for revising it. For their reviews of the second edition manuscript the author gratefully acknowledges the help of Debbie Purvis, Centennial College, Scarboro, Ontario, Canada; Steven Little, Hickman Mills High School, Kansas City, Missouri; Laura Kessler and Germaine Pino, Henri Bendel, New York; Gwen Reinauer, formerly with J.W. Robinson's, Los Angeles, California; and Judy Fox, May D and F, Englewood, Colorado.

Michael Buchman, Highland Park, New Jersey, provided much appreciated editorial assistance in preparing the projects for Units 1 and 4. The author also wishes to thank Malaine Hagemeier of Wentzville High School, Wentzville, Missouri, and Cindy Herman, Parkway South High School, Ballwin, Missouri, for their contributions to the objective test.

—Marilyn G. Winn

UNIT 1

Introduction to Apparel and Accessories

Welcome to the exciting and challenging world of apparel and accessories marketing! In this first unit, you will read about the many careers the fashion business offers and some important marketing principles. Chapter 1 should help you evaluate your personality and goals and decide if you want to pursue a fashion marketing career. Chapter 2 reviews the marketing concept, market segmentation, and market research.

These chapters will give you a solid foundation on which to build the skills needed for a successful career in this dynamic industry.

CHAPTER 1

The Scope of the Apparel and Accessories Industry

In 1967, Ralph Lauren went into the tie business and called his company Polo. He thought that the ties then in fashion were too thin and drab, so he designed some that were 4 inches wide and made of vibrant Italian silk fabrics. Later he created an entire menswear line, including wide-collar shirts and wide-lapel jackets to wear with the ties. His creations sparked the wide tie and lapel boom of the late 1960s and early 1970s.

In the spring of 1971, Lauren started to design shirts for women, and soon afterward he designed a women's sportswear collection. Today the Polo/Ralph Lauren name appears on children's wear, fragrances, menswear, women's wear, and home furnishings.

As of 1986, Lauren sold his products to 48 franchised Polo shops and 132 department store boutiques. His merchandise is also available in 16 factory outlets. In April of 1986, he opened his own store in a renovated mansion on Madison Avenue in Manhattan. His own store gives him complete creative control over marketing and a much higher profit margin than does selling to franchises and department stores. Although licensees make most of his products, Lauren manufactures the expensive Polo line of menswear at his own factory in Massachusetts. (A **licensee** is a company that has expertise in making a particular generic item. It pays the designer for the design, packaging, promotional ideas, and use of the designer's name.)

Referred to as the king of American sportswear, Lauren had retail sales approximating $1.3 billion in 1986. This was one-third higher than Calvin Klein's retail sales. If Lauren's company were a public corporation, it would rank 257th in the Fortune 500, a listing of the 500 largest corporations in the United States.[1]

Polo/Ralph Lauren is involved in designing, manufacturing, wholesaling, and retailing clothing and accessories. This successful company has a part in the full scope of apparel and accessory marketing activities.

■ READY-TO-WEAR

Less than a century ago, most Americans wore clothing made at home. Today, the majority of people wear **ready-to-wear,** clothing made in factories in standard sizes.

Every clothing and accessory item sold has passed through the stages of design, manufacturing, and distribution. It may take up to 18 months from the time a ready-to-wear item is designed until it is available for sale in stores or catalogs. (See Figure 1-1.)

Designing Clothing and Accessories

In fashion centers throughout the world, fashion designers create original designs. Using the designers' sketches and input, sample makers cut and sew the selected fabrics into finished samples.

The designers feature their designs at **couture** and **prêt-à-porter** showings for private customers, retail store buyers, and ready-to-wear and pattern manufacturers. (*Couture,* pronounced "koo-tour," is a French term that means "fashion." *Prêt-à-porter,* pronounced "pret-ah-portay," is the French term for "ready-to-wear.") At these fashion shows, private customers buy particular designs or "style numbers" which may be made to the customer's specific measurements. These designs are very expensive, costing many thousands of dollars. Retail store buyers

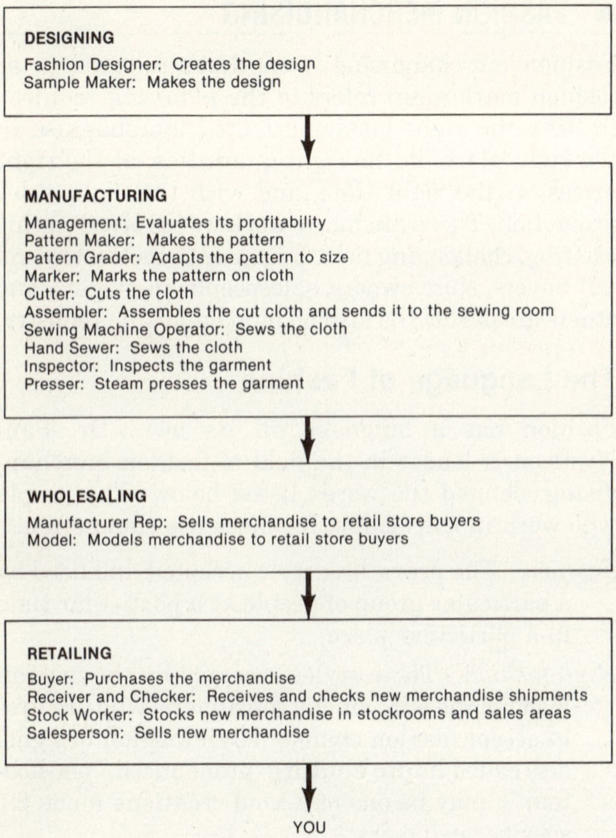

Figure 1-1 Before you buy an item of merchandise, it has gone through these four stages.

purchase particular style numbers or "numbers" for resale to their own customers. Ready-to-wear manufacturers, pattern manufacturers, and retail buyers also buy original designs that they copy and adapt for their own line. (A **line** is an assortment of new designs offered by a manufacturer to its customers, usually on a seasonal basis.) Manufacturers and retailers produce **"knock-offs,"** or mass-produced copies of the original designs in less expensive fabrics. Because production is on a larger scale, the knock-offs are priced much lower than the originals.

Some of the most famous designers whose original designs have shaped the course of fashion are Hubert de Givenchy, Marc Bohan of the House of Dior, and Yves St. Laurent. (See Figure 1-2.) Italian designers Georgio Armani, Claude Montana, and Mario Valentino and American designers Donald Brooks, Donna Karan, and Calvin Klein are other designers who have inspired fashion trends. Most famous designers live and work in one of the international fashion centers: Paris, Milan, New York, London, and Tokyo.

Manufacturing Clothing and Accessories

Before ready-to-wear manufacturers decide to produce a particular design, they evaluate its profitability. If a sufficient number of orders are received, and the company feels it can make a profit, the style is sent to a pattern maker and pattern grader. Using paper or fiberboard, the pattern maker constructs a master pattern for cutting out the clothing or accessory item. The master pattern is then sent to a pattern grader, who measures the pieces that make up the master pattern and changes them to fit many size ranges.

The marker arranges the pattern pieces on top of the many layers of fabric, leather, or other materials laid out on the cutting table and traces the pattern. Cutters cut the pattern pieces from the material. Assemblers bundle the pieces and trims needed to make a complete garment or accessory and send them to the sewing room.

Sewing machine operators work on an assembly line, sewing one or two parts of the garment. Sections of better quality items may be sewn by hand sewers, who do lapel basting and lining stitching.

Throughout the entire sewing process, inspectors and checkers examine the items and mark defects. (Figure 1-3) After the clothing or accessories are sewn together, they are sent to pressers. These employees use steam pressing machines to flatten seams and shape parts of the finished clothing and accessory items.

Although clothing and accessories may be produced in the United States, a large amount is manufactured abroad in countries such as Taiwan, Korea, and Brazil. Many American apparel companies are involved in **off-shore production.** Off-shore production means the clothing or accessories are manufactured in foreign countries where the cost of labor is much lower than in the United States. The foreign manufacturers use production specifications furnished by the American companies.

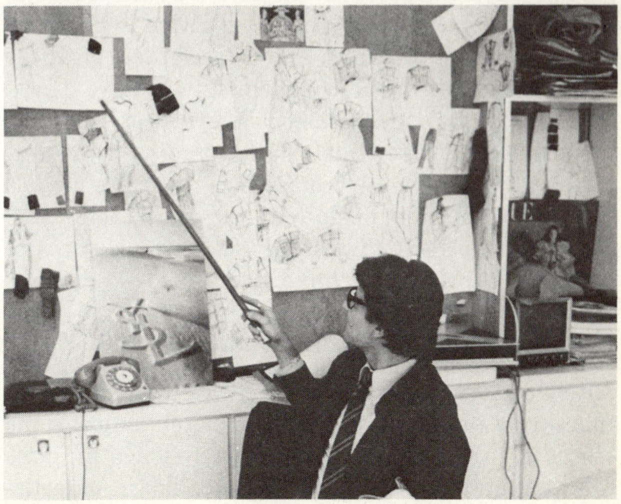

Figure 1-2 Yves St. Laurent is one of the best know designers of high fashion clothing for men and women. © Habans/Sygma.

Figure 1-3 Throughout the manufacturing process, garments must be checked to be sure that they match the pattern and size specifications. © Gabe Palmer/The Stock Market.

According to the Apparel Manufacturers Association, about 26 percent of boys' and men's knit shirts sold in the United States are made abroad. Over 17.6 million dozen imported sweaters entered the United States in 1986; 11.4 million dozen were produced domestically. The major apparel exporters to the United States between July 1985 and July 1986 were Taiwan, Hong Kong, South Korea, China, and the Philippines.[2]

Wholesaling Clothing and Accessories

The newly made clothing and accessories may be sold, or wholesaled, to retail stores. (**Vertical retailers,** those who design and manufacture the products they sell, do not use wholesalers or middlemen.) Approximately six months before the merchandise is available to the customer, retail store **buyers** travel to the showrooms in fashion-center cities to examine the manufacturers' lines. They place orders for those items they believe their customers will buy.

Companies employ salespeople, also called **manufacturer representatives** or reps, to sell the new styles to retail buyers. Sometimes there are models to display the new designs to buyers in the showrooms. Many reps call on buyers at their stores to check on how the merchandise is selling.

Retailing Clothing and Accessories

After retail store buyers place their orders, the manufacturers produce and ship the merchandise to the stores. At the store, the items are checked in, price marked, and displayed on the sales floor. Advertising encourages customers to purchase new merchandise; salespeople are hired to sell it.

■ FASHION MERCHANDISING

Fashion merchandising (sometimes referred to as fashion marketing) refers to the *planning* required to have the right fashion-oriented merchandise in the right place, in the right quantities, at the right prices, at the right time, and with the right sales promotion. There are many workers involved in this exciting, challenging field. Designers, wholesalers, retail buyers, store owners, salespeople, managers, and others are part of the fashion merchandising industry.

The Language of Fashion

Fashion has a language all its own. Dr. Paul Nystrom, a leader in the field of fashion merchandising, defined the words listed below. The people who work in the fashion business use them often.

Fashion The prevailing style accepted and used by a particular group of people at a particular time in a particular place.

High fashion Those styles accepted by the fashion leaders, the elite among consumers, who are first to accept fashion change. High fashion designs, also called **haute couture** (pronounced, "oat-koo-tour"), may be one-of-a-kind creations made for specific customers.

Mass fashion (or volume fashion) Styles widely accepted. These styles are manufactured and sold in large quantities at moderate to low prices.

Apparel Everything that a person wears or carries to complete an outfit: undergarments, outerwear, jewelry, handbags, gloves, hats, shoes, and stockings. In the fashion industry, *apparel* refers only to coats, suits, sportswear, and dresses. The industry uses different terms to describe other apparel items, such as footwear, loungewear, sleepwear, casual wear, accessories, jewelry, men's furnishings (shirts and ties), and so on.

Style The characteristic or distinctive way a garment looks—the combination of features that makes it different from other clothing items, including the silhouette (or shape), texture, or details. For example, the dirndl, wraparound, and slim skirt are three different skirt styles.

Fad A short-lived fashion that may include exaggerated details or loud colors and prints. Fads can come and go in a single season and usually affect a narrow group within a total population. For example, college students wearing bicycle racing shorts, called "bikers," is an example of a fad during the late 1980s.

Classic A style or design that remains "in fashion" over a long time. The cardigan sweater is an example of a classic.

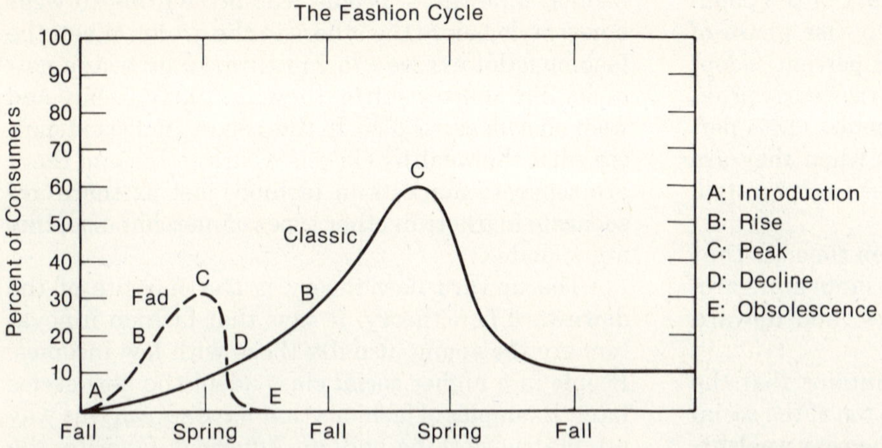

Figure 1-4 Compare the short cycle of a fad with that of a classic.

Fashion Principles

The fashion business is unlike any other business that markets products and services. This business is based on fashion, and to be successful in the fashion business, you must understand how fashion works. For example, you must be aware that most customers will reject a perfectly useful clothing item when it is no longer fashionable. You have to realize that today's biggest seller will become a poor seller (or "dog") when it is "out of fashion."

Consumers determine fashion. Designers, manufacturers, or retailers cannot dictate what clothing or accessories will become fashion. Consumers decide what styles they will buy and wear. For example, for many years, a large number of working women chose to wear the man-tailored suit, shirt, and silk tie, even though designers were creating and promoting many other looks.

Consumers' decisions regarding what clothing they buy and wear are influenced by a variety of factors, including the economy, current events, world leaders, movies, art, and the media—TV, newspapers, and magazines. Some clothing and accessory trends have been the result of the women's liberation movement, exposure to other cultures and ethnic influences, and sports and TV celebrities. The environment and weather also affect what people wear. For example, women in expensive, well-tailored business suits wearing sneakers are a common sight in New York City but very uncommon in cities where walking to work is less popular.

Fashions are not based on price. The price ticket on a clothing or accessory item is no indication of whether that item is currently fashionable. A new, very expensive designer style that is accepted by customers will be quickly copied by many companies and made available in a wide assortment of prices. Although the quality of the fabric and workmanship will be different, the style will be essentially the same.

Fashion changes. The look of clothing and accessories changes from season to season. Usually this change is gradual, or evolutionary. For example, men's lapels may vary a half inch from one season to another. Rarely is change revolutionary or abrupt. One reason for this is that customers buy new clothing and accessories to update their existing wardrobes, which may have been purchased the previous year or years before that. If current fashions are too dramatically different from those of the past, they won't fit in with what customers already own.

Fashions move in cycles. Fashions are introduced, gain acceptance, become popular, and go out of fashion. The length of a fashion cycle varies. A fad has a very short cycle, whereas a classic has a long fashion cycle. (See Figure 1-4.)

No amount of sales promotion can change the direction in which fashions are moving. Although sales promotion may persuade customers to look at a particular item, it cannot force people to buy what they don't want. When consumers reject a style, no amount of advertising or sales promotion can change their minds. That is one of the reasons why stores have markdown or clearance sales when the sales of a particular style start to drop off. Stores know that they must clear out the remaining merchandise as fast as possible and are willing to accept much lower prices to do so. This clearing out of old stock at end of month (EOM), end of season, and before inventory makes room for the next styles that customers have accepted. Companies that are not efficient at clearing their stocks are left with inventory that no one wants and serious cash flow problems.

Fashion Adoption

As shown in the fashion cycle, new styles are accepted by a very few people before they are accepted and worn by the majority. In fact, studies show that about 2½ percent of people are fashion innovators, those most likely to wear a new style. Then there are early adopters (about 13½ percent of the population), who wear items in the early-rise phase of the fashion cycle. The majority (68 percent) adopt new styles when they reach the late rise, early peak, and peak of the cycle. The late adopters (13½ percent) buy and wear the new styles when they are past their peak. Laggards (2½ percent) adopt a style when it's in decline.

There are three fashion adoption theories that attempt to explain who the fashion innovators are: the downward flow, horizontal flow, and upward flow.

The downward flow theory maintains that the fashion innovators are people at the top of the social pyramid, such as world leaders or the very wealthy. Over time, the style is accepted at progressively lower social levels. For example, there was great interest in what first ladies Nancy Reagan and Raisa Gorbachev wore to the summit meetings between the United States and the U.S.S.R. Princess Diana of England continues to set fashion trends, just as she did a few years ago with her hats, ruffles, and big white collars.

According to the horizontal flow (or mass market) theory, fashions move horizontally between groups on similar social levels, not vertically from one social class to another. In every group there are fashion innovators, people who are willing to wear a new style before the others in the group. When the fashion followers see a fashion innovator wear a new style, like shorter skirts, they are likely to buy and wear shorter skirts also. In this theory, it doesn't matter what the wealthy class is wearing. Instead there are separate markets in fashion, just as there are separate markets in other types of merchandise, like automobiles.

The upward flow theory is the opposite of the downward flow theory. It says that fashion innovators are the young, usually those with low incomes. People in a higher social class adopt the new styles later. Examples of fashions that have spread this way are prairie clothing and, an American favorite, the T-shirt. These items were first worn by young people with limited money; now they are included in the expensive lines of famous designers.

Although none of the theories completely explains how all fashions are adopted, each theory recognizes there are fashion innovators and fashion followers. Successful designers, manufacturers, and retailers understand and use all three theories to predict what styles, colors, and textures consumers prefer to wear and, hence, what styles become fashion.

Experience 1-1 ■ Fad or Classic?

Examine the following list of clothing styles. Circle the word to indicate whether the item is a classic or a fad.

1. Shirtwaist dress	**classic**	**fad**
2. Hosiery with seams	**classic**	**fad**
3. Kilt skirt	**classic**	**fad**
4. Stirrup pants	**classic**	**fad**
5. Pearl necklace	**classic**	**fad**
6. Overalls for children	**classic**	**fad**
7. Bare-midriff tops	**classic**	**fad**
8. Jeans	**classic**	**fad**
9. Knickers	**classic**	**fad**
10. Leather miniskirts	**classic**	**fad**

In the space below, identify three current clothing or accessory fads.

1. ______________________________
____
2. ______________________________
____
3. ______________________________
____

■ CAREER OPPORTUNITIES IN APPAREL AND ACCESSORY MARKETING

The fashion business offers vast career opportunities. There are designer jobs, which require an ability to create new clothing and accessory designs. Manufacturing jobs extend from producing cloth to cutting and sewing a garment. Advertising or publishing careers are available to people who photograph, write, or sketch. Fashion promotion jobs involve preparing fashion shows or modeling the latest styles. Wholesale trade salesworkers or manufacturer reps show clothing and accessory merchandise samples to retail store buyers. Wholesale trade salesworkers work for a company or are self-employed; they distribute products to retailers. Manufacturer reps work for a particular manufacturer and only sell that manufacturer's products.

Rather than choose a position in one of those areas, you may decide to begin your clothing and accessory career in a retail store. In a retailing career, positions are available for salespeople, buyer clericals, stock persons, and cashiers. Higher-level positions include department managers, buyers, merchandise managers, and store managers. Retail stores employ people in their sales promotion departments to advertise or display merchandise or plan fashion shows. Employees also work in service departments, such as gift wrap, customer credit, and merchandise delivery. After gaining experience in clothing and accessory retailing, you may decide to open your own store.

This manual provides the information you need to begin or advance your career in a clothing and accessory store. Unit 1 presents the skills and aptitudes needed to work as a salesperson, manager, or buyer. It also provides an overview of apparel and accessory marketing. Unit 2 discusses the basic and social skills needed for an apparel and accessory retailing career. There is advice on how to get along with people and how to communicate. You will use your mathematics skills as retail clothing and accessory employees do every workday.

Unit 3 includes information about the many kinds of women's, men's, and children's clothing and accessories sold. Unit 4 discusses how clothing and accessories are purchased, received at the store, promoted, displayed, and sold. Also in this unit, you will learn how to manage yourself and others.

The Clothing and Accessory Salesperson

Salespeople work with many different kinds of merchandise and customers. They maintain the merchandise on the sales floor, present it to customers, and answer their questions. They also ring up purchases on the cash register, handle money, and make change. Depending on the size of the store, salespeople may order merchandise and make merchandise displays, as well as receive, check, and ticket new merchandise. They may also take inventory and help keep their work area neat.

Successful salespeople enjoy working with the merchandise they sell. They get excited about new fashions and convey that excitement to customers. They know about the merchandise they sell—its construction, style details, and care. To help their customers choose attractive clothing and accessories, these salespeople have an eye for color, line, and design.

Successful salespeople are pleasant and like people. They make customers feel welcome and comfortable in the store. They have a service attitude and focus on the customer's wants and needs rather than on their own taste in clothing. And they are able to remain calm when unpleasant situations occur with customers or co-workers.

Although full-time clothing and accessory salespeople may work a 37½-hour, 5-day work week, they don't work regular 9-to-5 hours. Downtown stores are often open from 10 a.m. to 6 p.m. on Tuesday, Wednesday, and Saturday and from 10 a.m. to 9 p.m. on Monday, Thursday, and Friday. Suburban stores are usually open Monday through Friday from 10 a.m. to 9 p.m. and Saturday 10 a.m. to 6 p.m. Depending on the area, stores may also be open on Sunday. During the weeks before Christmas, the busiest time of the retailing year, most stores are open from 10 a.m. to 9 p.m. Sunday through Saturday. Some stores open as early as 8 a.m. and close as late as midnight to accommodate Christmas shoppers.

The working conditions in most clothing and accessory stores are very good. To attract customers, most stores are nicely decorated, well lighted, clean, and temperature-controlled. Some stores provide cafeterias for employees and/or customers.

Salespeople spend the majority of their day standing and walking. They need stamina and lots of energy. Good grooming is important for clothing and accessory salespeople, and most stores have written or unwritten dress codes that must be followed. Usually salespeople are encouraged to wear clothing similar to that sold by the store. (Stores provide attractive employee discounts so that their employees will purchase the clothing they sell.) Whatever clothing is worn should allow the salesperson to bend over, lift, and stand comfortably for long periods.

Clothing and accessory employers prefer to hire people with at least a high school diploma for full-time sales jobs. They look for people who have had training in business mathematics, merchandising, and retail selling. Many employers prefer to hire people who have taken marketing education classes because these employees have a wide background in merchandising, physical distribution, advertising, display, sales, management, human relations, business communications, and mathematics. Employers are looking for new employees who have a neat appearance and are able to communicate clearly.

Beginning clothing and accessory salespeople may earn minimum or higher wages or a **commission,** a percentage of their sales. Other benefits include an employee discount, health and life insurance, vacation and sick leave, retirement, and profit sharing.

Successful salespeople may be promoted to department managers or assistant buyers in 1 to 2 years. Usually, salespeople with college training are promoted faster than those without it. The employment outlook for salespeople is expected to grow about as fast as the average for all workers through the mid-1990s. People well trained in fashion and selling should be able to find good positions.

The Clothing and Accessory Buyer

Buyers are responsible for the profitability of the merchandise they select for resale in their stores. They evaluate the styles offered by vendors, order items that will meet their customers' wants and needs, and become involved in the promotional activities required to attract customers to purchase the merchandise.

Successful buyers are familiar with the manufacturers that make the merchandise they need. They know about changes in products and the development of new ones. To learn about merchandise, buyers talk with manufacturer representatives in their office or in the manufacturers' showrooms when they go to market. (**Market** refers to the central place or city where retail buyers go to examine manufacturers' or wholesalers' goods and place orders. Major U.S. market centers are New York, Dallas, and Los Angeles.) Buyers also regularly read trade journals, such as *Women's Wear Daily,* and fashion magazines, such as *Gentlemen's Quarterly,* to stay informed of the trends. (See Figure 1-5.)

Figure 1-5 Buyers order the latest apparel and accessory items when they visit manufacturers' showrooms. © NYT Pictures.

Buyers have busy schedules. During a day, they may work with salespeople, manufacturer reps, store executives, and customers. They spend time in their offices, on the telephone, and on the sales floor. A buyer's job is fast-paced and full of constant pressure. Buyers may be away from home an average of one week a month, and they usually work more than 40 hours a week.

Buyers use their mathematical skills in analyzing sales records, preparing merchandise buying plans, and computing their **open-to-buy** (the amount of money available to buy merchandise). They also need to be good at planning and decision making. A thorough knowledge of the merchandise they buy, leadership ability, and good communication skills are essential too.

Most buyers begin their careers in the executive training program or as assistant buyers. Promising salespeople may be promoted to assistant buyers. Stores usually prefer college or community college graduates. These employees often have a liberal arts, business, retailing, marketing education, or fashion merchandising background. After training in the various operations of a store, assistant buyers are promoted to buying positions.

Assistant buyers usually earn between $16,000 and $24,000 a year. Salaries for buyers are in the range of $30,000 to $70,000 (at the larger companies). Often buyers are paid a bonus in addition to their regular salaries if they reach their store's profit objectives for the year.

The employment outlook for buyers is expected to grow about as fast as average for all occupations through the mid-1990s. Competition for these jobs is keen. Clothing and accessory stores choose well-qualified people who enjoy the competitiveness of retailing and work best in a demanding, fast-paced job. Although there are buying positions in all parts of the country, most are in the major metropolitan areas where there are many retail stores.[3]

The Clothing and Accessory Manager

Store managers have the total responsibility for the profitability of a particular retail unit. Depending on the store's size and sales volume, they may be assisted by department managers, merchandise managers, and an operations manager. Department managers oversee the operations in a particular area, such as accessories. They make sure merchandise is received, checked in, priced, and displayed properly on racks and fixtures. They track the sales of their merchandise and make future buying recommendations to their manager. They are also responsible for staffing the department with trained, motivated salespeople and for controlling shortage in their area. Merchandise managers plan and coordinate the store's purchases. They communicate their customers' needs and wants to the buyers. The operations manager is responsible for keeping the physical plant functioning properly, including the heating, air conditioning, elevators, and escalators. They are also in charge of the supply, credit, housekeeping, and receiving departments and for maintaining the store's budget.

Although management jobs vary, all management positions require the ability to plan and organize, select and supervise employees, and control particular business activities. A manager has many responsibilities, works long hours, and makes important decisions. Managers may also be required to relocate, as are, for instance, chain store managers.

The National Retail Merchants Association reports that top executives in retailing management are younger and earn more than top executives in other fields. Persistence, ability, and hard work advance people into clothing and accessory management positions. For the right person, this job offers a satisfying career.

Owning Your Own Store

There are many opportunities to open your own clothing and accessory store. Before opening your own store, though, you should be familiar with how a store operates, how merchandise is purchased, inventoried, and sold, as well as the ins and outs of the fashion marketing industry. Working as a salesperson, department manager, or buyer is almost a necessity before starting a store of your own. Many new apparel businesses fail due to the owner's lack of retailing experience.

■ YOUR FUTURE IN APPAREL AND ACCESSORY RETAILING

Apparel and accessory retailing involves working in a fast-paced, dynamic environment. It's fun and exciting, but it's also hard work. Good mathematic skills, good communication skills, and the ability to get along with others are necessary. To be successful, you need to know the basic facts about your merchandise—the fabrics, style names, size ranges, and care. Familiarity with buying, receiving, advertising, display, and selling techniques is also important. As you advance on the career ladder, you also need to acquire management skills.

Here are a few suggestions to prepare yourself for a successful career in apparel and accessory retailing:

- If you are in high school and do not plan to attend college, start investigating the clothing and accessory stores in your area. Find out from managers which stores will have openings now or when you finish school. Ask about the kind of jobs in which you would start. Your first job may be as a stockperson, cashier, or salesperson. The pay may not be as much as in beginning jobs in other fields, but the opportunities to advance will be greater. You will also need this experience before you can be considered for promotion.
- If you are in high school and plan to go to college, try to get a summer or part-time job in a clothing and accessory store. This will let you see if you like the work.
- While in high school and college, take some basic courses in retailing, selling, and business. Try to get as much on-the-job experience as possible while still in school. If you enroll in the marketing education program at your school, you may be able to "earn and learn" through on-the-job cooperative education and classroom training.
- If you are graduating from a community college or four-year college, be sure to see the representatives of retail companies who visit your college placement office. Write to stores that do not visit your college and that you would like to know more about, and ask for an employment interview. Campus recruiters hire college students for executive training programs. These are excellent programs if you want to become a buyer or manager.

Your work will be challenging, fun, and interesting if you choose a retailing career in clothing and accessory stores. This is one field where you can test your knowledge daily and where you can tell almost immediately if you have made the right choice. Good luck with your career decision and welcome to the fascinating world of apparel and accessory retailing.

Experience 1-2 ■ Is It Right For You?

Only you can tell if you will be happy in an apparel and accessory retailing career. Carefully examine your own personality and the requirements of the job. The following questions may help you decide if this career really interests you. A "yes" answer to most of them may indicate that you would probably enjoy a career in the fascinating world of retailing.

1. Do you like to do your own shopping? **yes no**
2. Do you like to browse through stores? **yes no**
3. Are you interested in what's in fashion? **yes no**
4. Do you notice what others are wearing? **yes no**
5. Do you like to read fashion magazines? **yes no**
6. Do you remember the prices of clothing you buy? **yes no**
7. Do you compare prices? **yes no**
8. Do you enjoy meeting people? **yes no**
9. Do you like being around a lot of people? **yes no**
10. Do you try to figure out people? **yes no**
11. Is is easy for you to talk with strangers? **yes no**
12. Do you like change? **yes no**
13. Do you like your days to be very busy? **yes no**
14. Do you have a high energy level? **yes no**
15. Do you like to compete with others? **yes no**
16. Do you enjoy making decisions? **yes no**
17. Would you take a risk for a big gain? **yes no**

Chapter Notes

1. *Time,* September 1, 1986, pp. 54–61, and *New York,* October 25, 1985, pp. 40–47.
2. *The New York Times,* December 14, 1986, p. 6F.
3. *Occupational Outlook Handbook, 1986–87,* U.S. Department of Labor, April 1986, pp. 54, 262.

CHAPTER 2

Marketing: An Introduction

In the early 1970s, The Gap was the place teenagers shopped for clothes. The specialty chain, which began in 1969, became a quick success mostly because of the huge number of American teenagers who loved wearing denim jeans and cords, the store's biggest products. Another reason for its success was the increase in the number of malls during the 1970s, which is where most of the approximately 700 Gap stores are located. Also, there were relatively few competitors offering such a great selection of denim pants.

As time went on, however, the teenagers who used to shop at The Gap grew up. Although they still wore jeans, they wanted more sophisticated fashions. The competition became tougher also. Stores specializing in jeans were popping up all over. And, to make the competition even more intense, Levi Strauss, The Gap's major supplier, started selling the same styles to other stores, like Sears Roebuck and Co. and J.C. Penney. By 1983, sales at The Gap slowed to less than 1 percent of the total of all other comparable stores (which doesn't include sales at new stores opened that year). Gap founder and chairman, Donald Fisher, decided something had to be done to turn the company around.

In late 1983, Fisher hired a new president, Mickey Drexler, who changed the way The Gap does business. Drexler redefined The Gap's customer, the type of merchandise it sells, its pricing policy, the store interiors and displays, and its advertising. (See Figure 2-1.) His goal was to make The Gap a "purveyor of fashionable sportswear," selling casual clothing, from socks to sweaters, to American men and women. Gap Kids, which was started in 1986, is a version of The Gap for children up to the age of 12. Mickey Drexler prepared a new marketing plan for The Gap that paid off in increased sales and profits.[1]

Every clothing and accessory store has its own marketing plan or way of providing products and services to customers. This chapter will discuss marketing, the marketing concept, the marketing mix, and market research. These concepts are important in understanding how to operate a successful clothing and accessory store, like The Gap.

The American Marketing Association defines **marketing** as the process of planning and executing the conception, pricing, promotion, and distribution of ideas, goods, and services to create exchanges that satisfy individual and organizational objectives.[2] In other words, marketing includes those activities that businesses perform to satisfy current and potential customers and earn a profit.

In the past, *marketing* was defined more narrowly as a series of business activities that moved goods and services from producers to consumers. Although marketing continues to fulfill those functions, today it also focuses on nonbusiness services and marketing activities for businesses and organizations. For example, the Crafted with Pride in U.S.A. Council, Inc., a nonprofit organization that was founded in 1984 by American fiber, textile, and apparel companies, developed a marketing plan to increase awareness and convince consumers to purchase American-made apparel and textiles. One of the major elements of this marketing plan was network television commercials that featured celebrities such as Bob Hope and Don Johnson wearing American-made clothing. These marketing efforts have resulted in a significant increase in the number of consumers who prefer to "buy American."[3]

Figure 2-1 The Gap's advertising is designed to appeal to fashion-conscious, young adult sportswear customers. Courtesy of The Gap.

■ KEY MARKETING FUNCTIONS

Marketing is a big business in America and getting bigger. Approximately one-third of the labor force is engaged in performing marketing functions and about 50 percent of every buyer's dollar goes for marketing costs.[4] The persons employed in marketing perform a wide range of functions.

Product Planning

One of the key marketing functions is product planning, that is, developing products in response to market opportunities. At The Gap, the design staff devotes its time to selecting the exact quality, styles, and colors of their private-label merchandise, including such labels as Gap Sport and Gap. (**Private label** merchandise has met specific standards identified by a retail organization. It is merchandise which belongs exclusively to the retailer who developed it.) Retailers develop private label merchandise to ensure consistent quality of product, meet price competition, provide customers with unique items that no other store has, and increase the store's profit margin.

Buying and Pricing

Two of the key marketing functions, buying and pricing, are performed by retail buyers. At The Gap, buyers determine how many units (or items) to purchase and send to the more than 700 retail stores. Due to the large volume of clothing and accessories purchased, The Gap usually gets the best possible price.

In addition to obtaining merchandise for resale, Gap buyers determine the price that consumers will pay for their sweaters, jeans, sweats, and other items in the merchandise assortment. Customers must perceive the price charged as fair, or they will not buy. The price charged must also provide the company with a profit.

Distribution

Distribution includes those activities associated with the physical movement or transfer of ownership of products from producer to consumer. Some of The Gap's merchandise is manufactured outside the United States where labor costs are less expensive. This merchandise is transported to the United States by ship and then by truck to its two distribution centers in California and Kentucky. Other merchandise, such as the Levi products sold by The Gap, are manufactured in the United States and sent to the east and west coast distribution centers by truck. At both distribution centers, the shipments are quickly processed and sent to the individual stores, where the merchandise is made ready for sale.

The Gap keeps select merchandise in storage at a warehouse. This stored merchandise consists of basic stock items, such as The Gap jeans or sweats. Basic stock items are automatically reordered through the computerized point-of-sale system, in which the cash registers are actually computer terminals linked together. The goods are sent to the store by truck with the next shipment of fashion items. Because The Gap specializes in fashion goods that have a limited selling period, storing merchandise is not common.

After merchandise has been on the selling floor for awhile, The Gap consolidates the remaining styles, colors, and sizes to select stores. Specific stores are instructed to transfer their "broken assortment" to other Gap stores. Consolidating and transferring merchandise are two other distribution activities performed by multiunit clothing and accessory stores.

Promotion

The Gap uses a variety of promotion methods to communicate information about its products, services, and image. For example, it places advertisements in fashion magazines such as *Glamour*. It designs the stores in bright colors with attractive fixtures that place the focus on the merchandise. And the company requires employees to follow specific folding guidelines to keep the merchandise displayed on the tables looking fresh and appealing.

Selling

Selling, another key marketing function, means responding to consumer needs and wants through planned *personalized* communications. The purpose of selling is to influence purchase decisions and ensure customer satisfaction. The Gap staffs each store with salespeople to personally help customers coordinate fashion looks. It schedules its employees so that the most salespeople are working during the store's busiest customer periods. It provides product training to increase the staff's knowledge of how to coordinate and care for the merchandise.

Financing

When The Gap purchases merchandise from its suppliers, it does not pay the bill immediately. In fact, the bill may not be paid for 30 or more days. Likewise, customers may use their MasterCard, Visa, or American Express credit card when purchasing clothing and delay payment. In both instances, The Gap is involved in the financing of goods through the extension of credit. Financing is another important marketing function.

Risk Management

Risk management means being prepared for the possible losses a business may sustain, including theft of merchandise by customers or employees, markdowns due to poor buying decisions, and bad checks and credit card losses. The Gap has a series of procedures that are designed to minimize the losses it may incur, such as plans for handling shoplifting, check acceptance, and markdowns.

Marketing Information Management

The continuous gathering, analyzing, and disseminating of information that is needed when making marketing decisions is called **marketing information management,** or market research. Accurate, timely marketing information greatly increases the probability that the company will deliver products that customers want, when they want them, and at a price they are willing to pay. Market research will be discussed in more detail later in this chapter.

■ THE MARKETING CONCEPT

Successful businesses follow a philosophy called the marketing concept. The **marketing concept** is a belief that the company's total efforts should be directed at satisfying customers' needs and wants in order to earn a profit. A company that follows the marketing concept does what its customers want.

Nordstrom, the west coast specialty chain retailer, is a specific example of a store that is committed to satisfying customer needs. Salespeople approach shoppers within two minutes. They accompany shoppers to well-appointed, spotless dressing rooms and return to offer to get additional items. They accept returns politely and gift wrap merchandise at the same counter where it is purchased.[5] The Nordstrom employee handbook is a small card. It states, "Our number one goal is to provide outstanding customer service." It also lists the only Nordstrom rule, "Use your good judgment in all situations."

Some clothing stores stay open late on weeknights and weekends to make it easier for their 9-to-5 working customers and families to shop. Even though it may be difficult to find salespeople to work these hours, these stores stay open late because it's what the customer wants. Many clothing and accessory stores increase their sales staff during the busy 11-to-2 lunchtime hours so that customers who work in nearby offices can shop and get back to work quickly. In return for meeting their needs, customers reward these stores by purchasing merchandise and, ultimately, helping the stores to earn a profit.

Unfortunately, not all companies follow the marketing concept. You probably have your own examples of times when you were shopping and didn't get the service you wanted. Perhaps you wanted a salesperson to help you coordinate a new fashion look or find you a different size when you were in the fitting room. You may have become frustrated or angry when no one was available to help you. You may have left the store without making any purchase.

That's what happens when companies don't follow the marketing concept. They lose sales...they lose customers...they lose their chance to earn a profit...and when that continues, they eventually go out of business.

■ MARKET SEGMENTATION

A market consists of all the people who are willing and financially able to buy products and services. Because no clothing and accessory business can possibly meet the needs and wants of the entire market, companies divide or segment the total market into groups of customers who have identifiable common characteristics. This is called **market segmentation.** A group of customers with similar characteristics is called a **market segment.** By identifying smaller customer groups that exist in the total market, a company is better able to choose its **target market,** or the market segments it wants to serve.

The four most common ways to segment the total market are demographic segmentation, geographic segmentation, psychographic segmentation, and behavioral segmentation.

Demographic Segmentation

When the market is divided into groups according to population data, such as age, sex, race, income, nationality, occupation, and social class, this is called **demographic segmentation.** This information helps a company identify how large a market segment is and whether it would be a profitable group to target.

Age data may indicate whether a person might be interested in buying a particular clothing item. A children's shoe store, such as Stride Rite Shoes, uses age data to find out the number of infants and children living within a market area. Businesses selling trendy, junior-sized fashions, such as the junior chain Fashion Bug, want to know the number of young women aged 13 to 20.

Retailers such as K mart and Kinney Shoe Corporation use demographic data when determining their promotion plans. For example, K mart ads aimed at Hispanics can be quite different from ads directed at the Anglo market. Among the differences is a greater emphasis on children's apparel and on whole-family apparel. Kinney Shoe Corporation hired an advertising agency in Puerto Rico to handle its Hispanic advertising and public relations in Puerto Rico and U.S. Hispanic markets. Kinney allocated the lion's share of its Hispanic ad budget to the sponsorship of a dance segment on a television game show that had a large number of Hispanic viewers.[6]

Some companies segment the market by sex. There are probably many clothing stores in your local area that sell clothing just for men or just for women, such as Casual Corner.

Income can also be used to segment markets. Income alone is no assurance that people will buy expensive merchandise. It is, however, an indicator that people have the earning capacity to enable them to pay for a "designer original." For example, Barney's, an upscale retailer in New York City, appeals to the market segment that is willing to pay many hundreds of dollars for a blouse made by a popular designer.

Geographic Segmentation

Geographic segmentation involves dividing the total market into groups based on differences in geography. Clothing and accessory companies should be aware of the differences that exist among people living in different regions of the country, with a wide variety of climates and customs. (See Figure 2-2.) There are also differences that exist between people living in cities, suburbs, and rural areas.

FASHION

WHAT'S CHIC IN CHICAGO MAY BE TOO LOUD FOR LITTLE ROCK

Here, from impeccable on-the-spot sources: the most important things to know about business dressing in ten cities. Will your suit go day-into-evening in D.C.? Should you bother with a swimsuit in Miami? Do you need a down coat for Denver? Are pants okay in St. Louis? Look for the answers *you* need, *below*.

CITY	IMPORTANT DOS	DON'TS	F.Y.I.
ATLANTA	● Pack a good-looking raincoat in spring, fall, winter. ● Pants need a jacket to make them work-right.	● Don't overdress. Atlanta is a casual city, not the place for knock-'em-dead fashion (even at night). ● Don't bring winter boots.	● Darker colors and neutrals are favored for business. ● Skin-tone hosiery is your best choice for day; dark sheers are fine at night.
BOSTON	● You'll need an umbrella. ● Spring and fall, pack a zip-lined raincoat. ● Waterproof boots are handy winter, spring and fall.	● Don't wear high heels. Boston is a walking city. ● Don't overdo hair and makeup. Bostonians favor a natural look.	● The navy-suit-and-skinny-tie uniform isn't current here. Sophisticated looks prevail. ● Winters are fierce. You really need to dress warmly.
CHICAGO	● Warm waterproof boots are a must in winter. ● A warm coat counts for winter, spring and fall. ● Winter, pack earmuffs or a warm head cover.	● Don't wear light colors in winter; the slush will get you. ● Don't wear pants to meetings or other formal business occasions.	● Chicago is a city of conservatively stylish women. Avoid anything trendy. ● In winter, pack for cold blustery weather. In summer, it's hot and humid.
DALLAS	● You'll need a lined trench for every season but summer.	● Pants are not acceptable businesswear in most fields.	● Dallas is a fairly dressy city. ● Skin-tone hosiery is a good choice for day.
DENVER	● Bring a trench coat or equivalent year-round. In winter, it should be lined. ● Boots are key in winter.	● Don't bother with an umbrella. ● Don't wear an outfit that's all bright or pastel. Play colors off neutrals.	● Tailored dresses and pulled-together jersey looks are popular. And you don't need a business jacket.
LOS ANGELES	● Pack a jacket or sweater for winter evenings. ● Separates are the norm—more the L.A. look than severe suits.	● Don't wear summer clothes in February. L.A. natives change their look with the seasons. ● Silks are too hot in summer.	● Don't wear anything too bland if you want to impress the natives. ● Knits are popular—but must be well-accessorized.
MIAMI	● Bring a swimsuit with cover-up and a little black dress for evening. Business events are often held on a boat or at outdoor clubs.	● Don't dress more casually for business than you would in any other city. ● Don't wear flats; heels are the norm for business.	● Styles are conservative but up-to-date. Black is a popular daytime color choice. ● Lightweight wools work best in winter.
NEW YORK	● Pack something black. It's the favored color for business and casual wear. ● A good-looking tote is essential. You'll be walking a lot.	● Overly feminine clothes (small floral prints, puffed sleeves) are inappropriate for business. ● Don't skip accessories. New York is a put-together city.	● You can probably dress up a notch from your usual businesswear. When in doubt, wear the best of what you'd wear at home in a similar business situation.
ST. LOUIS	● Bring something navy blue—the business favorite here. ● Conservative, tailored looks are worn year-round, even through summer heat.	● Don't wear pants. ● Don't bare too much skin in summer. ● Don't wear more than a splash of pastels.	● Bold jewelry—one piece—is a popular accent. ● This is a conservative city. Noticeable shoulder pads still elicit comment, we're told.
WASHINGTON DC	● A dark knit suit, preferably black or navy, works best in most places. ● Bring an umbrella and raincoat, regardless of forecast.	● In political circles, avoid clothes that are ultra-feminine, including furs. ● Staid suits are wrong in less conservative fields.	● A work suit with P.M. accessories will see you through most dressy evenings. Otherwise, going-out clothes are fairly casual.

Figure 2-2 Fashion preferences vary from city to city. *Glamour*, Copyright © 1987 by the Conde Nast Publications Inc.

For example, customers living in Phoenix purchase a winter wardrobe that is very different from that of customers living in Denver. People who live near water have a higher demand for swimsuits; people who live near ski slopes buy more ski wear. People who live in the suburbs tend to purchase more casual clothing than their urban counterparts. People who live in rural areas do not have as large a selection of clothing stores as people in the city have. Rural customers may be more likely to purchase the clothing needed for a full season at one time rather than go shopping once or twice a month, as a person in the city can.

There are even differences in demand among customers who live within the same city. In fact, stores with multiple outlets in the same city often provide different merchandise assortments in each location to meet the needs of its customers.

Psychographic Segmentation

Dividing the market according to differences in lifestyle, self-concept, and personality is called **psychographic segmentation.** This method of segmenting the market suggests that people within the same demographic or geographic group will display many different characteristics.

Lifestyle research determines people's activities, interests, and opinions (AIOs). For example, not all 24-year-old women who live in Tucson, Arizona, have the same clothing and accessory needs or tastes. The lifestyle profile of a college student might reveal that she rides her bike to school, prefers natural fibers (even though they may take more time to care for), and enjoys browsing through the local specialty clothing shops near campus. A 24-year-old who works in an office and has a small child at home gears her wardrobe to business. When she takes the time to shop, she looks for synthetic fibers, which don't require special care.

When consumers are asked about their activities, interests, and opinions, lifestyle profiles are developed which identify similar characteristics of the groups. Although this information is difficult to collect, it is very helpful in segmenting markets for clothing and accessories. The retailer who has chosen to target the market to which the college student belongs would stock different kinds of clothing and accessories and promote them differently as compared with the retailer who is targeting the young working mother.

Behavioral Segmentation

In this method of market segmentation, consumers are grouped according to how they use the product. For example, a few years ago leg warmers, which had been worn primarily by professional dancers, became part of the fashionable exercise wear seen in exercise clubs. Later, women wore them on the street, over jeans and other casual pants. The characteristics of dancers, exercisers, and women who wore leg warmers on the street may be different. By knowing more about each segment, a company could fine-tune its marketing plan.

Another example of behavioral segmentation is how often a product is used. For example, women who buy many pairs of very sheer pantyhose, which snag and run easily, may have different characteristics from women who never or seldom buy this type of product.

■ THE MARKETING MIX

Once the target market is selected, the company should design a marketing mix that meets the needs of those customers. The marketing mix is the blend of product, price, place, promotion, and people (also called the 5 Ps) that best satisfies the needs of the target market. (See Figure 2-3.)

To create the right marketing mix, ask yourself the following questions:

What product will appeal to the target market?

What price would these customers be willing to pay?

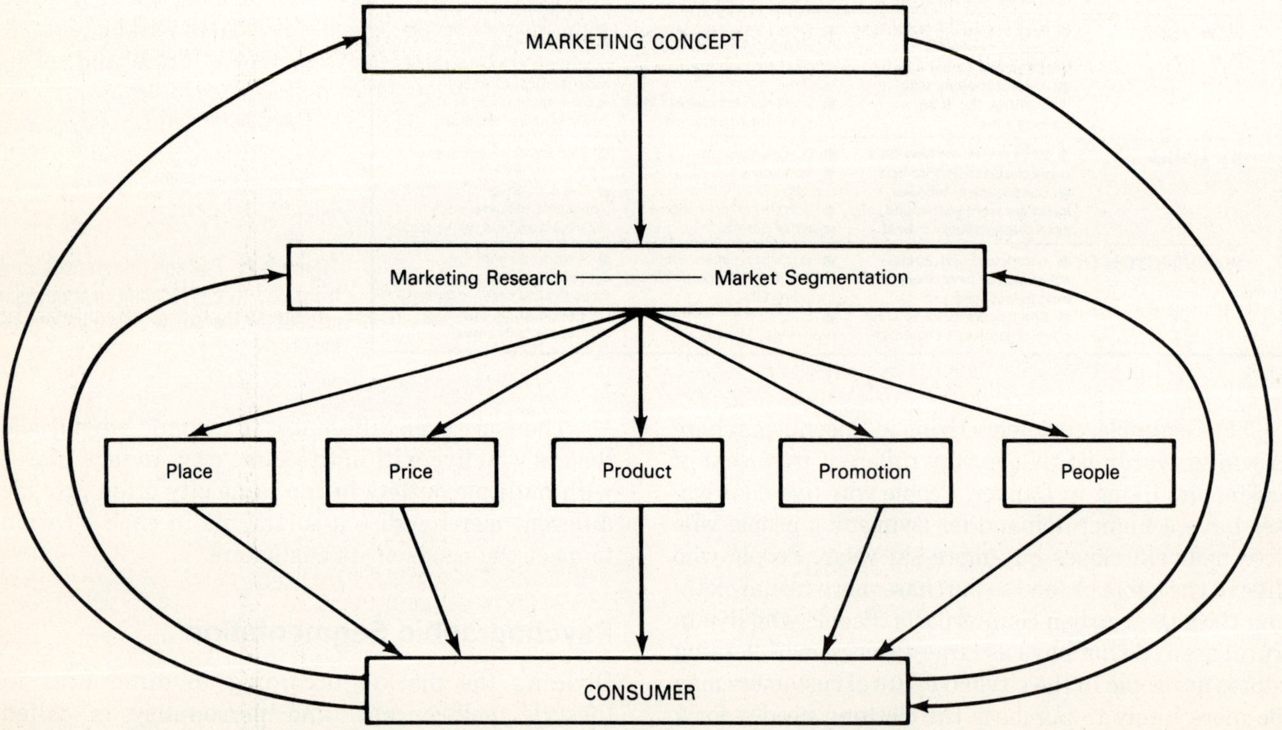

Figure 2-3 Clothing and accessories retailers reach their target customers through their marketing mix. Source: Richard L. Lynch, *Marketing Your Business Program*, New York: McGraw-Hill Book Co., 1986.

Where should the products be available for sale to customers?

What is the best way to tell customers about the products?

What types of salespeople would be best to guide and assist our customers in their clothing and accessory purchases?

Product

Clothing and accessory companies offer products or merchandise and services to their customers. Products can be defined as a combination of benefits for which a customer pays a price. The customer who purchases a dress, for example, is interested in its physical characteristics, including the style, fabric, color, workmanship, label, and size. The consumer is also interested in appropriateness of the dress for a specific occasion, how it fits, how it moves, how comfortable it is to wear, and its price. Just as important as these tangible factors is the psychological satisfaction received from buying the dress. Whether it makes her look and feel attractive are a big part of the buying decision.

A service is an intangible activity that is provided to satisfy customer needs but does not lead to the ownership of a tangible product. Some examples of services usually provided by clothing stores are alterations, delivery, gift wrapping or special packaging, accepting credit cards, and the advice of special wardrobe consultants. Stores offer these services to set themselves apart from the competition. They may be the deciding factors in why a customer chooses to purchase clothing from a particular company.

Some services are sold. For example, a clothing store may charge for alterations. For a tailor or seamstress, sewing or repair of clothing may be the entire source of income. Services may also be offered at no extra cost with the purchase of merchandise. For example, an exclusive designer salon may offer free alterations and the advice of a trained consultant. Such services become part of the whole collection of benefits a customer buys when shopping in a designer salon.

Price

The next of the 5 Ps, price, involves setting prices that are acceptable to customers in the target market and will provide a profit for the company. Rather than there being one "right" price for a particular product, there is usually a price range that is acceptable to the targeted group. The highest price charged should be the amount customers are willing to pay for the product. The lowest acceptable price should cover the company's costs and still provide a profit.

When setting prices, companies consider how much competitors are charging for the same or similar merchandise. They realize customers go from store to store comparing prices and looking for the best value. If another store has the identical item for less, customers will purchase from the competitor.

Pricing policies are concerned not only with setting the original price but also the markdown prices. The price of merchandise is lowered, or marked down, to move out slow sellers, reduce inventory, and stimulate sales.

In addition to the other parts of the marketing mix, price is one of the ways a company communicates its store image and the quality of its merchandise to customers. Although a high price does not guarantee quality nor a low price inferior workmanship, many customers judge a store and its merchandise by the prices charged.

Companies use advertising and display to communicate their pricing policy to customers. Some stores have a promotional image, meaning they frequently advertise sale merchandise. Stores with a high fashion, quality image, mark down merchandise less. They may have one big sale at the end of a season to clear out the remaining stock.

Place

Place is another part of the marketing mix. One of the marketing decisions regarding place is concerned with where the retail outlets will be located. Companies spend a great deal of time and effort determining where they should locate stores. Traffic access, the location of competitors, and the cost of leasing space are some of the factors they review. Another place consideration is how the retailer will transport merchandise from the manufacturers or suppliers to the stores. Although retailers use air and ship to transport clothing and accessory items, most merchandise is brought in by truck.

Where the merchandise will be located in the store is another place consideration. The **adjacencies,** or departments and items that are next to a merchandise group, are carefully thought out. Whether an item is placed on a fixture at the back of the store or department, inside a locked case, or on a self-service display are all questions the clothing and accessory retailer must answer.

Promotion

Promotion is the communication component of the marketing mix. It includes all the ways companies inform their target market about their products and services. Companies "tell and sell" their products and services through advertising, visual merchandising, personal selling, sales promotion, special events, and publicity. The Gap, for example, places advertisements in fashion magazines. Other stores, such as Macy's, send catalogs to customers' homes. Stores hire salespeople to help customers coordinate fashion looks. When a store sends a 20 percent coupon to its customers, it is using a sales promotion tool. In addition, some stores hold fashion shows to influence customers to purchase their merchandise.

People

Personnel, or the people working for a company, are often referred to as a fifth *P* of the marketing mix. They are responsible for planning, implementing, and controlling the marketing mix strategies. For example, it is the buyers who learn all they can about the target market in order to select the merchandise that best satisfies their customers. They establish the original selling price and determine when markdowns should be taken. The place strategy is executed by people such as the mail and phone room manager, who manages the staff that takes orders for merchandise presented in the company's catalogs. The salespeople are directly responsible for advising customers on fashion trends and assisting them with their purchases.

It's people who create a marketing mix and it is only through the cooperative, coordinated efforts of the employees that the marketing mix strategies are blended to meet the target market's needs.

Experience 2-1 ■ Analyze the Mix

Describe the marketing mix of a local clothing and accessory company. On a separate sheet of paper, describe the company's product, place, price, promotion, and people strategies. Next, briefly describe the target market you think this company is trying to serve. Finally, evaluate whether the mix is appropriate to serve the needs of the target market. Make recommendations on how the mix could be improved.

■ MARKET RESEARCH

To really satisfy a target market's needs, a company must know as much as possible about its customers. What merchandise do they need and want? How much will they spend? Where do they want to buy the merchandise? What is the best way to tell and sell the customers on the products? To answer these questions and others like them, companies conduct market research.

Market research is an organized way of finding answers or solutions to problems. It is the process of collecting, sorting, and analyzing market information. It is used by managers to make decisions about the marketing mix.

The focus of market research is always a specific problem. Therefore, the first step in conducting research is the careful definition of the problem. Once a problem is clearly defined, it is possible to determine what factors should be considered to solve the problem. For example, if a store's sales drop 10 percent in one month, the company will want to know why. Do competitors have the same merchandise at a lower price? Is road construction preventing customers from driving to the store? Is another store that recently opened taking customers away?

After a problem is defined, data must be collected. All data can be classified as either primary data or secondary data. **Primary data** is information collected for a specific research project, and it is collected directly from respondents, for instance, customer surveys. **Secondary data** is information someone else has collected for a specific purpose. Stock and sales records collected for financial and inventory control purposes are examples of secondary data.

When the problem is identified and the data is collected, the information must be analyzed and reported. Management studies the reports and decides what action it will take to remedy the problem.

Examples of some of the primary and secondary data most clothing and accessory managers use to solve marketing problems are customer surveys, comparison shopping reports, internal records, and external marketing information.

Customer Surveys

Sometimes companies face marketing issues that require them to go to their customers and ask for information. For example, a store may want to know what customers think of its service, merchandise, and store appearance. To collect the data, a manager may prepare a customer survey. Some stores hire researchers to ask customers if they would be willing to take a moment and answer the questions while in the store. Other stores place postage-paid questionnaires in the customers' shopping bags that request specific information. (See Figure 2-4.) After the data is collected, it is then organized into a usable format and analyzed. Management uses the data to make improvements to its marketing mix.

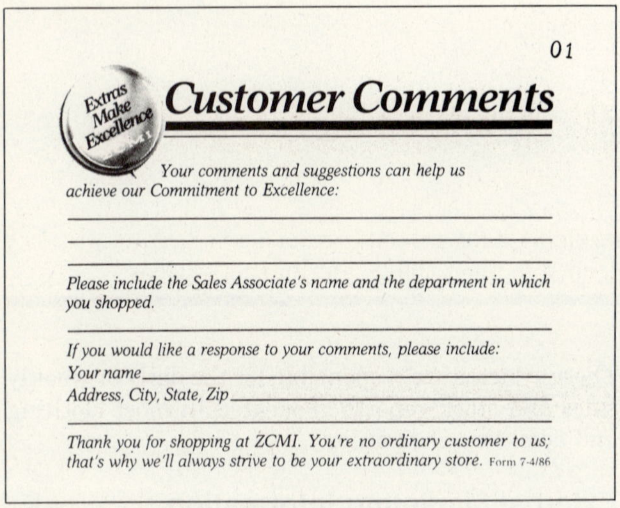
01

Extras Make Excellence

Customer Comments

Your comments and suggestions can help us achieve our Commitment to Excellence:

____

____

____

Please include the Sales Associate's name and the department in which you shopped.

____

If you would like a response to your comments, please include:

Your name ______________________

Address, City, State, Zip ______________________

____

Thank you for shopping at ZCMI. You're no ordinary customer to us; that's why we'll always strive to be your extraordinary store. Form 7-4/86

Figure 2-4 Customer surveys are a form of primary research used by many clothing and accessories retailers. Courtesy of ZCMI.

Comparison Shopping

Changes in the marketing mix of a competitor can seriously affect the sales and profits of a clothing and accessory company. For this reason, clothing and accessory managers should always be aware of the location of competitors; the products and services the competition offers; the prices charged; the competitor's advertising, displays, special events, and promotions; and the service level offered. Most managers often make it part of their weekly routine to visit competing stores and learn how they are trying to satisfy their target market. When collecting primary data, managers may prepare a comparison shopping report as well as keep a file of the competitors' ads. Knowing what the competition is doing is especially important in today's economy, when there are so many companies trying to serve the same target market.

Additional secondary data about the competition or the marketing practices of the major trend-setting stores, such as Bloomingdale's or The Limited, may be found in business publications. Many managers regularly read *Women's Wear Daily, Business Week,* and other business magazines and newspapers.

Experience 2-2 ■ Shop the Competition

Visit two full-service clothing and accessory stores that carry similar merchandise. Act like a customer and attempt to locate two identical items sold by both stores. After you leave the store, complete the shopping report below.

Comparison Shopping Report

	Store #1	Store #2
Merchandise (describe)		
Easy to find?		
Displayed neatly and attractively?		
Price		
Promotion		
Currently advertised item?		
Sales staff		
Sufficient staff for number of customers?		
Approached quickly?		
Friendly?		
Helpful?		
Polite?		
Knowledgeable about product?		
Dressed appropriately?		
Wearing name badge?		

(continued)

Comparison Shopping Report *(Continued)*

	Store #1	Store #2
Checkout lines		
Efficient?		
Accurate entry of data into register?		
Accept credit cards/checks?		
Returns/exchanges		
Generous policy?		
Cleanliness of store?		
Location of store		
Why would a customer choose one store over the other to buy the merchandise?		

Internal Records

Clothing and accessory companies have huge amounts of secondary data gathered from the company's own records. There are sales reports that show sales of individual items by size, color, and style; department sales reports; and total store sales reports. These reports are useful in determining what merchandise the customers liked. There are corresponding stock reports that show what products are available for sale. Markdown reports indicate prices that were too high and had to be lowered. Customer complaints and merchandise-return reports identify products that didn't meet the customers' expectations. Charge account records identify specific customers who have purchased merchandise in the past and would be likely buyers for similar items in the future.

Often, internal secondary data is ignored because there is so much data or it is not in a usable, organized format. Therefore, many companies have computerized marketing information systems that collect and organize internal data on a continual basis. These systems are responsible for the daily or weekly sales and stock reports generated in most clothing and accessory companies.

External Marketing Information

In addition to the internal data that marketing managers need, successful clothing and accessory companies also collect, organize, and analyze external marketing information. External data helps managers better understand the demographic, economic, legal, social, and cultural environment in which their company must operate. Most of this secondary data comes from government reports or trade publications. Examples of economic information that would be important for a company to know about are specific demographic changes in the trading area's population (from the Chamber of Commerce), reported sales and earnings of major companies in the apparel or retail industry (from business publications), and economic forecasts published by the U.S. Department of Commerce and U.S. Department of Labor.

Experience 2-3 ■ Research!

Locate two articles in periodicals such as *Business Week, Forbes, Fortune, Barrons,* or the *Wall Street Journal,* that discuss external marketing information that would be important for a local clothing and accessory company to know. On a separate sheet of paper, write a brief summary of each article. Also describe why this information would help a manager make better marketing decisions.

■ TRENDS

There are a number of trends that will affect the marketing mix of clothing and accessory companies. These trends, which represent directions that certain developments are taking, should be watched and analyzed. Although companies cannot control trends, they can often adapt their marketing mix to capitalize on the opportunities the trends create. Some of the trends that clothing and accessory companies are likely to find during their market research are described below.

Changes in Demographics

The baby boom generation, that huge number of people born between 1945 and 1962, is reaching middle age and becoming an even more important force in the market. Total personal income of this age group will climb by 90 percent between now and the year 2000. The baby boomers not only earn more than their parents did at the same age but many will also inherit substantial wealth from their parents. This great amount of buying power, coupled with the huge number of baby boomers, will result in an increased demand for clothing, especially upscale, quality items. (See Table 1.)

In 1986, *Mademoiselle* surveyed working women across the country with a median age of 30 years and average annual income of $27,000. Over 92 percent of the women reported they shop in traditional department stores. The average amount spent on clothing was $1,683 per year. Seventy-five percent of the women said they considered themselves style-conscious; 60 percent said they usually purchased better or designer clothing. To 50 percent of the respondents, the most important influence on their apparel purchases was the designer or manufacturer label. (See Table 2.) These statistics have major implications for clothing and accessory retailers. Over 60 percent of all women over the age of 18 are working, and this number is expected to rise. Working women spend about 35 percent more on apparel per capita than nonworking women. The demand for suits, dresses, and activewear should increase in the years ahead.

The birth rate is declining; however, more of those births, at least 42 percent, are first children. Since parents typically spend more on their first child than on the next, consumer spending for children's apparel should continue to increase. In addition, a large number of babies are born into affluent, two-income families—often with well-off grandparents, making the outlook for children's wear even better. (See Table 3.)

The Hispanic market is growing, already controlling as much as $70 billion worth of buying power. Retailers who want to appeal to this market will have to address their special language needs in terms of signs and language capabilities of salespeople.

Finally, the members of two-career families (those in which both partners are wage earners) have little time to shop. These consumers are willing to spend more if they can save time. Personal shoppers and stores that are clearly laid out will be important to this market, as will home shopping.[7]

Private-Label Merchandise

To improve their profit margins, stores are selling merchandise under their own labels. Sometimes stores use their own name on the label; sometimes they choose a special name that identifies their private label merchandise. For example, The Gap uses the label "Gap Sport." Federated Department Stores produces quality dress shirts under the Allen Solly label.

Private label merchandise offers stores a way of capitalizing on the quality reputation they have built with customers. Stores use private-label goods to separate themselves from the competition. Customers cannot find identical merchandise at any other location. Private-label merchandise has definitely changed the product assortment in stores today.

■ Table 1
Population by Age: 1985–2000 (in thousands)

	1985	1990	1995	2000	% change
All ages	238,631	249,657	259,559	267,955	12.3
Under 15	51,861	54,582	56,724	55,903	7.8
15–24	39,717	35,548	34,110	36,088	− 9.1
25–34	**41,788***	**43,529**	40,520	36,415	− 12.9
35–44	**32,004**	**37,847**	**41,997**	**43,743**	**36.7**
45–54	22,464	25,402	**31,397**	**37,119**	**65.2**
55–64	22,188	21,051	20,923	23,767	7.1
65 +	28,609	31,697	33,888	34,921	22.1

*Boldface = baby boom generation.

Sources: U.S. Census Bureau; "Who's Counting," by Penny Gill, *Stores,* May 1988, p. 35.

■ Table 2
Working Women's Buying Habits

Where Working Women Shop for Their Wardrobes	
Traditional department stores	92%
Specialty stores/boutiques	77
Discount stores	49
Mail order catalogs	46
Chain department stores	26
Department store catalogs	21
Mass merchandisers	8
Annual Amount Working Women Spend on Apparel	
$500 or less	14%
$500–$999	25
$1000–$1499	21
$1500–$1999	13
$2000–$2999	9
$3000–$4999	9
$5000 +	5
Median spent	$1,262
Average spent	$1,683
What Women's Career Wardrobes Consist Of	
Separates	53%
Suits	31
Dresses	11
Uniforms	5
In Which Department Working Women Usually Shop	
Misses	60%
Juniors	22
Petites	14
Half/Large sizes	4

Sources: Mademoiselle Career Marketing Board Survey on Fashion, 1986; "Who's Counting," by Penny Gill, *Stores,* May 1988, p. 43.

Home Shopping

One trend that affects the place strategy is home shopping. Americans are making more purchases from home than ever before. They are choosing products from direct mail catalogs, television shopping services on cable TV networks, and electronic shopping via computers.

According to *The New York Times,* home shopping sales totaled more than $15 billion in 1987 and have been growing at 12 percent each year. It is estimated that within a decade, they could rise to about one-third of all retail sales.[8]

Catalog buying was responsible for over 75 percent of the home shopping sales in 1987. The growth of catalog buying is due to such factors as consumer bank cards (like Visa and MasterCard), the 800 toll-free number, the use of computers to maintain mail-

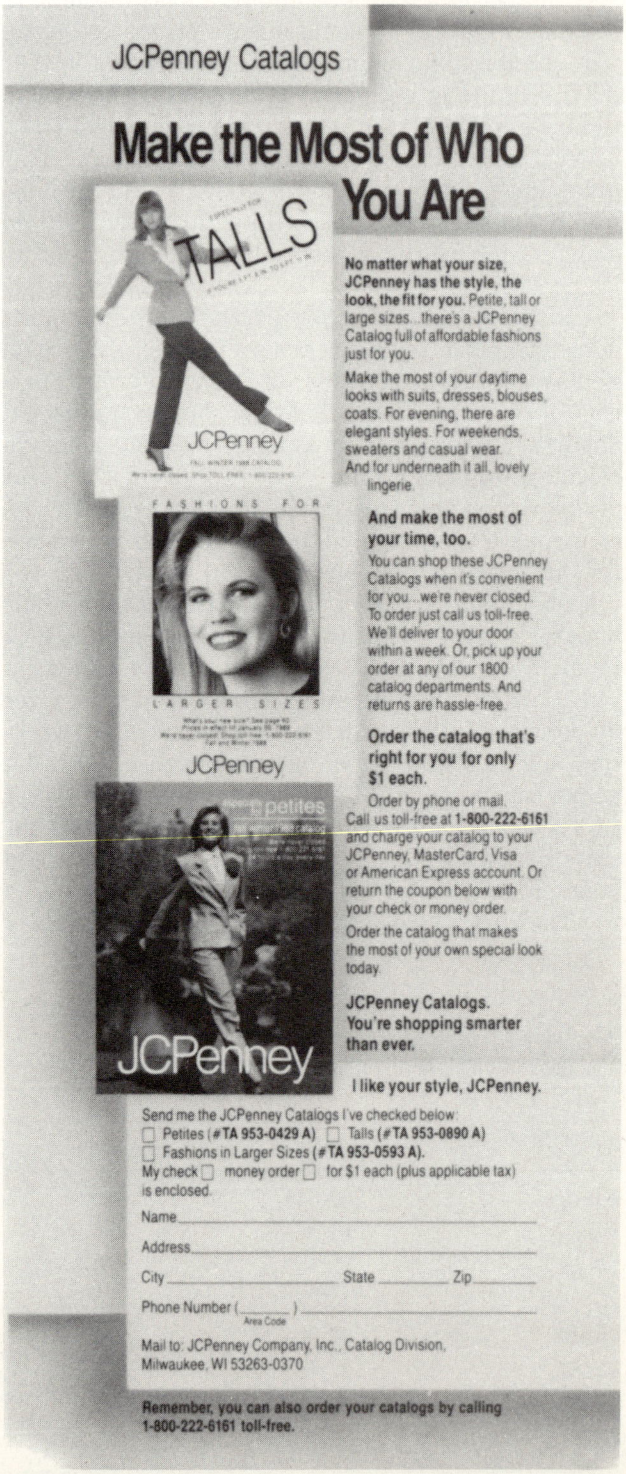

Figure 2-5 J.C. Penney uses specialized catalogs to reach different segments of its women's wear market. Courtesy of J.C. Penney Company, Inc.

■ **Table 3**
Child Population in the U.S.—Estimates and Projections (in thousands)

Age	1980	1987	1994	2000
Under 5	18,128	18,886	18,823	17,626
5–9	17,291	17,511	19,281	18,758
10–14	16,564	16,139	18,552	19,519
Birth to 14	52,073	52,516	56,656	55,953

Sources: U.S. Department of Commerce, Bureau of the Census, Population Estimates & Projections, Series P-25, No. 952, No. 1000; "Who's Counting," by Penny Gill, *Stores,* May 1988, p. 41.

ing lists, and the increase in the number of working women who don't have a lot of time for in-store shopping. (See Figure 2-5.)

There are more than 40 cable TV shopping programs selling merchandise to customers. When customers see a product they want to buy on the screen, they order the merchandise by telephone and then wait for home delivery. These programs have been called 24-hour commercials. Even Sears Roebuck and Co. and J.C. Penney are selling merchandise on cable TV.[9]

Professional Selling Staff

Small specialty shops are often known for their personalized customer service. They have knowledgeable salespeople eager to help customers coordinate clothing outfits and provide advice on how a clothing item can be accessorized. Other clothing and accessory stores are trying to duplicate that professional, friendly service. They are trying to improve their service by changing the salesperson's job, providing sales training, and paying their sales staff a commission on merchandise sold. For example, at Parisian, a company headquartered in Birmingham, Alabama, management has created career paths for salespeople and designed special training programs to help the sales staff meet the needs of customers. New employees are hired as sales assistants and required to meet 80 percent of the store's standards within the first 90 days before they are promoted to sales associate. If a sales associate has the drive and wants to succeed and receive additional financial gain and a different set of benefits, there is the possibility of promotion to senior sales associate.[10]

More companies are moving to paying retail salespeople a commission or percentage of their sales instead of a set hourly wage. With a commission, the people who sell the most, earn the most. At Nordstrom salespeople usually receive 7 percent of their sales. In 1987, the company expected them to sell as much as $125 per hour.[11]

Vertical Retailing

Another growing trend is vertical retailing. A clothing and accessory company that practices **vertical retailing** produces its own products and sells them directly to customers, without going through any middlemen. Vertical retailing allows a company total control from production of merchandise to the presentation of the items on the sales floor. It gives companies complete control over how its clothing will be marketed and the impression it makes on consumers. Examples of vertical retailers are Benetton, Esprit, and Banana Republic.

Even designers Ralph Lauren and Liz Claiborne, whose merchandise is also purchased for resale by department stores and specialty shops throughout the country, have their own stores.

Mergers

Between 1986 and 1988 the retail industry underwent the greatest change in ownership in its history. In 1986 May Department Stores purchased Associated Dry Goods, the company that owned such retail giants as Lord & Taylor, J.W. Robinsons, and L.S. Ayres and Company. The merger created a $10 billion complex under the May Company name. Campeau Corporation, a Canadian firm, purchased Allied Stores, which operated such stores as Ann Taylor, Brooks Bros., Jordan Marsh, and Sterns. In 1988, Campeau purchased Federated Department Stores. Federated owned more than 650 department stores, specialty stores, mass merchandising stores, and supermarkets in 36 states and had annual sales of more than $11 billion in 1987. They owned Bloomingdale's in New York, Bullock's on the west coast, Foley's in Texas, Rich's in Atlanta, and Lazarus in the Midwest.[12]

These major changes in ownership have dramatically increased the buying power of major retail companies. By ordering huge quantities of merchandise for all their stores, these companies have tremendous power with manufacturers. They are able to demand the merchandise they want at big quantity discounts.

■ THE MARKETING PLAN

The marketing plan is a road map that helps a clothing and accessory company achieve its twin goals of serving the target market well and earning a profit. (See Figure 2-6.) Marketing plans answer three important questions:

Where are we now?

Where do we want to be?

How are we going to get there?

Information about where we are now can be obtained through market research. The company's own records and reports as well as information about the environment in which the company is operating help managers determine how well the company satisfies the needs of the market. The Gap, for example, was a promotional store selling mostly jeans to teenagers. It had fierce competition from all the other companies selling denim pants. Although it had over 600 locations in 1983, mostly in malls, it was losing market share and sales were sluggish.

Where we want to be is a tough question. Companies may establish sales, expense, and profit goals, but achieving these goals is much more difficult. A company must clearly identify the market it wants to serve and understand what unfilled needs and wants its customers have. Only then can a company identify how it can satisfy its target market. At The Gap, management decided it wanted to be the store where men and women age 25 to 44 buy their casual fashion sportswear.

The final question, how can we get there, calls for an evaluation and, usually, a redesign of the marketing mix. The Gap focused on changing its product, promotion, and pricing strategy. It reduced the number of private labels from seven to one. It increased the assortment of tops and expanded its color assortment. It remodeled the stores and created a new advertising campaign. The Gap also changed from its promotional image of frequent markdowns and sales to everyday good values.

Marketing plans are needed by all businesses—large and small. They guide clothing and accessory managers in making marketing decisions that will satisfy the target market. In return, the target market rewards the company by helping it achieve its sales and profit goals.

AN OUTLINE FOR A MARKETING PLAN FOR A CLOTHING AND ACCESSORIES RETAILER

1. ***Type of clothing and accessories business.*** What is the name of the business? Describe carefully the clothing and accessories business for which this plan is developed.
2. ***Statement of philosophy.*** What is to be the philosophy of this clothing and accessories retailer? What are its reasons for being in business?
3. ***Target markets.*** To whom are the goods and services to be targeted? What are the demographic, geographic, psychographic, and/or behavioristic characteristics of the envisioned targeted market(s)? Prepare a typical customer profile for each of the targeted markets.
4. ***Marketing research.*** What information is needed to help this retailer succeed? How is the information to be used?
5. ***Product or service plan.*** Prepare a complete listing of products and services to be offered by this business. What are their costs, major features, and selling benefits relative to each targeted group?
6. ***Place marketing.*** How is the business's merchandise distributed? At what locations will the clothing and/or accessories be offered? What are the advantages (and disadvantages) of these locations? What are the costs associated with this location?
7. ***Pricing.*** How will prices for the goods or services be determined? How do they compare with those of the competition?
8. ***Promotion.*** Prepare a detailed description of the promotion plan for the clothing and accessories retailed. What products, services, or image aspects of the business are to be promoted and for what purposes? Include objectives, strategies, costs, and envisioned evaluation for advertising, visual merchandising sales promotion, personal selling, and public relations.
9. ***Personnel.*** Describe the type and number of people that need to be initially employed. What background experiences and characteristics should they possess? How should they be trained? What training is envisioned for current employees? Costs? How are personnel to be rewarded for exemplary service?
10. ***Assessment.*** How will the effectiveness of the marketing plan be assessed? At what periods of time will evaluative measures be taken? How will evaluative information be used to modify the marketing plan?

Figure 2-6 The marketing plan tells how to get from where your business is now to where you want it to be.

Chapter Notes

1. Jules Abend, "Widening the Gap," *Stores,* November 1985, pp. 95–99; Pamela Ellis-Simons, "A Cinderella Store," *Marketing and Media Decisions,* Winter 1986, pp. 46–50; Nancy R. Gibbs, "Falling into The Gap," *Time,* October 5, 1987, p. 54; Jean Sherman, "Bridging The Gap," *Working Woman,* June 1987, p. 58; and "Put the Zipper on the Back," *Forbes,* June 29, 1987, p. 89.

2. "AMA Board Approves New Marketing Definition," *Marketing News,* March 1, 1985, p. 1.

3. Interview with Robert Swift, Executive Director, Crafted with Pride in U.S.A. Council, Inc., New York, August 16, 1988.

4. William M. Pride and O. C. Ferrell, *Marketing,* 4th ed., Houghton Mifflin Company, Boston, 1985, p. 6.

5. Tom Peters, "The Store Is Where the Action Is," *U.S. News and World Report,* May 12, 1986, p. 58.

6. Marita Thomas, "The Hispanic Market," *Stores,* May 1988, pp. 53–56.

7. Penny Gill, "Who's Counting," *Stores,* May 1988, pp. 33–49.

8. Isadore Barmash, "Armchair Buying Becomes a Fast Growing Industry," *The New York Times,* February 19, 1987, pp. C1, C10.

9. "Home Shopping: Is It a Revolution in Retailing—or Just a Fad?" *Business Week,* December 15, 1986, pp. 62–69.

10. "Parisian: Steps to Professional Sales Staff," *Stores,* September 1986, p. 84.

11. "Nordstrom's High Style," *Newsweek,* January 5, 1987, p. 43.

12. "Federated Stores Mull New Purchase Offer," *Las Vegas Review-Journal,* March 1, 1988, p. 5F.

UNIT 1 PROJECT

Market Segmentation

In Unit 1 you have become acquainted with the marketing concept as it is practiced in the apparel and accessories industry. A crucial part of meeting the needs of your customers is knowing who your customers are.

In this project, you will assume the role of the manager of Neighbour's Shoes, a family shoe store in Sumpter County. You will ue market research to develop profiles of the market segments in Sumpter County, and you will attempt to identify a target market that your store can serve profitably.

As for the project for Unit 4, you will develop a marketing plan to launch a new product aimed at the needs of your selected target market.

You and Neighbour's Shoes

While you were working your way through Sumpter County Community College, the thought of running your own business was a distant dream. But when Uncle Joe and Aunt Sarah Neighbour retired, they left you in charge of their shoe store.

You had proved yourself to be a good worker, devoting evenings and weekends to Neighbour's Shoes, even before it moved to the mall. Now, with your degree in marketing, Joe and Sarah feel you are the best person to make the business grow. You have all agreed that you will buy the business over the next 10 years.

As manager of Neighbour's Shoes, you are studying the city of Central Beach and the surrounding Sumpter County based on data compiled in 19X0. You will also analyze data from a customer survey that you conducted throughout last year, 19X4. You believe that adding a new line of running shoes, Wings by Mercury, to your product mix, may be an opportunity for the store. You plan to segment the population and evaluate the sales potential of Wings within one subsegment. You hope to launch your new product this year, 19X5. Fortunately, the 19X0 figures include a population projection for 19X9.

Central Beach, Sumpter County

Central Beach started as a mill town on the Central River. As the center of Sumpter County and the county seat, the city has always enjoyed the bustle of commerce and government. The central business district is still around the courthouse at the corner of Tyler and 4th Avenue.

During the Great Depression, the Federal Government dammed the Central River, creating Central Reservoir and providing both energy for industrial growth and a recreational attraction. The county fair grounds were established near the reservoir, and soon a permanent amusement park and boardwalk complemented the beach and park land. Vacation cottages and, later, hotels and larger vacation homes surrounded River Road and the beachfront areas.

The main campus of Sumpter County Community College was established in the 1950s. In the 1960s, SCCC expanded its facilities from Oak Street north to Route 15. This expansion provided residential facilities for about half of its 7500 students, who come from outside the county.

The old clothing mills have long since been replaced with machine shops, which serve the automotive and small engine industries. New office parks support various white collar industries, including a major insurance company's regional office and the headquarters of a real estate development and construction firm. A shopping mall was built in the 1980s and draws people from the entire county.

Downtown shopping has remained strong with the growth of business in the city.

At the turn of the century, the town's business leaders built their houses in the northern hills, just beyond the college's current boundary. In those early days most working people lived north of 7th Street and in the "tree" streets. Now, housing is springing up across the river in the town of Sumpter Heights, where many two-career families are buying their first townhouse or detached home.

Together, Central Beach and Sumpter Heights comprise about 70 percent of the county's 100,000 population. Highway 56 divides Central Beach from Sumpter Heights and leads downstream to Metro City and upstream to the mountains and ski resorts. The other sections of the county are mostly rural but include a growing number of suburban developments and private estates. In many ways Central Beach and Sumpter County are an entrepreneur's dream, offering a diverse and growing economy halfway between the big city and the countryside.

Competition

Several other retailers sell shoes in Sumpter County. In the mall where you are located, Penn's Shoes specializes in women's shoes and carries a wide selection of expensive shoes. Good Sports carries athletic shoes as well as other sporting goods. K mart has a shoe department, and offers low prices and self-service. Orwycz Department Store offers men's, women's, and children's shoe departments, fair service, and high prices.

In Central Beach's central business district, Rand Shoes specializes in men's and women's shoes for office wear. Saveway Shoes has a branch in the retail triangle between Grant and Sumpter Avenues. They offer low prices and low quality goods.

With all the shoe retailers in the county, your marketing plan will be the key to Neighbour's continuing growth.

Segmentation

As discussed in Chapter 2, markets can be divided by applying principles of demographic, geographic, behavioristic, and psychographic segmentation in order to plan how best to serve the marketplace. A good place to start is with the characteristics of the general population.

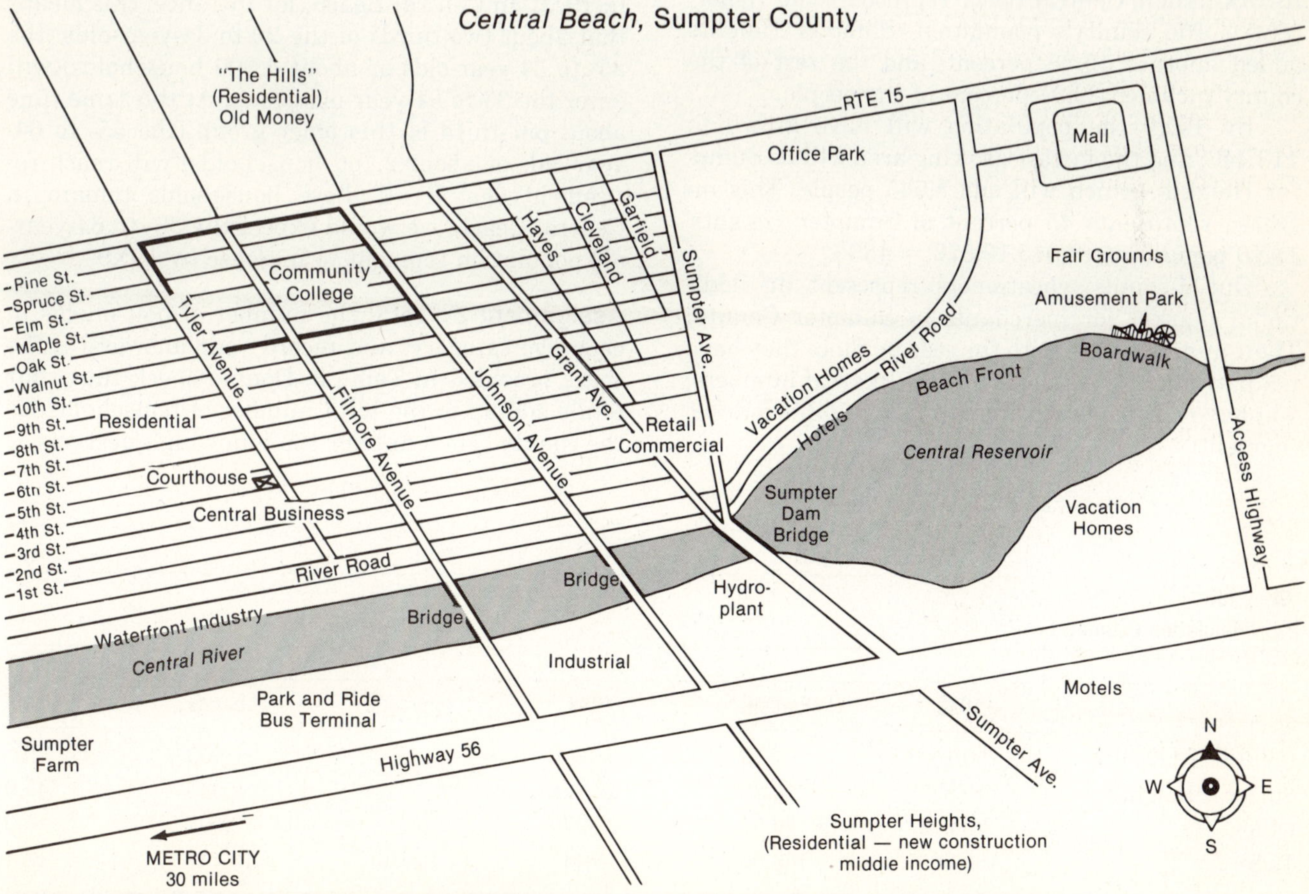

Figure 1 This map shows Central Beach and Sumpter County.

Characteristics of the General Population

The Chamber of Commerce of Sumpter County published selected census data from 19X0 to help merchants understand the region. These data give an overview of the demographics of the county and show the differences between its three geographic sections.

Population Size

Table 1 shows the population of Sumpter County broken down for Central Beach, Sumpter Heights, and the "other" sections of Sumpter County. The table also includes estimates for population growth expected by 19X9.

In 19X0 the total county population was 100,000. About half the people live in Central Beach (50,150). Roughly another 20,000 people lived in Sumpter Heights, and about 30,000 lived in the rest of the county. Table 1 shows the percent of the county represented by each section, which was calculated by dividing the population of the section by the population of the entire county.

Section/County × 100% = Section as a percent of the county

In 19X0, then, Central Beach represented 50.15 percent of the county's population. Sumpter Heights added another 19.88 percent, and the rest of the county included 29.97 percent of the people.

By 19X9 the population will have grown to 113,447, and the fastest-growing area will be Sumpter Heights, which will add 8,945 people. This increase represents 45 percent of Sumpter Heights' 19X0 population (8,945/19,877 = 45%).

Out-of-county vacationers represent an additional market for merchants in Sumpter County. Their numbers vary with the season. Since they have not patronized Neighbour's Shoes in great numbers, you have not made a point of analyzing this market segment.

Assignment 1 To see the relative size of each section of the county in 19X9, divide each section population by the population estimate for that year. Write your answers in Table 1.

Location, Age, and Income

Demographic data can be analyzed geographically by looking at differences between sections of the county. Table 2 shows the locations of households according to the age of the head of the household. Median household incomes are also given for each age range. Thus, we see that in Central Beach 230 households identified themselves as headed by a person 15 to 19 years of age. The median income for households headed by that age group in the county was $12,200. The richest age segment was the 35- to 64-year-olds.

The table also shows the population of each section and the average household size. We can see that households in the "other" sections of the county are a bit larger than average.

Studying the 19X0 age distribution can also help predict the future age distribution. In 19X9 (ten years after this information was collected), the 55-to 64-year-olds will have entered the 65+ group. They will be replaced with those who were 25 to 34 years old in 19X0. In Central Beach, for instance, this means that about two-thirds of the 20-to 34-year-olds (the 25- to 34-year-olds or about 4,300 households) will enter the 35-to 64-year-old group. At the same time about one-third of this older group (the 55- to 64-year-olds or about 2,700 households) will reach retirement age. If all these households remain in Central Beach, we would expect the 35- to 64-year-old population segment to increase by 19X9.

Assignment 2 Total the number of households in each age category, and record your answers in the space provided in Table 2. Double check that your totals add up to the total number of households for the county. Then answer the following questions.

■ Table 1
Population Estimates

	19X0	Percent	19X9	Percent	Change	Percent
Central Beach	50,150	50.15	51,655	_____	1,505	3.0
Sumpter Heights	19,877	19.88	28,822	_____	8,945	45.0
Other	29,973	29.97	32,970	_____	2,997	10.0
Total	100,000	100.00	113,447	100.00	13,447	13.4

■ **Table 2**
19X0 Age of Head of Household by Location and Income

Age Head of Household	Median Household Income	Central Beach	Sumpter Heights	Other	Total
15–19	$12,200	230	30	100	______
20–34	19,950	6,520	1,940	2,990	______
35–64	32,800	8,125	4,570	6,105	______
65 +	18,200	4,050	1,052	1,872	______
Total	28,100	18,925	7,592	11,067	37,584
Population		50,150	19,877	29,973	100,000
Household size		2.65	2.62	2.71	2.66
Median age of head of household		44.0	46.0	46.0	45.1
Median age of population		26.4	31.2	33.2	29.4
Median adult age		36.0	41.0	41.0	39.8

1. Which group has the most households? How many years does this age group span?

____

____

2. Which group has the second most households? How many years does this age group span?

____

____

3. Do you expect the 19X9 population of households headed by 35to 64-year-olds to increase from its 19X0 number? Explain.

____

____

4. Which is the youngest section of the county?

____

____

5. Why is the median age of the population much lower than the median age of heads of households?

____

____

6. Review Table 1. Notice the growth patterns in the three sections of the county. How will the general population change in terms of age and income? Explain.

____

____

Customer Characteristics

While you were studying the census data, you decided to run the store as it had been for the first year. After all, it was a successful business with a loyal following from its original neighborhood in Central Beach. For continuity, you kept Aunt Sarah and Uncle Joe's latest slogan in your advertising, "Neighbour's Shoes, Your Local Shoe Store in the Mall."

You need to find out more about your current customers, however, so you can compare them with the general population. Such a comparison will show you which population segments you serve best and which segments you do not serve well. You then will be able to develop strategies for extending (or changing) your customer base.

As part of your promotional program, you conducted four customer surveys, one during each season. People were encouraged to complete a questionnaire, which made them eligible for a drawing for a $100 gift certificate.

Using your sales figures and survey results, you projected the total number of customers you will serve. You hope to identify new opportunities as you analyze your information.

Household Income and Location

Income and location represent demographic and geographic variables. To compare your customers with the general population, you first put your survey results and census data in the same format.

Current Customers

Table 3 shows the number of your current customers by household income and location. You have also calculated the percent of all your customers represented

■ Table 3
Survey Results, 19X4
Customer Households by Income and Location

Income Range	Central Beach	Sumpter Heights	Other	Total
Lower	1,280	260	400	1,940
$0 to 24,999	34%	7%	10%	51%
Middle	740	250	270	1,260
$25,000 to 49,999	20%	7%	7%	33%
Higher	250	120	240	610
$50,000+	7%	3%	6%	16%
Total	2,270	630	910	3,810
	_____ %	_____ %	_____ %	_____ %
Median Income $	22,200	30,500	30,100	24,500

by each subgroup. For instance, in the Total column your survey indicated that you serve customers in 3,810 households. Of these, 1,260 customer households had incomes in the middle range. This represents 33 percent of all your customers (1,260/3,810 = .3307 = 33%).

Customers in subgroup ÷ Total customers × 100 = %
1,260 ÷ 3,810 × 100 = 33.07 rounds to 33%

Of these middle income customers, 740 were from Central Beach, 250 were from Sumpter Heights, and 270 were from other sections of the county.

The median income of your customers was $24,500: half of your customers had household incomes above $24,500 and half below. Your customers from Sumpter Heights and other sections are considerably more affluent than those from Central Beach.

Assignment 3 Calculate the percent of your customers from Central Beach, Sumpter Heights, and the other sections of the county. Write these percents in the spaces provided in Table 3. Now circle those values in the table that represent 20 percent or more of your customers.

General Population

Table 4 shows information parallel to Table 3 for the entire population of the county. The county is richer

■ Table 4
Census Data, 19X4
Households by Income and Location

Income Range	Central Beach	Sumpter Heights	Other	Total
Lower	10,220	2,668	4,196	17,084
$0 to 24,999	27%	7%	(11%)	45%
Middle	6,813	3,310	3,697	13,820
$25,000 to 49,999	18%	9%	10%	37%
Higher	1,893	1,580	3,208	6,681
$50,000+	5%	4%	9%	18%
Total	18,926	7,558	11,101	37,585
	50%		30%	100%
Median Income $	23,200	32,300	33,700	28,100

than your average customer. Only 45 percent of the households had "lower" range income in the county, while 51 percent of your customers fell into this range. Notice that the county median household income was $28,100, almost $4,000 higher than that of your average customer.

Furthermore, you note that while your survey is current, the census is four or five years old, and incomes have risen in recent years. A recent survey conducted by the mall's management estimated that the current median household income of a mall visitor was $29,000.

Assignment 4 Compare the figures in Tables 3 and 4. In Table 4 circle the subgroups with percents greater than those in Table 3; for instance, the "11 percent" for "Other" in the lower income range is circled, since the same segment in Table 3 was only 10 percent.

1. Name the subgroups that patronize your business *more* than the general population.

____

____

2. Name the subgroups that patronize your business *less* than the general population.

____

____

3. Compare the median incomes of your customers (Table 3) with the median incomes for the general population (Table 4).

   a. In which sections of Sumpter County are your customers richer than the general population? Explain.

   ____

   ____

   b. In which section of the county is the difference in income between your customers and the general population greatest? Explain.

   ____

   ____

Age

Census data show the age distribution in the county. Your customer survey lets you estimate the number of people in your customers' households. You can decide your relative strength with each age group by comparing these figures.

Table 5 shows age distribution for the county and for your customer households. Not all people in a customer household are customers, of course. Mom, Dad, or Uncle Ralph may shop elsewhere. The "people in customer housholds" are the highest estimate of your current share of each age market.

You will notice that in 19X0, 20,000 people in the county were 14 years old or younger (children). This represented 20 percent of the county population. In comparison, there were 1,800 children in the households of your customers, or 17 percent of the people represented by your survey. You can see that your customers have fewer children than the general population. Your current maximum market share is only 9 percent. (1,800 ÷ 20,000 = 9%.) Overall you now serve about 10 percent (10,388 of 100,000 = 10.39%) of the population.

■ Table 5
Population Ages

	19X0		19X4 Survey		
Age Range	**County**	**Percent of County**	**People in Customer Households**	**Percent of People in Customer Households**	**Current Maximum Share Percent**
0 to 14	20,000	20	1,800	17	9
15 to 19	8,887	9	1,155	11	13
20 to 34	24,457	24	1,957	19	8
35 to 64	34,416	____	4,130	____	____
65+	12,240	____	1,346	____	____
	100,000	100	10,388	100	10

On the other hand, your customers' households have more older teens (aged 15 to 19) than the general population. (Current maximum market share is 13 percent.) While you have not kept track of the number of sales to older teens, your strong impression is that sales to this group are much less than its 13 percent potential.

By studying this table you can characterize your current customers by age. You can then decide whether to try to specialize in your largest age groups, or to expand to age groups that are underrepresented in your current customer base.

Assignment 5 Calculate the "percent of county," "percent of people in customer households," and the "current maximum share percent" for the "35 to 64" and the "65+" age ranges. To find the percent of county, divide the age range's population by 100,000 (the county's total population) and multiply by 100 percent. For customer household age ranges, divide by 10,388 and multiply by 100 percent. For the current maximum share, divide the people in customer households by the county population for each group. Write the percents in the table. Then answer these questions:

1. Which are the largest and smallest age groups in the county? In your customers' households?

____

____

2. For which three age groups do you have the greatest "current maximum market" share?

____

____

3. Does the "current maximum share percent" represent actual sales? Explain.

____

____

4. Compared with the county population, do you attract more or fewer people aged 20 to 34 years?

____

____

5. Should you try to attract more people in this age group? Explain.

____

____

Spending Behavior

Marketers can use information about the way people buy and use products to segment the population. To get a picture of your customers' buying behavior you analyzed your sales figures.

Table 6 shows the number of shoes sold in each of three categories: children's, women's, and men's. The table also lists the 19X0 county population of the groups that would buy these products. We can see that Neighbour's sold 7,200 pairs of shoes last year. Half of those were women's shoes.

You have also noted that most shoe purchases involved women. Typically, women buy their own shoes and direct the purchases of their young children. They also advise men in their purchases in many cases.

The last column of the table shows the units sold divided by the county population of potential customers. Thus, you sold 1,152 pairs of children's shoes to a population of 14,000 children for a ratio of .082. In other words, on average you sold shoes worn by 8.2 percent of the children in the county.

You also analyzed your sales to two special populations: loyal customers and high spenders. You've defined loyal customers as those who spend half or more of their shoe budget in your store. High spenders are those who spend significantly more than average on shoes. To avoid counting a customer twice, you decided to categorize people who fit both definitions as loyalists. Since the median reported house-

Table 6
Sales by Shoe Category

	Units	Percent Units	County Population	Units/ Person
Children's (to age 10)	1,152	16	14,000	.082
Women's	3,600	50	44,600	____
Men's	2,448	____	41,400	____
	7,200	100	100,000	.072

Segment	Median Income	Median Adult Age	Primary Location (CB,SH,O,T)	Attitude	Households Total	Households Current	Households Percent
Current customers	$24,500	36.0	CB,O	C,D,$,Fr,P	37,584	3,810	10.1%
Loyalists	$27,000	38.0	CB,O	C,D,Fr,P	N/A	1,524	N/A
Sumpter Heights	(4)_____	(2)_____	SH_____	(8)_____	(1)_____	(3)_____	_____

3. How much larger is Sumpter Heights than your loyalist population?

____

4. What percent of Sumpter Heights do you now serve? How does this compare with the county in general?

____

5. Is Sumpter Heights an attractive target? Explain.

____

____

____

____

UNIT 2

Foundation Skills

To succeed at any apparel and accessories career, you need certain "foundation" skills. For example, you need to know how to get along with people—customers, co-workers, supervisors, and suppliers. You also must know how to communicate with others—by speaking, listening, reading, and writing. These skills are discussed in Chapter 3. The mathematics skills you will use in your apparel and accessories marketing career are covered in Chapter 4.

If you master these foundation skills, you can build a good career on them—a career that will give you financial and personal rewards.

CHAPTER 3

Human Relations and Communications

Since the late 1960s, most salespeople have been paid salaries only, and low ones at that. In only a few departments—shoes and men's suits, for instance—did the sales staff earn anything extra for making a sale. Generally, stores saw no need to provide incentives; retail sales were booming. But now that profits have begun to sag at many stores, retailers are adopting the commission compensation system.

Incentive plans vary from retailer to retailer. At Bloomingdale's, people who sell designer shoes earn a 10 percent commission and no salary. At Sears Roebuck and Co., where shoes cost much less, the sales staff gets a salary plus a 3 percent commission.[1]

The return of the commission system means a different relationship between the shopper and sales staff. For example, customers can expect salespeople to be more service-oriented, helping them to coordinate an outfit, since the salesperson's income will be directly affected by every sale.

A commission system may also create a different relationship between one salesperson and another, causing some salespeople to become pushy and uncooperative. This is only one of the many human relations problems that may result in unproductive behavior. In a successful clothing and accessories store, such problems must be prevented or, when they do occur, corrected. Human relations and communications must be directed toward meeting customer needs, maintaining a pleasant work environment, and running a profitable business.

HUMAN RELATIONS

Human relations concerns a person's behavior and how it affects others. People who practice good human relations skills are well-liked and get others to cooperate with them. They understand that every person is different because of unique personality traits. They understand that every person has a different intelligence level, ability to understand, and communication style. Each is also different because of his or her past experiences and cultural background.

People who realize that everyone is different vary how they talk and respond to others. They are more objective, not expecting others to act and react as they themselves do.

Personality Traits

Clothing and accessories employees display a variety of personality traits, such as interest in others, enthusiasm, initiative, flexibility, self-control, attention to appearance, flexibility, honesty, and trustworthiness. These traits affect the way they get along with each other, customers, and their supervisor.

Interest in Others An employee who is known for showing interest in others makes everyone feel special by really listening to what they are saying. The technique is to listen with your ears and watch with your eyes.

Listening is not a difficult skill to develop. You can improve your listening skills by:

- Being an active listener by giving full attention to what is being said
- Having an open mind by not making value judgments on what the speaker is saying
- Minimizing interruptions and distractions, such as loud noises or two people talking at the same time

Figure 3-1 Because employees have unique personality characteristics, they relate to others differently. © Jules Allen Photography/ U.S. Athletics.

- Being patient and not anticipating a speaker by interrupting or finishing sentences
- Giving verbal and nonverbal feedback by asking relevant questions, keeping eye contact, and nodding your head.

Enthusiasm It's easy to see when people are enthusiastic about their jobs; their eyes sparkle and they seem to enjoy what they are doing. The tone of voice is pleasant and sincere which shows that they honestly want to be of service. It's not a secret that customers enjoy dealing with enthusiastic people.

Initiative and Creativity Managers look for self-starters who can begin a job without being told what to do. As a new employee, you should find out all you can about your job and the place you work before putting your ideas into action. Once you learn the store's operation, you can make suggestions for improvement. These suggestions are usually appreciated, and your initiative and creativity may be rewarded with a promotion.

Flexibility At a clothing and accessory store, unplanned tasks occur all the time. For example, a customer may want to purchase the suit you just put on the window mannequin. Flexibility will help you adjust to unplanned events and to get along with your co-workers, supervisors, and customers.

Self-Control Salespeople have to work with many different customers. Some customers are very quick in their decisions, others are slow and seem to take forever. As a salesperson, you must learn patience and courtesy when helping all customers. It's easy for salespeople to tell customers what they want to hear in order to make sales. But many future sales are lost when customers lose trust in a store, product, or salesperson. Management holds salespeople responsible for knowing the facts and honestly communicating them to customers.

When customers become difficult, remember the following important points:

- Get the customer off the sales floor and to the manager's office or another part of the store where other customers will not be disturbed.
- Treat the customer with respect. Many stores want their salespeople to call their customers Mr., Mrs., or Ms. If you know the customer well, and if it's acceptable store policy, you may call him or her by the first name. Never insult the customer or call the customer rude names.
- Appeal to the customer's sense of fairness and explain the store policies and reasons for those policies.
- Don't be frightened by the customer.

Appearance An employee's clothing and personal appearance create an impression on customers and others who come into the store. A proper appearance establishes a good public image for the store. It's especially important in a store or department that sells clothing or accessories.

Many stores have guidelines or a dress code on what employees are to wear. (See Figure 3-2.) Stores without formal dress codes expect their employees to be clean and well-groomed. Be sure to check with

APPEARANCE STANDARDS

Rich's appearance standards reflect the good taste we practice in all aspects of our business. Accepting a job with Rich's means accepting responsibility for maintaining your appearance to the standards outlined here.

All employees of Rich's are expected to look well groomed, an essential mark of respect for both co-workers and customers. Good grooming begins with cleanliness, including clean, fresh clothes. Hair, too, should be clean and neatly styled. Cosmetics and fragrance should be tastefully applied for the most natural effect. Rich's I.D. name tags must be worn at lapel-level. Headgear is not permitted, unless part of a Rich's uniform.

Your clothes should reflect Rich's fashion image. Extreme or exotic looks are not appropriate in business. Your clothing should serve as a complement to the merchandise we sell. Also inappropriate are styles designed for sports or casual wear, loungewear or evening clothes. Most important, learn what looks good on you, realistically. Many styles offer a businesslike yet individual look. Choose carefully to look your best.

WOMEN-YES

Run-Free Hosiery
Foundations; Bra, Slip, etc.
Tailored, smooth-fitting pants
Skirts no shorter than 2'' above the knee; no longer than mid-calf
Dresses
*Blouses-cap sleeve minimum
Sweaters, Jackets, Vests
Tasteful accessories, cosmetics, hairstyles

MEN-YES

*Suit
Vest
*Conservative sport-coat
Dress shirt, long- or short-sleeve
Tie
Well-maintained shoes
Tasteful accessories, toiletries, hairstyles
Dress slacks
Socks

WOMEN-NO

Blue-jeans; denim or chambray in any form
No jean-cut pants in any fabric
Provocative looks: bare, brief or clinging
Headgear, including scarves
Apparel or accessories which violate safety standards, i.e., catch-on fixtures, etc.
Bedroom slippers, beach sandals, athletic shoes
Pants tucked into boots
Socks of any type (knee, anklet, or other)

MEN-NO

Blue-jeans
Denim in any form
Sandals or athletic shoes
Headgear
T-shirts, Turtlenecks

*The General Manager of each location may temporarily suspend these standards to permit male employees in select areas to remove their coats (but not to loosen their ties) and women to wear sleeveless tops (but not backless or low-cut).

Figure 3-2 Employees of an apparel and accessories store, especially the sales staff, contribute to the store's image. That is why many stores establish a dress code. Courtesy of Rich's.

the personnel department about dress standards before you start a new job because the standards vary from store to store.

Honesty Sometimes new employees are offended that their employer talks to them about honesty. But, after they work at the store for a while, they see the many opportunities for a dishonest person to steal from the store. For example, employees who fail to clock in and out on their time cards when they go to lunch are overpaid for the hours they actually worked.

An employee who has been convicted of theft may have a difficult time getting good jobs in the future, especially in the retail field.

Trustworthiness Clothing and accessory employees work with confidential, or private, information. Such information should not be discussed with anyone outside the store. It should only be discussed with other authorized store employees, such as your manager.

Trustworthy employees are exposed to the following types of confidential information: (1) credit status of customers, (2) wages and commissions earned by other employees, (3) markup and markdown percentages, (4) future store advertising or fashion promotions, and (5) store operation costs and budgets. As in all relationships where confidential communications exist, you are trusted to know the difference between public and private information. Some stores view telling confidential information to unauthorized persons as reason for dismissal, and it could also result in a lawsuit against the employer, the employee, or both.

Experience 3-1 ■ Taking Advantage

You work with three other part-time employees in the sportswear department of a medium-sized department other.

For the past three weeks, Chris, one of the other part-time employees, has asked you to work all his nights, and you agreed to the schedule change.

You are scheduled to work next Friday. But you want to attend a concert that night and ask Chris to work your shift. Chris says, "Me? I may have a date Friday night. Why don't you ask Marty or Les to work for you?"

What are you going to do?

On a separate sheet of paper, answer the following about the situation described.

1. Identify the human relations problem.
2. What are some important facts to be considered when evaluating this problem?
3. List several possible solutions to this problem.
4. Evaluate the results of each solution.
5. Which solution do you recommend and why?
6. How could this situation have been avoided by using better human relations skills?

■ COMMUNICATIONS

Communications deals with the way a message is transmitted to a person and the way the message is received. How we communicate affects how a person feels about us and responds to us. A small communication mix-up can cause an employee to quit, a customer to shop at another store, or a personal relationship to end.

The communications network allows a store's managers and employees to work together to achieve the goals of the store—profit through customer satisfaction. A store's communication network consists of many forms, reports, instructions, and discussions among management, employees, and customers. Some of the communications, such as the presentation of merchandise to customers, require good speaking and listening skills. Others, such as reading merchandise labels or writing reports, require reading and writing skills. These skills are important to a successful career in apparel and accessories marketing.

The Communication Model

All communications include five common elements:

1. A **sender,** the person who sends the message
2. A **message,** what is said and how it is said
3. A **medium,** or how the message travels to the receiver
4. A **receiver,** the person who receives the message
5. **Feedback,** the signals sent back to the original sender by the receiver

Let's examine these elements in a typical customer-salesperson communication. Suppose a salesperson is helping a customer choose a pair of slacks. The customer, who has tried on the slacks, is standing before a three-way mirror. The salesperson is checking the fit.

Customer: Karl, I don't feel comfortable with the waist so tight. Will you call the tailor?

Salesperson: Sure, Mr. Berkly. It will only take a minute to call him.

In this example, the sender is the customer and the receiver is the salesperson. The message is, "I don't feel comfortable with the waist so tight. Will you call the tailor?" If the customer said this message in an angry tone, the anger would also become part of the message. The medium is both vocal and visual; that is, the salesperson can both see and hear the customer. The feedback is the salesperson's reply to the customer "Sure, Mr. Berkly. It will only take a minute." This feedback helps the original sender, the customer, know that the message was received and correctly interpreted by the original receiver, the salesperson.

Communication Mix-Ups

Even in the most skillfully constructed communications, it is easy for misinterpretations to develop. Some usual reasons for communication mix-ups follow.

The receiver is not paying attention:

Salesperson: You said you wanted a size 14. Here they are.

Customer: I said I need a size 10.

The sender is not using words the receiver understands:

Buyer: Check that suit. What's the season?

Salesperson: It's wool...it must be for winter.

Buyer: I meant what was the season code on the price ticket. I was referring to the season we received the suit at the store.

The sender may say one thing, but the tone of voice and facial expression say another:

Manager: Linda, we need someone to work on Friday night. Would you be interested?

Salesperson: Uh...sure, Mrs. Naylor...I guess ...I mean well, maybe I can.

Effective Communications

In clothing stores, communications are often oral. At times, you will be the sender. When someone is speaking to you, you will be the receiver. In both situations, you will share responsibility in making communications effective and understood. When you are speaking, you want the receiver to understand what you mean, and it is your responsibility to use words the receiver understands. For example, when you talk with your buyer, you may use words like "open-to-buy" or "turnover." Buyers are familiar with these words. However, new salespeople may not understand these terms. Instead of "open-to-buy," you should say "the amount of money the buyer can use to purchase new merchandise." You are also responsible for speaking clearly in logical sentences. To be heard, you should face your listener.

To help your receiver understand what you mean, send a **consistent message.** This means that what you say should be the same as what you do. An example of a consistent message is saying, "I'm going to change the window display right now" and then changing the display. An example of an inconsistent message is saying, "I'm going to change the window display right now" and then going to lunch. In an **inconsistent message**—when you send one message verbally and then send a different message by your actions—the receiver becomes confused.

When you're the sender, try to get feedback from the receiver. This is a check to be sure that your message is being received the way you want it to be. You may say, "What do you think about that idea?" The person's reply, or feedback, lets you know if you both are talking about the same thing.

As a receiver, pay close attention to what a sender is saying. When you listen with "only one ear" or are thinking about something else, it is easy for you to miss the speaker's message. Some people are able to speak more clearly and more logically than others. If you are unable to understand a message, it is your responsibility to ask questions to clarify the subject. This will give the speaker feedback and will help the speaker clear up any misunderstandings.

Whether you are a sender or receiver, try to keep eye contact with the other person. When you keep your eyes focused on the other person, you show that you are paying attention and are interested in what is being said.

Experience 3-2 ■ What Did You Say?

Margie Hamilton is a salesperson at Creation's Best, a specialty clothing store. Although she has to work until 1:00 p.m., Margie has the afternoon off to move into a new apartment. Margie has been sizing clothing all morning, but her mind has been on her moving.

As Margie is resizing a group of slacks, her manager, Mrs. Wiley, comes up behind her and says, "I just received an important call from Mr. Henry, our buyer. He wants to send all the leather and suede skirts to our South Town store. While I go to lunch, will you fill out a transfer form and get these garments ready for transfer? The delivery truck will make a special stop at 2 p.m. to pick up the merchandise."

By the time Margie realizes her manager was talking to her, Mrs. Wiley has gone. "If it's important, Mrs. Wiley will mention it to me again. I can't wait to get home to start packing!"

Name ______________________________ Date ______________________

1. Identify the sender, receiver, message, medium, and feedback in this situation.

____

____

____

____

2. What could Mrs. Wiley and Margie have done to make this communication more effective?

____

____

____

____

Speaking

In a clothing store, the spoken word is an important "tool of the trade." Let's examine some situations in which the spoken word is important.

Presenting Merchandise Many factors influence customers' buying decisions. One of these is the courtesy the salesperson shows the customer. Another is the salesperson's enthusiasm for the merchandise. Other reasons include the salesperson's tone of voice, vocabulary, and grammar.

Poor grammar lowers the image of store personnel, the store, and the value of the merchandise. Customers are not impressed with a salesperson's slang, swearing, or improper grammar. When customers find a salesperson who does not speak correctly, they may doubt the salesperson's knowledgeability. Salespeople should remember that the way a person speaks is one of the factors that makes a first impression.

Handling Customer Inquiries Customers often ask salespeople for information. For example, they may want to know how late the store stays open. They may want directions to the mall entrance or another shop. Salespeople should be just as courteous to customers when handling inquiries as they are when making a sales presentation.

When answering customer questions, it is important to understand what the customer is asking. Try to answer the customer in the simplest way possible, and make your answers clear and concise. Sometimes, you will meet people who are hard of hearing or who are unfamiliar with English. In these special situations, you need to speak clearly and distinctly. At all times, avoid the use of slang.

Interpreting Store Policies Stores establish policies and procedures as guidelines to help employees respond in certain situations. For example, a store's policy might state that merchandise can be held in layaway for a maximum of 30 days. Many times, customers are confused about store policies and ask salespeople to explain the policies to them. They often will question why such a policy has been established. Customer problems occur when salespeople do not effectively communicate store policies.

Communicating Over the Phone Many customers call a store to find out whether the store sells a certain brand of clothing. Salespeople at branch stores telephone to see if a particular size or color of merchandise is in stock at the main store. Salespeople, especially those who work in catalog sales, use the telephone to sell customers new merchandise. For these reasons and many others, a salesperson should be able to communicate clearly and distinctly on the telephone.

When using the telephone, you are relying on your voice to communicate the message. Because the listener cannot see you, you need to speak clearly.

Writing

Most clothing and accessory employees must complete a number of forms. Some of these forms, such as charge slips, return slips, and markdown sheets, are for accounting purposes. Others are used to keep track of inventory or reorder merchandise items. Still others, such as layaway forms, are for the customer's benefit.

These forms are useless unless the written information is both correct and legible. Written mistakes can cause many problems. For example, an incorrect charge account number can delay a store in receiving payment for merchandise sold. The wrong street address or telephone number will prevent a customer from receiving the purchase when expected.

When completing forms, follow these two important guidelines:

Make Letters and Numbers Legible Fancy handwriting is hard to read. Remember to write clearly so that others can read the information. If your handwriting is poor, print.

Double-Check Written Information Reread what you have written to make sure it is accurate and understandable. Look for misspelled words and transposed figures (15 instead of 51). Give customers an opportunity to read sales slips to verify that the information is correct. Many forms have a special place for supervisors to sign after double-checking employees' work.

Reading

Even though salespeople are often trained by oral instructions from supervisors and co-workers, they also must read various materials. Many companies have pamphlets that explain fringe benefits, personnel policies, and employee regulations to newly hired workers. Stores also issue written materials about new store policies and procedures. Employees read fashion magazines and newspapers to be knowledgeable about the current merchandise styles that are being featured.

Sometimes employees may need only to skim written material, that is, read only the important facts. Often they may skim fashion magazines and newspapers to find out about new trends. Sometimes a person can **scan** the material, that is, read only specific information. Most employees scan sales charts or stock records because they may be interested in only a few items on the chart. Many times, however, an employee must read and understand nearly all of the written material. For example, a store policy on returns must be completely understood so as to be accurately explained to others. Although intensive reading takes more time than skimming and scanning, a careful reading of materials once can save rereading time in the future.

When you must remember information accurately, making notes on the important points is helpful. Always look up unfamiliar words in a dictionary.

Figure 3-3 Tags and labels on clothing provide important information that salespeople can explain to their customers. Courtesy of Mansanto Company.

Reading Merchandise Labels Some of the most important reading you will do is attached to the merchandise you sell, such as merchandise labels or hangtags. At some point, be sure to read the labels on new merchandise. If time does not permit you to read the tags while checking the merchandise, be sure to read them at a slower time—perhaps while you are straightening merchandise items. These tags often contain important selling points. (See Figure 3-3.) Use this information to describe the features and benefits of the merchandise to the customer or to answer customer questions.

Reading Mail Orders Some stores include merchandise order forms in their monthly credit statements or in newspaper ads. (See Figure 3-4.) Completed mail orders are distributed to particular departments and are filled by salespeople.

When filling these orders, you will have to read the customer's name, address, and merchandise selection carefully. You will have to locate the correct merchandise item and ring up the sale. Pay attention to how the customer wants to pay for the merchandise.

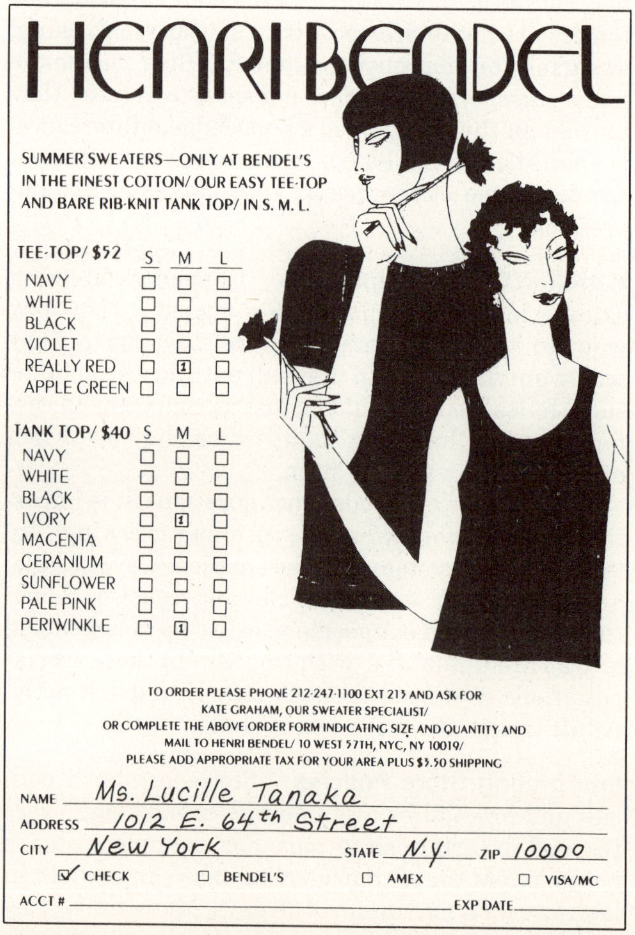

Figure 3-4 What information on a completed customer order form should you read with special attention? Courtesy of Henri Bendel; illustration by Albert Elia.

Name ______________________________ Date ______________________

Experience 3-3 ■ Sure You Read It, But Did You Read It?

Suppose your store is establishing a new policy on merchandise returns. This policy is shown to the right. Read the policy very carefully. After you are sure you understand the policy, answer the questions below.

1. What type of credit will customers who are returning merchandise without proof of purchase receive?

2. What type of credit will customers who are returning gifts receive? Assume the customers do not have a sales slip.

3. Are all sales really final? Explain your answer.

4. A customer, Mr. Malloy, purchased merchandise at full price, $24. The merchandise item was the wrong size. During the next week, when he returned the merchandise, which had not been worn and was in good condition, he found the merchandise was on sale for $18.99. The customer has his sales slip. How much will the customer receive back? What kind of credit will he receive?

5. Suppose it is October 7. A customer enters the store and wants to return a swimsuit cover-up that was purchased at full price in April. The customer has the sales ticket. How will you handle the return?

POLICY AND PROCEDURES

With few exceptions, a customer returning merchandise with which he or she is dissatisfied feels he or she has a justifiable complaint. The customer expects to encounter difficulty, so it is particularly important that he or she be put at ease and handled as courteously as possible. It is also extremely important that the matter be resolved quickly. The customer should not be passed from one person to another. With each delay or referral, the customer will become more upset. To the extent possible, adjustments should be handled promptly.

The following guidelines are set down in an effort to establish, as far as possible, a uniform adjustment policy at our store:

General: Merchandise that is current and salable is to be exchanged cheerfully, or, if necessary, a charge credit, merchandise credit, or cash credit is to be written.

Customers returning merchandise *without* proof of purchase will be given a charge credit or merchandise credit, but not a cash credit. The amount of the credit should be determined from the last retail price marked by the store.

Customers returning merchandise received as a gift will be given a charge credit or merchandise credit, but not a cash credit.

Sale Merchandise: Our policy regarding sale merchandise is that all sales are final; however, if, in your judgment, it will not harm the department and will accommodate the customer, an exchange of sale merchandise can be permitted.

Merchandise purchased recently but put on sale after the customer's purchase is to be redeemed at the customer's purchase price.

Fashion Merchandise: Merchandise that is completely out of season and not salable within the department is to be refused any allowance.

■ MANAGEMENT RESPONSIBILITIES

One of the most important jobs of clothing and accessory supervisors, department heads, or store managers is to see that employees work together effectively. If a manager's employees are not productive and efficient, then the manager is not successful, no matter how good a worker or technically proficient that manager is.

As an assistant department manager, merchandise manager, or store manager, you must organize your work and delegate it to the people who work for you. By training and encouraging your employees and monitoring their progress, you should achieve results. These results will be greater and more beneficial than if you did all the work yourself. If you find the productivity of your employees is disrupted because they do not work well together, you need to concentrate on improving the human relations skills of your staff and your management communications.

The Manager's Role

Managers are responsible for the work and actions of the people they supervise. They help new employees adjust to their jobs and see that they are acquainted with their fellow workers. Effective managers set good examples for their employees. In setting a good example, you become a role model for your employees. For instance, if you have a positive attitude toward work, enjoy doing a good job, and try to get along with others, so will your employees. If you display good work habits and ethical character, your staff will model their work behavior after yours. Your positive behavior may help your employees to change their actions and characteristics in a positive way.

Suppose, for example, that you worked for a person who was often late or who dressed sloppily. If this supervisor ever commented on your punctuality or dress, how would you react? You would probably think to yourself, "Why should I listen? I don't want to change my personality or behavior to become more like my supervisor!"

Managers must accept responsibility for influencing the attitudes of employees who work for them. Would you try to work hard and excel at your job if your manager did just enough to get by? Poor employee morale can often be caused by a manager who just doesn't care. Good employee morale can result from a manager who sets an example of personal work habits and character that people can admire and imitate.

As a manager, you are a vital link in the store's communication process. It is your responsibility to give instructions and training to employees who work for you. You must make sure these employees understand the store's policies, procedures, and plans. You will also have to prepare reports on the progress of your department or store. You will need to keep stockholders or other management persons informed of aspects of your business. As a manager, you must make your communications a two-way process, whether they are upward (to your boss), downward (to your employees), or lateral (to your peers).

Managers come in contact with much more confidential information than salespeople. They know about their salespeople's past work history and salary. They know the amount of markup their store charges and if it makes a profit. Managers should never violate the trust a store places in them. They should never talk about confidential information to other unauthorized employees, such as a salesperson, or to anyone outside the store. They should set a good example for employees and show that confidential information really is private.

Experience 3-4 ■ What's the Problem?

Lily's Dress Salon recently added two new departments: accessories and shoes. To fill one of the new positions, Lily promoted her nephew, David, from receiving and stock clerk to manager of the Shoe Department. Although he has never worked on the sales floor before, David is responsible for training and supervising four sales-people.

Since David has become manager, he takes an hour for lunch rather than the usual half hour. When new shipments of shoes arrive, he tells his workers to put away the stock. He also gets the employees to prepare his merchandise and sales reports for his aunt. David spends lots of time on the phone with his friends. Many of his friends also drop in the store just to chat with him.

In the three months that David has been department manager, the store has hired nine salespersons for his department. The average time the employees stay at the store before quitting is 4½ weeks. The sales for the department are 35 percent below plan.

On a separate sheet of paper, answer the following:

1. Identify the real problem.
2. What are the important facts to consider when solving this problem?
3. What are possible solutions to this problem?
4. Evaluate the results of each one of your possible solutions.
5. What is the best solution? Give reasons to support your choice.

Motivation

Because managers are responsible for the work of others, they spend much of their time motivating people to do a job and do it well. **Motivation** is the reason that each of us acts the way we do. Motivated workers have a goal they want to achieve. Their motivations, or reasons why they want to accomplish a goal, influence their behavior. Some of the ways clothing and accessory managers motivate their employees are:

Praise When a manager praises an employee for doing a good job, the manager increases the chances of that employee doing a good job in the future. To be effective, praise, or positive reinforcement, should be specific, timely, and based on facts.

Negative Reinforcement Telling employees gently what they did wrong is another way to motivate employees to do a better job in the future. Like positive reinforcement, or praise, negative reinforcement is most effective when it is specific, timely, and based on facts. Negative reinforcement should also let employees know what it is you want them to do. A manager should never correct an employee, or give negative reinforcement, in front of others.

Public Recognition Public recognition can be a form of motivation, for example, when the manager tells co-workers and other people who know the employee what a good job has been done. The store manager may sponsor an awards banquet and present outstanding workers with a plaque or trophy.

Competition Clothing stores sponsor sales contests to increase employee productivity. Prizes are usually money, trips, or merchandise. Many clothing stores publish weekly sales reports that list the sales per hour of each salesperson. People on the top of the list get a feeling of satisfaction, and sometimes the list spurs other employees to do a better job.

Money Outstanding employees may receive bonuses for doing a good job. Many salespeople earn a commission and can increase their pay by selling more. Some stores give the salesperson who received the most customer service commendations during the month a $100 bonus.

Promotion Companies usually promote the most productive and effective salespeople into positions of increased responsibility, such as department supervisor or sales manager. Managers can be promoted to buying positions or supervisor of larger areas or stores. Promotions give workers increased salaries and additional responsibility and authority.

Other Factors Setting a goal, telling yourself you can do it, and then satisfying yourself by meeting your goal is another motivational tool. Some employees get satisfaction from increased job responsibility or more decision-making ability on the job. Increased responsibility means that management recognizes their ability. This feeling of being in charge makes them feel important and needed. Sometimes physical surroundings, such as working in a plush, more expensive selling area, motivates workers.

Everybody is motivated by different things. What is important to one person is not important to another. As a manager, you need to realize that some motivators have more effect on some workers than on others. To achieve the most productive results from your workers, you must determine what incentives strongly motivate these employees.

You may wonder how you are going to find out what motivates your workers. When asked, most people can't say what really makes them work hard. People's motivations can only be estimated by observing and analyzing their behavior. For example, if a salesperson's sales increase by 25 percent every time there is a sales contest, it may be assumed that the employee is motivated by competition. During contests, other employees may be so "turned off" to the increased competition that their sales decrease. For these employees, contests are a negative motivation.

The only way managers can understand how to motivate others for best performance is to show an interest in the employees. By practicing effective human relations skills, managers should be able to get close to employees and find out "what really makes them tick."

Experience 3-5 ■ Motivate!

Suppose you are the owner of a clothing store that employs all the students in your class as salespeople. On a separate sheet of paper, describe in detail three ways you would motivate these salespeople to be productive and earn a profit for your store. Be specific. Describe the way each motivator would be set up.

Developing a Pleasant Work Environment

Some factors that make up the environment of a clothing and accessory store are the building, the fixtures, the merchandise, and the other employees. Other factors are the manager's leadership style and employee morale.

Store Environment The store environment affects the way employees feel about their job. When employees feel comfortable in their environment, they will work to meet the goals of the store. The environment is an important part of the store image conveyed to employees as well as to shoppers. If employees are dissatisfied with the work environment, they may spend their time complaining rather than working.

The Manager's Leadership Style How a manager plans the work, gives orders, and makes decisions also affect the working environment. For example, a manager who plans in advance eliminates the pressure that builds up when everything is left to the last minute. Managers who assign tasks by "ordering" rather than "asking" are a negative influence on their workers. No one likes being told what to do; people like to think that they make their own decisions.

How a manager makes decisions influences the work environment. In a democratic leadership style, managers ask employees for their opinion before making a decision, which makes employees feel needed and a part of things. Managers who make all the decisions themselves, without consulting their employees, are called autocratic managers. These managers create an environment that limits employee creativity and initiative.

Employee Morale Employee morale is one of the most important factors in a pleasant work environment. When employees have a bad attitude about their job, their co-workers, and their supervisors, that attitude spreads to others. As a manager, you should watch for signs of poor employee morale: absenteeism, gossip, and tardiness. High turnover rate of employees is another sign of very poor employee morale. Look for employee morale problems and correct them before they spread.

Training

As a manager, you are responsible for the work of other employees. You train employees to do various tasks, assign work duties, and follow up to see that the work is done properly. These tasks require you to be a good communicator.

To help in training, many clothing and accessory managers use training manuals that explain how to complete various tasks. Instructions in the manual are written in a logical, step-by-step way. Managers also use follow-up sheets in order to be certain that various points have been covered during training. (See Figure 3-5.)

DEPARTMENT SPONSOR MUST COMPLETE CHECKLIST AND RETURN TO THE STORE MANAGER WITHIN 3 DAYS FROM DATE OF ISSUE.

- ☐ Show where to put personal belongings.
- ☐ Show where nearest restrooms and water fountains are.
- ☐ Introduce to department personnel and Floor Supervisor.
- ☐ Show Department Manager's office location.
- ☐ Explain lunch time and location of weekly schedule.
- ☐ Give a tour of your department(s) to show stock by type, brands, sizes, reserve stock location.
- ☐ Ask Buyer or Sales Manager to assign specific stock responsibility.
- ☐ Explain PRICE tickets in detail and, if you STUB, go over this thoroughly.
- ☐ Show where supplies are kept and how they may be ordered.
- ☐ Show location of department bulletin board.
- ☐ Get selling supplies needed.
- ☐ Explain delivery procedure.
- ☐ Show how to handle gift wraps.
- ☐ Tell how to answer department phones, take messages and transfer calls.
- ☐ Review how to call Security.
- ☐ Explain alteration procedures if applicable.
- ☐ Show location of fitting rooms.
- ☐ Explain telephone and mail order procedures.
- ☐ Show how to handle co-worker package.
- ☐ Take person to location for depositing register media, keys, etc.
- ☐ Go over dress standards, grooming, selling, floor conduct, etc.
- ☐ Review co-worker's part in accepting bank checks.
- ☐ Show location of numbers to call for charge authorization.
- ☐ Show list of wanted charge plates and list of checks which are not to be accepted.

Figure 3-5 This example of a checklist for training new employees lists the many activities involved.

Policy manuals that give guidelines for employees are also part of the training program. These manuals include guidelines which state how a store employee is to respond in certain situations. For example, a policy manual may state how a salesperson should handle a customer's request to hold merchandise (that is, take merchandise off the selling floor for a day so the customer can decide whether to buy it).

You may develop new training or policy manuals. Or you may be asked to revise outdated manuals. To write these materials, you will need to plan carefully what should be included in the manuals and decide how to organize it. You will need to write the information in an interesting and logical way in order to hold the reader's attention.

Conducting Meetings

As a salesperson, you will attend many meetings. As a manager, you will attend even more meetings. Often, you will be responsible for conducting the meetings. Your organization and speaking skills will be evident to all who attend your meetings. Many managers rely on a written agenda to keep the meeting progressing. Usually, the most important matters are put on the top of the agenda, to make sure they are discussed before time runs out. Less important items are covered later in the meeting. The manager is responsible for keeping the discussion focused on the items listed on the agenda.

Often, visual aids help managers clarify a particular point. Graphs, charts, overhead transparen cies, or handouts help your audience understand what you are trying to say and make listening to you more interesting.

When conducting department meetings, not all your time should be spent talking. You should also set aside some time to listen. Your employees will have questions. By encouraging employees to ask questions or discuss their concerns, you will promote a feeling of friendship and cooperation among your staff. Also encourage employees to discuss their concerns.

Experience 3-6 ■ Training Meeting

Suppose you are the manager of a medium-sized junior specialty chain. Choose a particular task that new employees would need to perform to be successful at your store. (For example, coordinating new fashions, accepting customer checks, or arranging new stock on the selling floor.) Prepare a 5- to 10-minute training session for the new employees. Design your presentation so that you can rate a "yes" to the questions 1 to 7 on the evaluation (to the right). You may use any visual aids or handouts you design.

When your presentation is ready, train two to five of your classmates in the task you have chosen. After the presentation, ask your classmates to complete the training evaluation.

(*Note:* If you need to become more familiar with the training topic you have chosen, visit two or three junior specialty stores. Explain to the manager that you are studying apparel and accessory marketing in school. Interview the manager about the topic.)

Training Evaluation

1. Did the instructor make you feel comfortable? **yes no**
2. Did the instructor begin the session by explaining what you will be learning? **yes no**
3. Did the instructor clearly explain the information? **yes no**
4. Did the instructor present the information in a logical sequence? **yes no**
5. Did the instructor ask for questions? **yes no**
6. Did the instructor satisfactorily answer any questions? **yes no**
7. Based on the information presented, could you perform the task discussed? **yes no**
8. On a separate sheet of paper, briefly explain how the training session could be improved.

Employee Counseling

After a manager gives employees their initial training, the manager must follow up on the work of the employees to be sure they are able to perform their job duties as required. It's the manager's responsibility to give employees feedback regarding their work—either to praise them for doing a good job or to correct them when they are not meeting the job standards. Sometimes an employee's poor performance does not improve because the employee does not want to improve or doesn't have the skills required to do the job. In these situations, the manager should conduct an employee counseling session.

Employee counseling is the first formal step in the disciplinary process. Employees whose performance does not improve to meet the job standard may eventually be fired. Because employee counseling is very serious and can be the first step leading to dismissal, all counseling interviews should be documented, that is, a written record should be made of what was said by both manager and employee.

When counseling employees for poor performance, remember these steps:

Privacy Discuss the situation with the involved parties in private so that the other employees or customers are not listening and will not interrupt. Your job is not to embarrass an employee; it is to solve a problem.

Clarity Clearly explain to employees why their actions are causing a problem in the store or department. Be sure the employees realize that what they do affects the total operations of the store or department.

Fairness Ask the employees if the facts you've presented are true. Ask for their side of the story. Carefully listen to them and reevaluate if you are being fair.

Shared Responsibility Ask the employees how you can help them solve the problem. Let the employees see that you want to help them improve.

When handling employee corrections or complaints, you have to be sensitive to employees' feelings. You shouldn't attack a person's intelligence, attitude or character. Instead, you should talk about the employee's behavior and how it can be changed.

Chapter Note

1. Leonard Sloane, "Stores Offer Staff Incentive as Sales Commissions Return," *The New York Times,* August 6, 1988, p. 16.

CHAPTER 4

Mathematics

In November 1986, *Stores* magazine reported the results of a 1986 survey on point-of-sale (POS) systems and electronic cash registers (ECRs). Seventy-two percent of the companies that responded said they have POS systems at all their locations. They said the greatest use of the POS was stockkeeping unit (SKU) data capture.[1]

A **stockkeeping unit** is an identification or code number that is assigned to a single item of merchandise or a group of similar items for the purpose of keeping separate sales and stock records. The SKU number is printed on a merchandise price ticket.

When a customer buys, returns, or exchanges merchandise, the salesperson enters the SKU data into the cash register either with a wand or by hand. This data is then transmitted to a central computer that collects and summarizes what was entered. Reports are then generated that show exactly what merchandise was sold and returned and when each transaction was made. The reports are used by the buyers who purchase merchandise and the managers who run the stores. Because of POS systems, they receive faster and more accurate information. They are better able to determine what merchandise is selling and should be reordered and what merchandise should be marked down and moved out of the store. Of course, the reports are only as good as the information entered into the electronic cash register. That's why the salesperson must be accurate when working with the register.

WORKING WITH THE CASH REGISTER

As a clothing and accessory salesperson, you may use various cash registers that record a wide variety of business transactions. Some models may be similar to an adding machine; they are unable to send the information entered to a computer. The majority of the cash registers will be electronic. They are part of a POS system similar to the one described above. (See Figure 4-1.)

Opening the Register

At the beginning of each day, the cashier, salesperson, or manager must prepare the cash register to handle the day's business. One of the duties of the person in charge of "opening the register" is to put money in the cash drawer. This money is to make

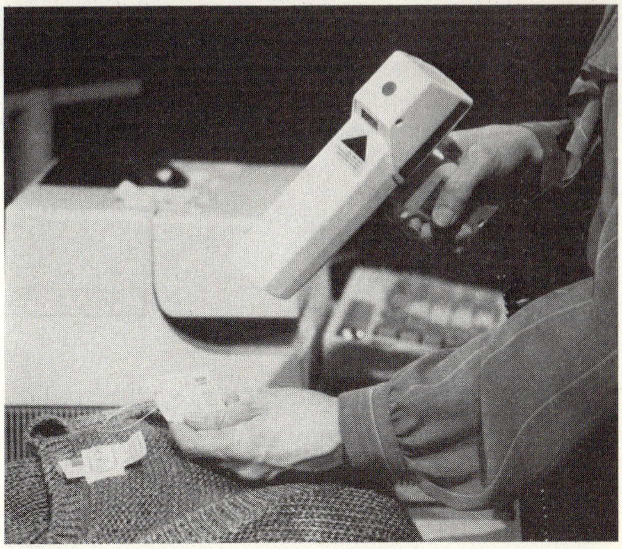

Figure 4-1 Electronic cash register systems have been designed especially for apparel and accessory stores. The register is actually a terminal connected to a computer network. © IBM.

change for the day's customers and is called the change fund or opening bank.

Many apparel and accessory stores start the day with a $50 bank. The actual breakdown of the bank into bills and coins of different denominations varies depending on the change that was in the register at the end of the previous day. For example, the men's furnishings department may have the following breakdown in its $50 bank: one $10, four $5, twelve $1, sixteen quarters, thirty dimes, fifteen nickels, and twenty-five pennies. This money is arranged in the cash drawer in numerical order with pennies and dollar bills on the right. Care is taken to keep all the pennies together, all the nickels together, and so on. (See Figure 4-2.)

The employee in charge of opening the register must get the bank from the store's credit office or manager before the start of business. This person must count the opening bank to be sure the exact amount that is supposed to be in the bank is really there. If the amounts do not agree, the employee immediately reports the disagreement to the manager or the person in charge. Some apparel and accessory stores require that the person who opens a register complete a special form when the opening bank is over or short. (See Figure 4-3.)

Ringing up Sales

After a customer has decided to buy merchandise, the price of the item or items is entered into the cash register. The procedure used in ringing up sales will vary depending on the size of the store and the type of cash register used. Let's examine the cash register procedures at a fictional children's wear specialty chain called Jasper's. Although this chain has six stores, it does not use electronic cash registers. In fact, Jasper's salespeople write up sales checks for each customer purchase. When a sales check is completed, the salesperson gives this check to a cashier who rings up the total price on the cash register. (See Figure 4-4.)

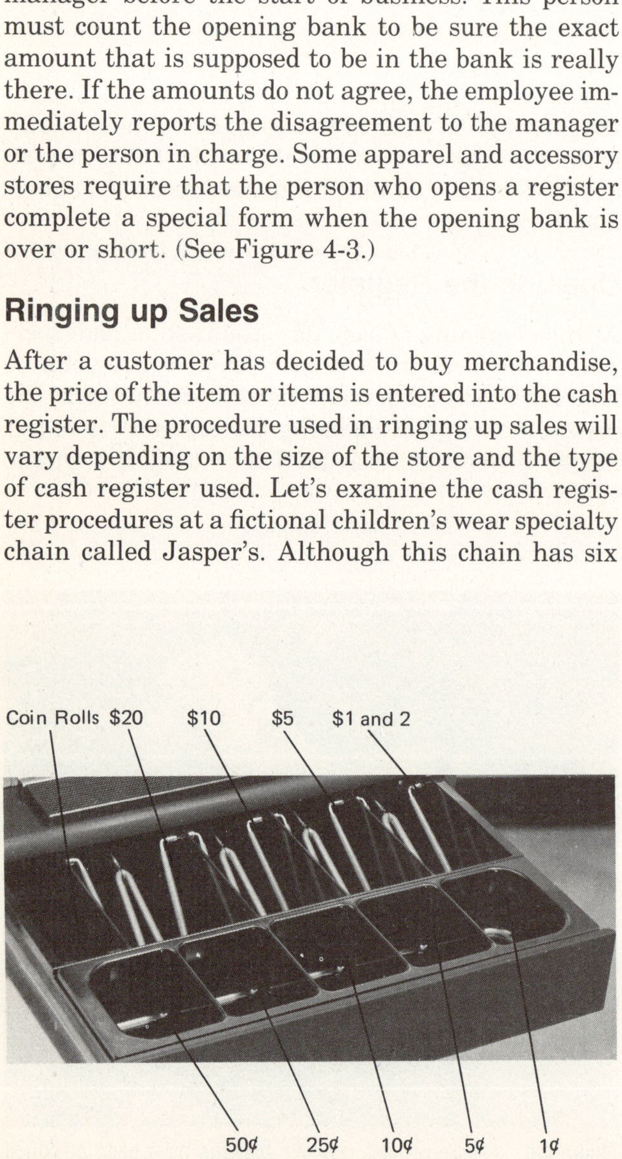

Figure 4-2 Notice where places are provided in this cash drawer for bills and change of various denominations.

Lee Muller, manager of a Jasper's outlet, explains, "For a cash sale, a salesperson should fill in the date, store number, his or her initials, merchandise quantity and description, the individual price of the merchandise items, and the amount. Accuracy is very important when completing the sales check. The prices on the sales check should correspond with the price tags on the merchandise. Numbers should be written clearly and in their correct columns. When customers purchase multiple quantities of merchandise, such as several pairs of socks, the salesperson should carefully count all items to be sure each has been included on the sales check.

At the bottom of the sales check, the salesperson subtotals the amount column, adds tax, and computes the total amount due. Each salesperson is familiar with state tax charts, which are taped to the cashier counter. To use these charts, the salesperson locates the subtotal at the left of each column and reads across to find the corresponding tax amount. For example, on a purchase of $34.75, the tax is $2.09. (See Figure 4-5.)

"In our state the tax is 6 percent on all apparel and accessory items. A salesperson could also compute the correct tax amount by multiplying the subtotal amount by 0.06 (6 percent). We try to discourage salespeople from doing their own tax computations. They are more accurate when they use the tax chart.

"All of our sales checks are written in duplicate. One copy is given to the customer as a receipt. The other copy stays at the store and is later used to compute the daily sales figures for each department."

Another system of recording sales is used at Weaver's Department Store. Each salesperson is responsible for ringing up his or her own sales on the cash register. The cash register prints a customer receipt that shows the price and total of the merchandise purchased. Manager Mike Garcia explains his store's system, "The salesperson helps the customer choose merchandise. The items purchased are entered into a register. Our registers require the salesperson to enter his or her salesperson number, the department and class numbers of each merchandise item, the merchandise manufacturer and style number, and the price. All this merchandise information is printed on the price ticket. The salesperson then depresses the subtotal key and the register automatically prints the correct tax on the receipt. Finally, the salesperson enters how the customer is paying for the merchandise.

OVER/SHORT REGISTER CHANGE FUND

Report of discrepancy in funds at beginning of day — Date 7/11/--

Change fund was:

☑ Over $ 26¢ — Register No. 21

☐ Short $ — Salesperson #3583

Errors in amount of change left in register must be recounted and approved by authorized O.K. Enclose this form in register detail envelope.

Store No. #12 — Authorized O.K. K. Ralston, mgr.

Figure 4-3 When the change fund is over or short, it is reported on a form similar to this.

205397 — DATE Feb. 14 19 —

NAME

STREET

CITY

SOLD BY	CASH	C.O.D.	CHARGE	ON ACCT	MDSE RETD	PAID OUT	STORE
TK	✓						#3

QUAN		DESCRIPTION	PRICE	AMOUNT	
6	1	Socks	4.00	24	00
1	2	Blouse	19.99	19	99
1	3	Pant	29.99	29	99
	4				
	5				
	6				
	7				
	8				
	9				
	10				
	11				
	12				
	13				
	14				
	15				
	16				
	17		Tax	4	44
	18		Total	78	42

CUSTOMER'S ORDER NO. — REC'D BY — 78.42

KEEP THIS SLIP FOR REFERENCE

519

Figure 4-4 This is an example of a sales check that must be written up by a salesperson. Tax was computed at 6 percent.

6% TAX SCHEDULE

SALE TAX	SALE TAX	SALE TAX	SALE TAX
.11 — .01	8.92 — .54	17.75 — 1.07	26.59 — 1.60
.23 — .02	9.09 — .55	17.92 — 1.08	26.75 — 1.61
.40 — .03	9.25 — .56	18.09 — 1.09	26.92 — 1.62
.57 — .04	9.42 — .57	18.25 — 1.10	27.09 — 1.63
.74 — .05	9.59 — .58	18.42 — 1.11	27.25 — 1.64
.91 — .06	9.75 — .59	18.59 — 1.12	27.42 — 1.65
1.09 — .07	9.92 — .60	18.75 — 1.13	27.59 — 1.66
1.25 — .08	10.09 — .61	18.92 — 1.14	27.75 — 1.67
1.42 — .09	10.25 — .62	19.09 — 1.15	27.92 — 1.68
1.59 — .10	10.42 — .63	19.25 — 1.16	28.09 — 1.69
1.75 — .11	10.59 — .64	19.42 — 1.17	28.25 — 1.70
1.92 — .12	10.75 — .65	19.59 — 1.18	28.42 — 1.71
2.09 — .13	10.92 — .66	19.75 — 1.19	20.59 — 1.72
2.25 — .14	11.09 — .67	19.92 — 1.20	28.75 — 1.73
2.42 — .15	11.25 — .68	20.09 — 1.21	28.92 — 1.74
2.59 — .16	11.42 — .69	20.25 — 1.22	29.09 — 1.75
2.75 — .17	11.59 — .70	20.42 — 1.23	29.25 — 1.76
2.92 — .18	11.75 — .71	20.59 — 1.24	29.42 — 1.77
3.09 — .19	11.92 — .72	20.75 — 1.25	29.59 — 1.78
3.25 — .20	12.09 — .73	20.92 — 1.26	29.75 — 1.79
3.42 — .21	12.25 — .74	21.09 — 1.27	29.92 — 1.80
3.59 — .22	12.42 — .75	21.25 — 1.28	30.09 — 1.81
3.75 — .23	12.59 — .76	21.42 — 1.29	30.25 — 1.82
3.92 — .24	12.75 — .77	21.59 — 1.30	30.42 — 1.83
4.09 — .25	12.92 — .78	21.75 — 1.31	30.59 — 1.84
4.25 — .26	13.09 — .79	21.92 — 1.32	30.75 — 1.85
4.42 — .27	13.25 — .80	22.09 — 1.33	30.92 — 1.86
4.59 — .28	13.42 — .81	22.25 — 1.34	31.09 — 1.87
4.75 — .29	13.59 — .82	22.42 — 1.35	31.25 — 1.88
4.92 — .30	13.75 — .83	22.59 — 1.36	31.42 — 1.89
5.09 — .31	13.92 — .84	22.75 — 1.37	31.59 — 1.90
5.25 — .32	14.09 — .85	22.92 — 1.38	31.75 — 1.91
5.42 — .33	14.25 — .86	23.09 — 1.39	31.92 — 1.92
5.59 — .34	14.42 — .87	23.25 — 1.40	32.09 — 1.93
5.75 — .35	14.59 — .88	23.42 — 1.41	32.25 — 1.94
5.92 — .36	14.75 — .89	23.59 — 1.42	32.42 — 1.95
6.09 — .37	14.92 — .90	23.75 — 1.43	32.59 — 1.96
6.25 — .38	15.09 — .91	23.92 — 1.44	32.75 — 1.97
6.42 — .39	15.25 — .92	24.09 — 1.45	32.92 — 1.98
6.59 — .40	15.42 — .93	24.25 — 1.46	33.09 — 1.99
6.75 — .41	15.59 — .94	24.42 — 1.47	33.25 — 2.00
6.92 — .42	15.75 — .95	24.59 — 1.48	33.42 — 2.01
7.09 — .43	15.92 — .96	24.75 — 1.49	33.59 — 2.02
7.25 — .44	16.09 — .97	24.92 — 1.50	33.75 — 2.03
7.42 — .45	16.25 — .98	25.09 — 1.51	33.92 — 2.04
7.59 — .46	16.42 — .99	25.25 — 1.52	34.09 — 2.05
7.75 — .47	16.59 — 1.00	25.42 — 1.53	34.25 — 2.06
7.92 — .48	16.75 — 1.01	25.59 — 1.54	34.42 — 2.07
8.09 — .49	16.92 — 1.02	25.75 — 1.55	34.59 — 2.08
8.25 — .50	17.09 — 1.03	25.92 — 1.56	34.75 — 2.09
8.42 — .51	17.25 — 1.04	26.09 — 1.57	34.92 — 2.10
8.59 — .52	17.42 — 1.05	26.26 — 1.58	35.09 — 2.11
8.75 — .53	17.59 — 1.06	26.42 — 1.59	35.25 — 2.12

Figure 4-5 If a sales total does not appear on this sales tax chart, charge the tax for the closest lower total that does appear. Courtesy Magenetax, Bellevue, WA

"Our store, like most other clothing and accessory stores, honors bank credit cards, such as Visa and MasterCard. When customers present these cards for payment, the salesperson must complete a bank credit card sales slip. The salesperson must be careful that all information written on the slip is accurate and correctly added." (See Figure 4-6.)

"The salesperson must also check the card number in the "Lost or Stolen Credit Card" booklet to be sure that the card is valid and compare the customer's signature on the card with the signature written on the slip."

High Castle is a chain of discount clothing stores that uses electronic cash registers and a computerized inventory system. Store owner Dorothy Lewis explains, "Our salespeople don't have to worry about adding the sales checks. Each merchandise item is assigned a special number that is coded onto its price ticket. When the item is purchased, the salesperson runs a wand over the price ticket. The computer finds the price of that item in its memory bank and records the price on the sales check. The computer automatically totals the items purchased, adds tax, and computes the change due. This system eliminates many human errors and has saved our store money. It also provides our buyers with reports of what merchandise has been sold, returned, or is still in stock."

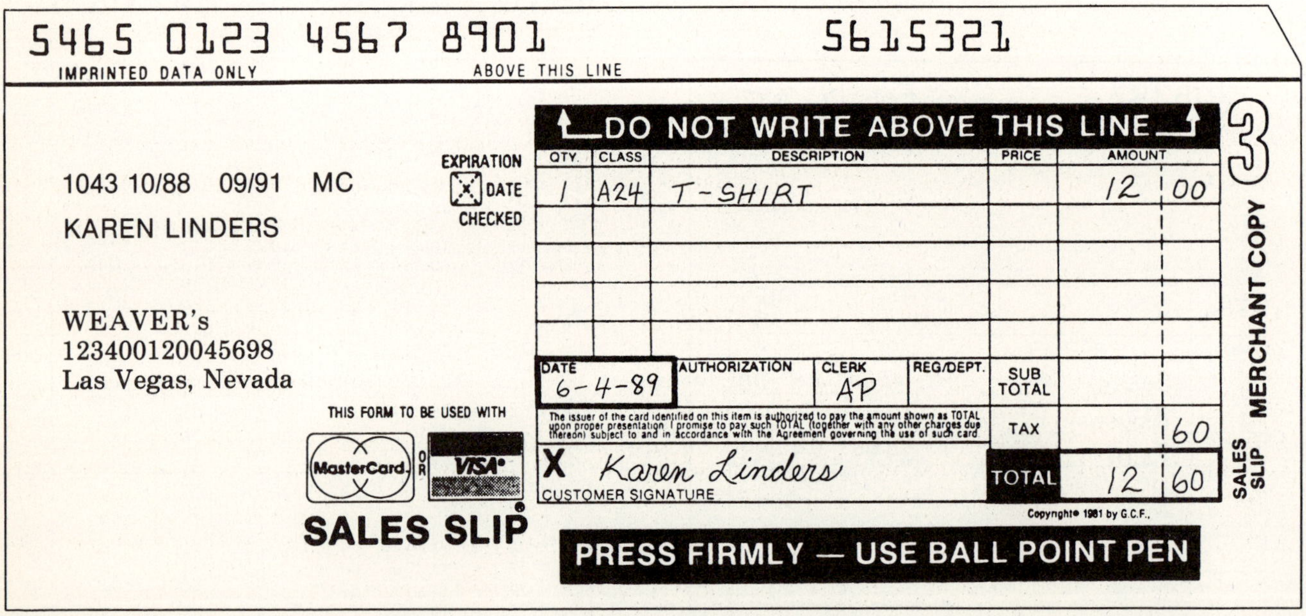
5465 0123 4567 8901 5615321
IMPRINTED DATA ONLY ABOVE THIS LINE
1043 10/88 09/91 MC
KAREN LINDERS
WEAVER's
123400120045698
Las Vegas, Nevada
EXPIRATION DATE CHECKED
DO NOT WRITE ABOVE THIS LINE

QTY.	CLASS	DESCRIPTION	PRICE	AMOUNT
1	A24	T-SHIRT		12 00

DATE 6-4-89 | AUTHORIZATION | CLERK AP | REG/DEPT. | SUB TOTAL
The issuer of the card identified on this item is authorized to pay the amount shown as TOTAL upon proper presentation. I promise to pay such TOTAL (together with any other charges due thereon) subject to and in accordance with the Agreement governing the use of such card
TAX 60
X Karen Linders
CUSTOMER SIGNATURE
TOTAL 12 60
THIS FORM TO BE USED WITH MasterCard OR VISA
SALES SLIP
Copyright© 1981 by G.C.F.
PRESS FIRMLY — USE BALL POINT PEN
3 MERCHANT COPY SALES SLIP

Figure 4-6 Here's a typical sales slip written for a bank credit card customer. Note that 5 percent sales tax is charged. © Visa/MasterCard.

Experience 4-1 ■ The Extra Cost: Tax

1. Using the tax chart on page 53, determine the tax on the following customer purchases.
   a. $12.89 ________
   b. $20.40 ________
   c. $28.15 ________
   d. $31.52 ________
   e. $33.72 ________

2. Suppose the state sales tax chart you are working with only goes up to $50. Compute the sales tax, at 6½ percent, for the following customer purchases.
   a. $59.87 ________
   b. $71.90 ________
   c. $151.49 ________
   d. $187.79 ________

Experience 4-2 ■ What's the Total?

Suppose you are a salesperson at Jasper's, store no. 2. Complete each sales check below for the two customer purchases that follow. Because both sales are cash, it is not necessary for you to write the customer's name and address on each sales check. Compute the sales tax at 6 percent.

1. The customer has purchased four items: two pair jeans at $34 each, one sweatshirt at $34.99, and one knit tank at $22.50.
2. The customer has purchased two items: a wool sweater and flannel slacks. The price tickets are shown to the right.

JASPER'S	
DEPT.	CLASS
8	10
SIZE	PRICE
8	$68.00

Sweater

JASPER'S	
DEPT.	CLASS
8	20
SIZE	PRICE
8	$58.00

Slacks

QUAN.	DESCRIPTION	PRICE	AMOUNT	

QUAN.	DESCRIPTION	PRICE	AMOUNT	

Ringing Up Returns and Exchanges

Sometimes customers find that their purchases are the wrong size, color, or style. Most clothing and accessory stores allow customers who have their receipt to bring back unsatisfactory merchandise and receive money or other merchandise in return. In a return or exchange, salespeople usually complete a credit slip or refund check. (See Figure 4-7.) The return or exchange must be entered into the cash register in order to keep the records correct.

Andy Willis works at Becker's Shoes. Recently, Mr. Kramer returned to the store with a pair of shoes he had purchased. Andy explains the situation as follows, "It seems as though the shoes, which were purchased a week ago, had a split seam on the instep and the customer wanted to exchange them. The customer decided on a pair of $44.95 shoes. He had his sales slip for the $65 shoes he wanted to return. Because the shoes he decided on were less expensive, I had to compute how much money the store owed Mr. Kramer.

"Our state has a 4¾ percent sales tax, so the total amount of the second shoe sale, with tax, came to $47.09. I figured I owed him $21 by working out this table. I realize this is a simple computation. But when I am under pressure, I get nervous and make mistakes. Since the store doesn't have calculators on the sales floor, I have to rely on my arithmetic skills."

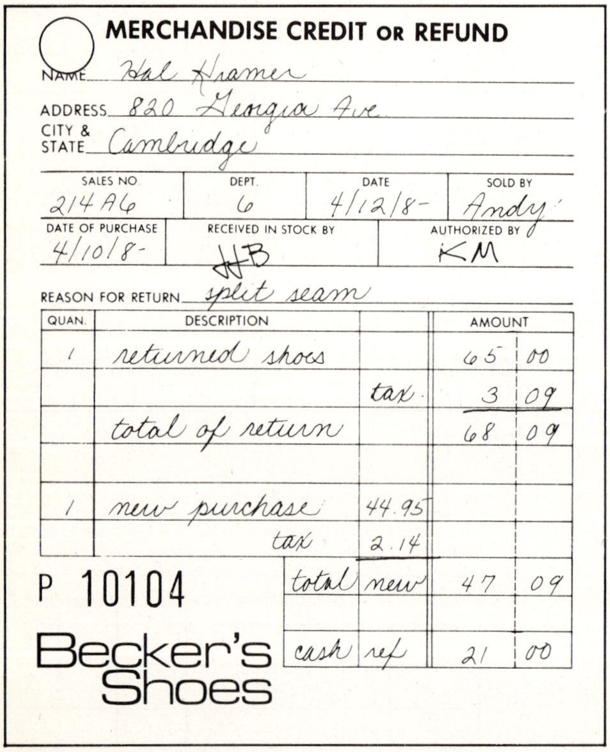
MERCHANDISE CREDIT OR REFUND

NAME Hal Kramer
ADDRESS 820 Georgia Ave
CITY & STATE Cambridge

SALES NO.	DEPT.	DATE	SOLD BY
214A6	6	4/12/8-	Andy

DATE OF PURCHASE	RECEIVED IN STOCK BY	AUTHORIZED BY
4/10/8-	HB	KM

REASON FOR RETURN split seam

QUAN.	DESCRIPTION		AMOUNT	
1	returned shoes		65	00
		tax	3	09
	total of return		68	09
1	new purchase	44.95		
	tax	2.14		
	total	new	47	09
	cash	ref	21	00

P 10104

Becker's Shoes

Figure 4-7 A refund slip such as this is used for a return or an exchange.

Returned Merchandise

$65.00 Retail price
+ 3.09 Tax
$68.09 Total

Exchange Merchandise

$44.95 Retail price
+ 2.14 Tax
$47.09 Total

Amount of Refund

$68.09 Return total
−47.09 Exchange Total
$21.00 Refund

Experience 4-3 ■ I Owe You

Mrs. Henricks, a regular customer, says she wants to exchange two of the four pairs of slacks she purchased. The slacks she is returning cost $60 each without tax. The slacks she wants in exchange cost $39.95. Compute the amount of the refund due this customer. Complete the quantity, description, and amount columns on the refund check in the opposite column. Compute the sales tax at 6¼ percent.

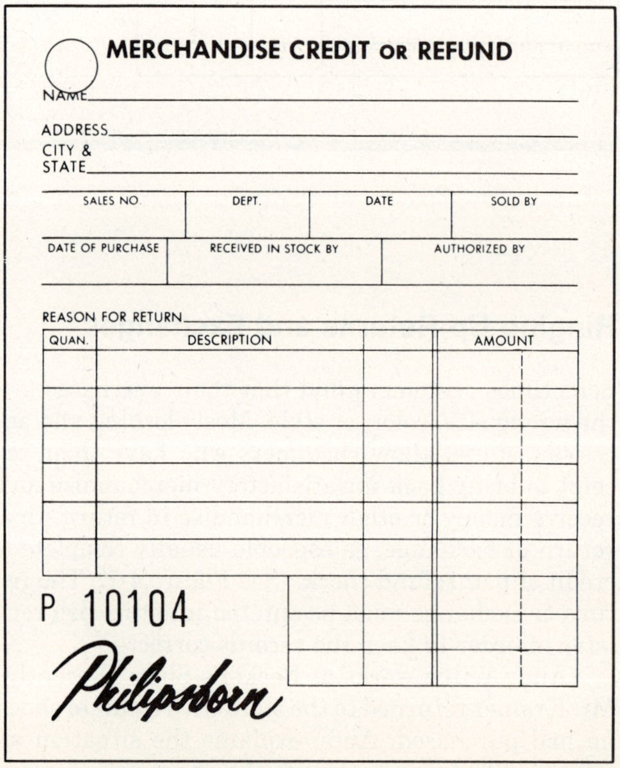
MERCHANDISE CREDIT OR REFUND

NAME
ADDRESS
CITY & STATE

SALES NO.	DEPT.	DATE	SOLD BY

DATE OF PURCHASE	RECEIVED IN STOCK BY	AUTHORIZED BY

REASON FOR RETURN

QUAN.	DESCRIPTION		AMOUNT

P 10104

Philipsborn

Ringing up Layaways

Many stores offer customers a service called layaway. With this service, customers can pay a small deposit to have merchandise put aside for them. The customers have a certain amount of time to pay the total price, plus tax, of the merchandise item. When the merchandise is completely paid for, the merchandise belongs to the customer. Although most stores ask for a small service charge for placing merchandise in layaway (usually $1 to $3), there is no finance charge for layaway.

Customers who take advantage of a store's layaway service must pay a percentage of the total cost of the merchandise item at the time of the layaway, usually 20 percent. The salesperson must compute the layaway deposit and balance due on the account.

For example, suppose a customer wants to layaway a $175 coat and the store requires a 20 percent layaway deposit. The salesperson would have to add tax to the purchase and compute the total amount of the sale. Assume that the tax is 8 percent, or $14. A 20 percent deposit on a total purchase of $189 ($175 + $14) is $37.80. The balance due is $151.20. The salesperson records all this information, including the customer's name, address, and the merchandise description on the layaway form. Every time the customer pays on the account, the salesperson records the payment on a special form and computes a new balance.

Experience 4-4 ■ You Say You Want This on Layaway?

1. A customer wants to layaway the following merchandise items: a silk crepe de chine dress for $165, an embossed calfskin belt for $44, and a pair of stacked heel pumps for $68. Your store requires a 25 percent deposit on all layaway items. Compute the tax at 5½ percent. Complete the layaway form, using your own information and the information provided here. Compute the balance due after the deposit is paid.
2. On April 4, the customer paid $75 on the layaway balance. On April 15, the customer paid the balance. How much was the April 15 payment? Record both payments on the layaway form below.

LAY-AWAY

NAME

ADDRESS

LAY-AWAY NUMBER 14576

CITY PHONE

☐ BOX ☐ HANGER | SOLD BY | DATE / / 19 | TIME LIMIT

DATE	PAYMENTS	BALANCE DUE

Ringing Up Employee Discounts

One of the benefits of working for an apparel and accessory store is that employees usually receive a discount on the store's merchandise. Typical employee discounts range from 20 to 40 percent.

If the cash register does not automatically compute the employee cost of the merchandise, the salesperson must multiply the merchandise price by the amount of the discount. For example, suppose a customer purchased a shirt ($34.50) and atie ($22) and received a 20 percent discount. The employee cost for these items would be $42.50 plus tax. (See Figure 4-8 for an example of the sales check.)

Many stores issue their employees and their employees' dependents a shopping pass called a discount card. When completing an employee purchase, the salesperson usually writes the employee's name and number on the sales check. Most stores ask employees to sign the sales check, even when it is a cash sale.

Computing Delivery Charges

Clothing stores usually charge their customers a small fee to send merchandise to their homes or any other destination. Many stores charge the same delivery fee for any merchandise that weighs less than a certain amount and is sent anywhere in the continental United States. These stores often use a package delivery service, such as United Parcel Service (UPS) or the U.S. Postal Service. When merchandise is being sent, the salesperson must add the delivery fee to the total sales price.

SOLD TO Philip Bell - Employee #403

CLERK F.B. | DATE 11/5 | Cash | CHARGE | REC'D ON ACCT

DEPT	QUAN	ARTICLE	AMOUNT	
	1	Tie	22	00
	1	Shirt	34	50
			56	50
		Employee Discount	11	30
			45	20
		TAX	2	71
		TOTAL	47	91

All claims and returned goods must be accompanied by this bill

Philip Bell
CUSTOMER'S SIGNATURE

Thank You!

Figure 4-8 Here's an employee discount on a sales check.

Correcting Cash Register Errors

Even careful employees occasionally make mistakes. These mistakes may involve entering the wrong department, class, SKU, or price into the register. Two common errors are overrings and underrings. A salesperson makes an overring error when the amount entered is more than the actual price of the merchandise, for example, $63.95 rather than $53.95. When the amount entered into the register is less than the price of the merchandise, the salesperson has made an underring error, for example $60 rather than $90.

The total daily sales rung up on a cash register are recorded on the cash register detail or journal tape. These tapes can be added together to determine the total sales of a department or store. During inventory, the store compares the merchandise that has moved through the department with the total sales figures. Cash register underrings and overrings cause the store management to have an inaccurate picture of the store's inventory levels and shoplifting problem. Therefore, underrings and overrings must be corrected as soon as they are discovered.

How a cash register error is handled depends on store policy. Usually, if a mistake is made before a total or enter key is depressed, the salesperson can correct the error by pushing the clear key or correction lever. The salesperson can then enter the correct data into the register. If the error is recorded on the sales slip, the salesperson may have to void the slip with the incorrect information and enter the information again. Many stores give their salesperson a void form on which the salesperson records each void (mistake). The voided sales slips are attached to this form and turned in at the end of the day to the credit office or manager. Some stores require a lead salesperson or supervisor to approve each void.

Other reasons for a salesperson to void a sales slip are (1) the customer decides to charge a purchase after the salesperson has rung up a cash sale and (2) the customer decides not to purchase an item after the sale is recorded on the cash register.

Making Change

Although all electronic and many mechanical cash registers compute the change due customers, some salespeople may be required to calculate change. For example, if a customer's purchase totals $45.39 and the customer offers a $50 bill, the salesperson would have to subtract $45.39 from $50 to determine the customer's change, $4.61.

When making change, a salesperson should (1) tell the customer the total amount due, including tax. (For example, "That will be $45.39, please.") And (2)

take the customer's money, announce the amount, and place it on the change plate of the register. (For example, "Out of $50.")

Counting Up Calculate the amount of change due by using the count up method. For example, the salesperson counts up from the total, $45.39, to the amount given, $50. The salesperson takes a penny out of the drawer, for $45.40, then a dime, for $45.50, then two quarters or a half dollar, for $46. The salesperson then takes four dollar bills out of the drawer for $50.

Notice that when making change, the salesperson uses the least number of bills and coins possible. Instead of giving the customer a penny and six dimes or sixty-one pennies, the salesperson gives the customer a penny, one dime, and two quarters or one half dollar. By using the fewest bills and coins, the salesperson is usually able to keep a good supply of coins and bills to make change.

Count the change back to the customer while placing the bills and coins in the customer's hand. For example, say "$45.39 from $50." While putting the change in the customer's hand, the salesperson should count up to the amount tendered like this:

Say to Customer	Give to Customer
$45.39	1 penny
$45.40	1 dime
$45.50	1 half dollar
$46.00	1 dollar
$47.00	1 dollar
$48.00	1 dollar
$49.00	1 dollar
$50.00	

Counting Change Due When the cash register computes the amount of change due, the salesperson does not have to subtract the amount of the sale from the amount tendered. After the sale is totaled, the salesperson enters the amount tendered into the register. Change due the customer appears on the cash register read-out.

For example, suppose the customer pays for a purchase costing $17.11 with a $20 bill. The read-out on the cash register says $2.89 CHG DUE. The salesperson takes two $1 bills, a half dollar and one quarter or three quarters, one dime, and four pennies out of the register. Notice that the salesperson uses the fewest coins and bills to make change.

When giving change back to the customer, the salesperson would say, "Your purchase totaled $17.11, out of $20, your change is $2.89." The salesperson then gives change by following this procedure shown below:

Say to Customer	Give to Customer
$2.00	two $1 bills
.50	one half dollar
.75	one quarter
.85	one dime
.89	four pennies

The salesperson always thanks the customer for shopping at the store. Showing appreciation for the customer's business is a courteous way to complete any sales transaction.

Experience 4-5 ■ That Was $38.50 Out of $40

Have a classmate role play a customer. You play the part of a salesperson. Assume your cash register does not tell you how much change is due. Using play money, demonstrate how you would make change using the count up method. Be sure to use the smallest number of coins and bills possible in making change for each of the situations in the opposite column.

Situation	Total Customer Purchase	Customer Pays
1	$17.63	$20.03
2	$24.61	$25.01
3	$37.57	$40.00
4	$54.89	$60.00
5	$93.43	$100.00

Replenishing the Change Fund

During the day, salespeople may find that they are almost out of a particular coin or currency in their change fund. Suppose, for example, that there are no $1 bills in the cash drawer. Because $1 bills are used so frequently in making change, the salesperson must replenish the supply. This process is called "buying change" because a salesperson must use other money in the cash drawer to buy the needed bills or coins. The salesperson could take a $20 bill, two $10 bills, or any other combination totaling $20 out of the cash drawer and buy twenty $1 bills. Change may be bought from the credit office or from another cash register, depending on store policy. Notice that money is never taken out of one cash register and put into another cash register without an equal transfer of funds. When buying change, some stores require a change wanted form to be completed. (See Figure 4-9.)

Closing out the Register

At the end of the day, every apparel and accessory store follows certain closing procedures. The money from the day's sales and the charge and cash receipts are taken out of the cash drawer and turned into the credit office or manager. How each store "closes out" a register varies, but the general principles are the same.

Let's watch how the manager, Randy Tucker, explains the closing procedure to a new employee, Emily Hardy. Randy says, "Emily, you have worked here for a month and are fairly familiar with how the store operates. Next week you will work nights and have to close out the cash registers. We have two cash registers in the department and both must be closed out. You can close out one register about 8:30 p.m. if it is slow. Close out the other cash register after 9:00 p.m. when all customers have left the department and the store is closed.

"To close out a register, you:

1. Open the cash register drawer.
2. Count the coins and bills that will make up the next day's change fund. Because our change fund equals $50, you must count out $50 in coins and bills every night. Include as many small coins and small bills, such as ones and fives, as possible.
3. Place the next day's change fund into the special money bag provided by the credit office.
4. Count the remaining coins and bills and record them on the daily cash report. (See Figure 4-10, A.)
5. Add the checks and record them on the daily cash report. (See Figure 4-10, B.)
6. Add the bank credit card sales slips, such as MasterCard and Visa, and record them on the daily cash report. (See Figure 4-10, C.)
7. Total the coins, bills, checks, and bank credit card sales slips and record the total on the daily cash report. (See Figure 4-10, D.)
8. Add the paid-out vouchers (receipts for money that has been taken out of the cash drawer during the day) and record them on the daily cash report. (See Figure 4-10, E.)
9. Subtract the paid-out vouchers from the total cash in drawer and record the answer on the daily cash report. (See Figure 4-10, F.)

CHANGE WANTED

Date 9/17/8-

Amount sent for change: 20.00

Please send change as follows:

Dollars and Cents		Bills	
Dollars		Twenties	
Halves		Tens	
Quarters		Fives	10.00
Dimes		Twos	
Nickels		Ones	10.00
Pennies			

John Doe

Cashier (Please sign in ink)

Figure 4-9 Here's a typical form used for buying change.

Daily Cash Report

No. 448

Register No. 16 Date 11/7/8-

Pennies		63	A
Nickels		75	A
Dimes		90	A
Quarters	1	25	A
Halves		50	A
Bills	104	00	A
Checks	74	38	B
Bank Card Credit Sales Slips	95	40	C
Total Cash in Drawer	277	81	D
Paid-Outs	12	00	E
Total Cash Received from Sales	265	81	F
Refunds on Cash Sales	14	73	G
NET CASH SALES	251	08	H

Salesperson Emily Hardy

Figure 4-10 Here's a typical example of a daily cash report form.

10. Add all the refunds made during the day and record the total on the daily cash report. (See Figure 4-10, G.)

11. Subtract the refunds from the total cash received from sales and record the net cash sales on the daily cash report. (See Figure 4-10, H.)

12. Place the coins, bills, checks, bank credit card sales slips, paid-out vouchers, refund slips, and daily cash report in another money bag. Then deliver both money bags to the credit office."

Experience 4-6 ■ Closing Time

Assume you are responsible for closing out cash register 14 in Department 12. Your salesperson number is 318. The change fund needs to be returned to $50.

In the cash drawer, you find the following. Complete the daily cash report below.

Coins	Bills	Checks	Bank Credit Card Sales Slips	Refunds
34 pennies	Fourteen $1	$89.92	$82.18	$33.40
17 nickels	Seventeen $5	$67.85	$111.75	$77.38
15 dimes	Seven $10	$121.40		
8 quarters	Four $20			
1 half dollar	Five $50			
	One $100			

Daily Cash Report

No. 448

Register No. _______ Date _______________

Pennies		
Nickels		
Dimes		
Quarters		
Halves		
Bills		
Checks		
Bank Card Credit Sales Slips		
Total Cash in Drawer		
Paid-Outs		
Total Cash Received from Sales		
Refunds on Cash Sales		
NET CASH SALES		

Salesperson ______________________

■ MERCHANDISE COUNTS, REORDERS, AND MARKDOWNS

In addition to entering sales into the cash register and computing returns, layaways, employee discounts, taxes, and delivery fees, clothing and accessory employees use their math skills when taking inventory counts, determining how much stock to reorder, and computing markdowns. Each of these tasks involves basic arithmetic: adding, subtracting, multiplying, and dividing. For example, although the manager or buyer usually determines the percentage of the retail price by which merchandise items are reduced, the salesperson often computes the exact markdown reduction.

To compute a markdown price, the salesperson multiplies the markdown percentage by the original sales price and subtracts that amount from the original price. Most stores have a specific amount that the marked-down price is rounded to. For example, if the marked-down prices of the store always end in 99, \$26.08 would be rounded to \$25.99.

Counting, reordering, and calculating prices of merchandise are discussed in detail in Chapter 11, "Buying and Pricing."

Example of Markdown

Markdown percentage = 33⅓ percent
Original price = \$224

Markdown percentage × Original price
= Amount of markdown

33⅓ % × \$224 = \$73.92

Original price − Amount of markdown
= New sales price

\$224 − \$73.92 = \$150.08, rounded to \$149.99

Experience 4-7 ■ What's This Selling For?

Your buyer has just handed you a list of merchandise markdowns. Compute the new sales price of each item. Assume your store rounds any markdown price to the nearest 99¢.

Item	Markdown Percent	Regular Price	Reduced Price
1. Hooded sweatshirts	25	\$34	______
2. Cotton sweaters	33	\$68	______
3. Wool scarfs	50	\$48	______
4. Acrylic pullovers	20	\$27.99	______
5. Knit turtlenecks	25	\$24.99	______

Chapter Note

1. "Point-of-Sale '86: A New Study," *Stores,* November 1986, pp. 88–89.

UNIT 2 CASE

Selling on Commission

At A&S, salespeople are in charge of their income and additional benefits. Every women's shoe salesperson earns a straight comission on all sales at this chain of department stores with headquarters in Brooklyn, New York. The more they sell, the bigger the paycheck.

The pay system at A&S works like this. The employee writes his or her ID number on the box of the shoes the customer is buying. A cashier rings up the sale (so as to not take the salesperson's valuable selling time away from working with other customers) and gives the salesperson a copy of the sales receipt.

The salesperson keeps a copy of his or her receipts, and once a week adds up the sales, and compares the net sales against those listed on a printout. The printout shows how much was sold according to the company's records. (See Figure 1, letter A.) The printout also lists, under the heading, "credits," the returns charged against the salesperson (see letter B). An 8 percent commission is earned on net sales (gross sales − credits = net sales). (See letter C.)

For example, see the printout for Jeff Samuel, who works 37.5 hours per week and earns an 8 percent commission. Jeff sold $7,400 of shoes, had returns of $465, and had earnings of $554.80 ($7,400 − $465 = $6,935 × .08 = $554.80). During this one week, he earned over $12 per hour!

Like all commission salespeople, Jeff is paid a guaranteed weekly draw based on anticipated commission earnings. Jeff's draw is computed by multiplying the number of hours worked by a preagreed hourly rate, such as $5.00 per hour. As shown on the printout, Jeff's draw is $187.50. The balance column represents commission earned on sales minus his draw ($554.80 − $187.50 = $367.30). Because Jeff's commission earned on sales has been consistently greater than his draw, he does not have anything in

DATE		HOURS WORKED ST / OT	NET SALES PRIOR TO THIS PERIOD
1/15/--	123-45-6789 JEFF SAMUEL 37.5 8%	37.5/00	195012.75

THIS PERIOD			THIS PERIOD					
GROSS SALES (Excl. Sales Tax)	CREDITS	NET SALES (Excl. Sales Tax)	COMM EARNED ON SALES Not Incl. Tax	DRAW	BALANCE	BAL. UNEARNED PREVIOUS PER	TOTAL UNEARNED	AMOUNT ENCLOSED
Ⓐ 7400 00	Ⓑ 465 00	Ⓒ 6935 00	554 80	187 50	367 30	00	00	367 30

DATE		HOURS WORKED ST / OT	NET SALES PRIOR TO THIS PERIOD
1/15/--	012-34-5678 JENNIFER SCOTT 37.5 8%	37.5/00	11430.16

THIS PERIOD			THIS PERIOD					
GROSS SALES (Excl. Sales Tax)	CREDITS	NET SALES (Excl. Sales Tax)	COMM EARNED ON SALES Not Incl. Tax	DRAW	BALANCE	BAL. UNEARNED PREVIOUS PER	TOTAL UNEARNED	AMOUNT ENCLOSED
Ⓐ 1946 00	Ⓑ 92 00	Ⓒ 1872 00	149 76	187 50	-37 74	738 72	776 46	00

Figure 1 The printout shows how much each commissioned salesperson is paid and how the pay was calculated.

the balance unearned previous period or total unearned columns. On payday he receives his draw, $187.50 plus his commission, $367.30.

The example of Jennifer Scott is very different. Jennifer also works full time, 37.5 hours per week, and receives 8 percent straight commission and a guaranteed $5.00 per hour draw. For the week shown on the printout, Jennifer had gross sales of $1,946, returns of $74.00, and net sales of $1,872. The commission earned on sales equals $149.76 ($1,872 × .08). Jennifer's draw, which happens to be the same as Jeff's, is more than the commission earned. Therefore, she has a negative balance of $37.74. In prior weeks, Jennifer has had other negative balances that total $738.72.

Her total unearned balance is $776.46 ($37.74 plus $738.72). For this week, Jennifer will receive a check for her draw only. She will have to pay back the total unearned amount with commission earned on sales in future weeks before she will receive any more than her draw.

By providing commission salespeople with a guaranteed draw, A&S is smoothing out some of the wide swings in earnings that often occur because of fluctuations in sales. This kind of commission system helps employees to feel more secure because they know their paycheck will never be below a certain amount. However, they are required to pay back the draw from future commission earnings.

In addition to the opportunity to set their own earnings, A&S rewards outstanding performers with additional benefits and recognition. Salespeople who maintain a specified rate of sales over time receive cash bonuses, time off with pay, and recognition at store meetings.

To help employees earn as much income as possible, A&S gives each shoe salesperson a preferred customer record book, complete with postcards and envelopes. Salespeople are encouraged to send customers thank-you notes and to inform customers of upcoming promotional or special events. This customer file system enables a salesperson to build a clientele who will come back to him or her for all their shoe purchases. All shoe salespeople also receive business cards, which are to be given to customers at the close of each sale, with a sincere "Thank you for shopping at A&S."

Although the commission system motivates employees to sell, it can also put an added strain on the employees' human relation skills. For example, sometimes situations occur when two salespeople help a customer, and only one receives credit for making the sale, causing hurt feelings between co-workers.

Another issue that the A&S shoe department managers have to deal with is motivating salespeople to perform their nonselling duties, such as floor and stockroom maintenance. When new employees are hired, the manager explains that they are required to spend at least 20 minutes of each shift keeping the selling floor neat or making sure that shoes have been returned to their proper location in the stockroom. Nevertheless, salespeople are not eager to spend the time away from the customers, because time spent off the floor impacts earnings. It is also difficult for managers to motivate commissioned employees to attend any meetings that take them off the floor, which makes it difficult to find the time to discuss new products or review selling techniques or policy changes.

Even with these minor concerns, A&S is committed to motivating employees through a commission pay system. It provides customers and employees with the best situation—quality service and the opportunity to earn a good income.

1. A full-time, 37.5 hour per week, shoe salesperson has gross sales of $5,472 and returns of $316. This salesperson is paid an 8 percent commission and receives a guaranteed draw of $6 per hour. What was this salesperson's commission earned on sales?

____

____

2. What recommendations can you make concerning this commission system that would motivate salespeople to perform willingly their nonselling duties?

____

____

____

3. One of the ways the A&S shoes salespeople build a customer following is by writing thank-you notes to customers who have made a purchase. Suppose you are a shoe salesperson, and write an effective customer thank-you note in the space provided.

____

____

____

____

____

Source: Based on personal interview with Mindy Fabricant, Employment Manager (A&S), Paramus, January 5, 1989.

UNIT 3

Product and Service Technology

Anyone who works in a marketing job must know about the products or services being sold. This unit discusses the products you may be working with in your clothing and accessories marketing career. First, in Chapter 5, you'll learn about the fabrics used to make today's clothing. In Chapters 6, 7, and 8, you will find out about the sizes and styles of women's, men's, and children's clothing. You will also learn how to recognize apparel that fits properly and how to coordinate clothing and accessories to make a child, man, or woman look his or her very best. Chapter 9 presents information about shoes. Other accessories, including handbags, belts, and costume jewelry, are discussed in Chapter 10.

With this product information, you should be able to begin and advance in your career.

CHAPTER 5

Fabrics

Glamour magazine featured the following advice in an article about choosing the right workout clothes at the right price.[1]

> When it comes to every day clothes, few materials surpass the comfort and beauty of natural fibers. But for the slopes, track, pool, and gym, synthetics are the hands-down favorite. Light, strong, and washable, they keep you cool when it's hot, warm when it's cold, and dry inside and out.... Your choice depends on your needs.... If you are a cold-weather runner, a suit of sturdy, waterproof Gore-Tex is the best buy. (Gore-Tex is a thin, plastic membrane laminated to other fabrics, such as nylon, to produce windproof, waterproof cloth. It's breathable because microscopic openings are big enough to let body vapor escape but are too small for raindrops to enter....) If you are a fair-weather runner, a nylon suit will do the job. And if you only run indoors or in the summer, nylon or cotton blend shorts and a top are all you need.

When you work in the clothing industry, you will find that customers want information about the clothing they purchase, similar to the information above. They'll want to know what certain terms, such as Gore-Tex, mean. They'll want to know about a garment's durability and care, whether it will shrink, wear well, or fade. And they'll ask about how to launder the clothing and whether ironing is required. To answer their questions, you need to be able to identify fabrics and know something about how they are made. Your knowledge about fabrics will increase your customers' confidence in the clothing and accessory suggestions you make. Information about fabrics will benefit your customers by helping them select the right merchandise to fit their needs. Knowing about fabrics will also benefit you through increased sales and a loyal clientele.

■ THE LANGUAGE OF FABRICS

In dealing with fabrics, there are many new words you will need to learn and understand:

Fiber is the smallest unit in a textile fabric.

Yarn is a continuous thread formed by twisting or spinning fibers together.

Finish is something done to a fabric, yarn, or fiber to change its appearance, performance, or feel.

Hand describes the texture or feel of a fabric, such as its softness or crispness.

Textiles are any products made from fibers. These fibers can be woven, knitted, felted, braided, or bonded into cloth.

Most fabrics are a combination of four major items: fiber, yarn, construction, and finish. (See Figure 5-1.)

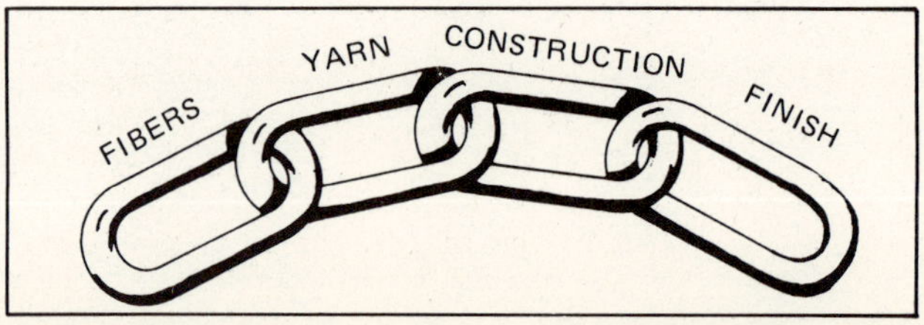

Figure 5-1 Most fabrics are made up of a combination of the four items shown here.

■ FIBERS

Up to the early 1900s, clothing was made only from natural fibers, such as cotton, linen, wool, and silk. Natural fibers come from plants or animals. Beginning in the 1920s, man-made fibers were developed and used in clothing. Examples of man-made fibers, which are produced in chemical factories, are rayon, nylon, and polyester. Man-made fibers can look and feel like natural fibers. For example, a 100 percent polyester dress can look like silk. Many customers, however, prefer to wear 100 percent natural fibers because of their comfort.

There are also blends of man-made fibers and natural fibers. A **blend** is a combination of two or more fibers, natural and/or man-made, before they are spun into yarn. Blending combines the best features of each fiber. For example, polyester added to cotton or wool gives strength, crease resistance, and shrink resistance. Rayon added to cotton gives luster. Any fabric may be made up of different fibers. For example, flannel can be blended from acetate and rayon, or acrylic and wool. The fibers used in a fabric affect its care and durability.

Natural Fibers

Cotton, linen, wool, and silk are the most common natural fibers. Cotton and linen come from plants; wool and silk come from animals. Table 1 provides specific characteristics and care details for cotton, linen, wool, and silk.

■ Table 1
Natural Fibers

Characteristics	Properties	Care Instructions[1]
	Cotton	
Source: Boll of cotton plant. The world's major textile fiber. Cotton is used alone or in blends. New wash-and-wear finishes have made cotton easier to care for. Two of the finest quality cottons are Sea Island cotton and Pima cotton.	Is absorbent and colorfast (colors don't fade), is strong when wet or dry; takes dye nicely; is comfortable; shrinks unless treated with special finish; wrinkles, and needs pressing; has poor crease retention (cannot hold creases); does not melt, shrink, or stretch when ironed.	Usually, cotton can be machine-washed and tumble-dried. Sheer fabrics, or those with "doubtful" colorfastness should be hand-washed. Cotton can be bleached with chlorine or peroxide bleaches. However, cotton with a resin finish cannot be chlorine bleached (a resin finish makes a garment shrink resistant, and easy to care for). Some finishes may cause the fabric to yellow when exposed to chlorine bleach. Medium- and heavy-weight cotton can be pressed with a hot iron. Use warm iron for thin or sheer cottons. Remove stains before washing. May mildew.
	Linen	
Source: Flax plant. Flax that is processed into yarn or fabric is called linen. Because of its limited production and costly hand labor involved in producing it, linen is imported into the United States. Linen is more expensive than cotton.	Is highly absorbent, cool, and comfortable in summer; is very durable (stronger than cotton); resists shrinking; soils less easily than cotton and is colorfast; wrinkles easily unless treated.	Some linens are machine-washable and can be tumble-dried. It is best to iron linen when damp; first on outside and then on inside. May mildew.
	Wool	
Source: The hair covering a sheep, lamb, Angora or Cashmere goat, camel, alpaca, vicuna, or llama. Wool can be sheared from the living animal (called fleece or clipped wool) or pulled from the hide after the	Insulates wearer against heat and cold; is absorbent; is colorfast; possesses elasticity (can stretch up to 30% of its length without strain); takes dye easily; resists wrinkles; is resilient but will pill (pills are balls	Most wool garments should be dry-cleaned. However, if the label says hand-wash, use cool water, mild detergent, and gentle action. Never soak, rub, or twist. Handle the garments carefully when wet or damp.

(continued)

Table 1 *(Continued)*

Characteristics	Properties	Care Instructions[1]
	Wool	
animal has been slaughtered for its meat (called pulled wool). Pulled wool is of inferior quality to fleece wool. The Wool Products Labeling Act classifies wool into three groups: (1) *Wool or virgin wool:* fiber being used for the first time in the complete manufacture of a wool product. This is the strongest and most resilient (able to spring back into shape after being crushed or wrinkled). (2) *Reprocessed wool:* wool that has been made into fabric before but never worn or used; such as mill end pieces. (3) *Reused wool:* wool that has been made into fabric before and worn or used. This wool is "reclaimed" and reprocessed. These classifications will appear on the label.	of tangled fibers on the fabric's surface); poor resistance to bleach, friction, moths, perspiration, and strong soaps.	Dry on a flat surface. Press with a cool iron and steam. By hanging wool garments in a steam-filled room, most set-in wrinkles can be removed. After wearing, wool garments should be brushed gently. Garments should be allowed to rest 24 hours before they are worn again. Knitted sweaters should be folded and stored in drawers to prevent sagging and misshaping from hangers.
	Silk	
Source: Cocoon of silkworm. Silk may be weighted or nonweighted. In their original state, silk filaments are covered by a gummy substance. When the gum is removed, the silk loses about 25% of its weight. This weight can be replaced by putting the silk in a solution of metallic salts. *Pure silk, all silk, pure silk-dye,* and *silk* are terms that describe an all-silk fabric that contains no metallic weightings and contains no more than 10% (by weight) of dyes or finished materials. Pure silk holds shape better and wrinkles less than weighted silk. Weighted silk has appearance and feel of pure silk, but is less expensive. Weighted silk is more sensitive to sunlight, air, and perspiration damage.	Is luxurious in appearance and feel; takes dye beautifully; is the strongest of the natural fibers; wears well and resists strain; is absorbent and warm; does not soil easily; resists wrinkles and is colorfast; has dimensional stability (ability of fabric to hold its shape and size after being worn, washed, and/or dry-cleaned); drapes nicely.	Depending on the care label, silk should be hand-laundered or dry-cleaned. Iron with a warm iron on the wrong side. Use a press cloth to finish pressing seams, collars, and hems on the right side.

[1]These care instructions are general ones for the fibers listed. The care instructions on the garment's label should be followed closely.

Man-Made Fibers

Man-made fibers are made from thick liquids that pass through tiny holes of a mechanical device called a spinnerette. (See Figure 5-2.) The long, continuous fibers that emerge from the spinnerette are called filaments. Filaments may be cut into short lengths and twisted together to form yarn or may be made into yarn without being cut. The most common man-made fibers used for apparel are acetate, acrylic, modacrylic, nylon, polyester, rayon, spandex, and triacetate.

There are two categories of man-made fibers: cellulose and noncellulose. Rayon, acetate, and triacetate are cellulose or modified cellulose fibers and are made from natural substances, such as cotton linters or wood pulp. (Cotton linters are the cotton fibers that are too short for yarn or fabric

Figure 5-2 Man-made fibers are passed through a spinnerette like the one pictured above. Courtesy Celanese Fibers Marketing Co.

manufacturing.) Cellulose fibers have characteristics similar to those of cotton and linen. Generally, these fibers are absorbent, easy to dye, have a high natural luster, and are economical. Table 2 details characteristics and care instructions for garments made of cellulose fibers.

Noncellulose fibers generally are thermoplastic (that is, able to be molded or shaped at high temperatures), able to withstand surface wear and rubbing, nonabsorbent, and resilient (able to spring back into shape). They resist sunlight, mildew, and moths. These fibers are as strong when wet as they are when dry and require little or no ironing after laundering. Table 3 details characteristics and care instructions for garments made of noncellulose fibers.

Today any garment can be made of either cellulose or noncellulose fibers. For example, many slips are made from rayon, triacetate, or acetate, but many slips are also made from nylon and nylon/polyester blends.

■ Table 2
Cellulose and Modified Cellulose Fibers

Principal Uses	Properties	Care Instructions for Cellulose Fibers	Trade Names
		Acetate—A lustrous, silky fiber	
Dresses, blouses, foundation garments, lingerie, fashion fabrics (satin, jersey, tafetta, lace, faille, brocade, tricot, crepe)	Luxurious feel and silklike appearance; shrink-resistant; able to retain crispness; able to take beautiful color; drapes well; resists mildew and moths; fairly absorbent; wears out easily and not self-extinguishing; can be made flame resistant. Acetate melts at 500°F. Nail polish, paint remover, and some perfumes dissolve acetate fibers.	Some acetate fabrics must be drycleaned. Others can be home laundered. If home laundered, hand-wash using warm water, mild soap, gentle agitation. Do not soak, wring out, or twist. Iron with cool iron; a safe ironing temperature is 250°–300°F. Press garment while damp on wrong side, use a press cloth. If iron is too warm, may put a shine on the fabric. Let dry on nonrust hanger.	Ariloft Celanese Chromspun Estron Loftura
		Rayon—Combines well with many other fibers (used extensively in blends)	
Dresses, blouses, suits, lingerie, coats, slacks, ties, lining fabrics	Very absorbent (cool); comfortable; drapes well; easy to dye; durable; not elastic (does not go back into shape when stretched); wrinkles easily unless it has resin finish; brushed or napped fabrics are	Rayon can be washed by hand, machine-washed and tumble-dried, or dry-cleaned, depending on care label. If not handled carefully, may stretch when wet. Do not wring or twist when wet. May shrink or	Avril Coloray Enkrome Fibro Zantrel

(continued)

Table 2 *(Continued)*

Principal Uses	Properties	Care Instructions for Cellulose Fibers	Trade Names
	flammable; can be made flame-resistant; will shrink unless has shrink-resistant finish.	lose finish if washed or dried at high temperatures. Rayon will scorch but not melt if iron is too hot.	
Triacetate—Similar to acetate but can withstand higher temperatures			
Bonded and tricot fabrics, permanently pleated garments	Resists heat and wrinkles and shrinking; holds pleats; wears out easily; easily washed; stronger than acetate; can withstand higher temperatures.	Most garments made of 100% triacetate can be machine-washed and dried. If ironing is necessary, a hot iron can be used (450°).	Arnel

■ **Table 3**
Noncellulose Fibers

Principal Uses	Properties	Care Instructions for Noncellulose Fibers	Trade Names
Acrylic—A fiber with wool-like and easy-care qualities			
Dresses, sweaters, skirts, suits, socks, fleece and pile fabrics	Soft; provides warmth without weight; shrink-resistant; strong; wrinkle-resistant; low absorbency; quick drying; subject to static buildup; ignites and burns easily; subject to pilling (little balls forming on surface).	Can be machine-washed and tumble-dried at low temperature. Remove garments from dryer as soon as tumbling is done. If ironing is necessary, use a moderately warm iron, not hot iron. Sweaters should be dried on a flat surface. Acrylic fabrics should be brushed if they get fuzzy.	Acrilan Creslan Orlon Zefran
Modacrylic—A modified acrylic fiber with many properties similar to those of acrylic			
Deep pile coats, trims and linings, fleece fabrics	Soft and warm; long wearing; flame-resistant (generally self-extinguishing); quick drying; colorfast; softens at low temperatures; shrinks at 260°F.	Machine-wash in warm water and tumble-dry at low temperatures. Remove from dryer as soon as tumbling stops. Use low setting if ironing is necessary. Deep pile should be dry-cleaned. No ironing for 100% modacrylic.	Dynel Elura SEF Verel
Nylon—A very versatile and strong fiber			
Hosiery, dresses, lingerie, underwear, suits, stretch fabrics	Exceptionally strong; keeps its shape; elastic; easily dyes; smooth; durable; low in moisture absorbency; dries fast; lustrous; easy to wash; woven fabrics sometimes hot and uncomfortable to wear; does not ignite easily but when ignited will burn, melt, and drip.	Most items made from nylon can be machine-washed and tumble- dried at low temperature (fabric softener will prevent static cling); can be bleached with chlorine; requires little or no ironing; wash delicate items by hand in warm water and mild soap; wash whites separately.	Anso Antron Cantrece Caprolan Celanese

(continued)

Table 3 *(Continued)*

Principal Uses	Properties	Care Instructions for Noncellulose Fibers	Trade Names
Polyester—Used extensively in blends with other fibers (especially wool and cotton)			
Permanent press apparel	Resists wrinkles, shrinkage, stretching, surface wear, and pilling; strong; holds pleats and creases well; easily washed; quick drying; colorfast; needs little or no ironing; does not ignite easily; but when ignited will burn, melt and drip. Can look like wool, silk, cotton.	Wash delicate garments by hand in warm water and mild soap. Gently squeeze out water, smooth or shake out garment, and let dry on nonrust hanger; most polyester items can be machine-washed. Use warm water and add fabric softener to rinse cycle. Set dryer low and remove as soon as tumbling is done. Use warm iron, if ironing is necessary; may be dry-cleaned.	Avlin Dacron Encron Fortrel Kodel Trevira
Spandex—Few garments are made of 100% spandex; only small amount is needed to give garment necessary holding power			
Girdles, bras, swimsuits, ski pants, support and surgical hose, elastic waist bands	Elasticity (ability to recover shape after being stretched); extremely strong; lightweight; yellows with age at temperatures above 300°F.	Can be hand- or machine-washed in warm water. Drip dry or tumble-dry low. Can be ironed at low temperature. Do not use chlorine bleach on any fabric containing spandex.	Lycra

Government Regulation of Fiber Identification

Until recently, it was difficult to determine the fibers used in a clothing item. That's why the government requires clothing manufacturers to attach labels that state the fiber content in a garment. The Textile Fiber Products Identification Act (TFPIA), effective since 1960, requires all textile merchandise to carry a label that shows:

1. The generic names of the fibers in the fabric in the order of their proportion. (A **generic name** is a standard term describing the fiber.) Polyester (a man-made fiber) and cotton (a natural fiber) are examples of generic names. The companies that produce these fibers are also allowed to include their trade name on the label. For example, Dacron polyester is a polyester fiber made by Du Pont. Fortrel polyester is made by the Celanese Corporation. Dacron and Fortrel are trade names.
2. The percentage of each fiber in the fabric.
3. The garment manufacturer's name, address, or registered identification number.
4. The country of origin of imported fibers.

The Federal Trade Commission's Trade Regulation Rule on Care Labeling requires apparel to have care and maintenance labels. (See Figure 5-3 and Table 4.)

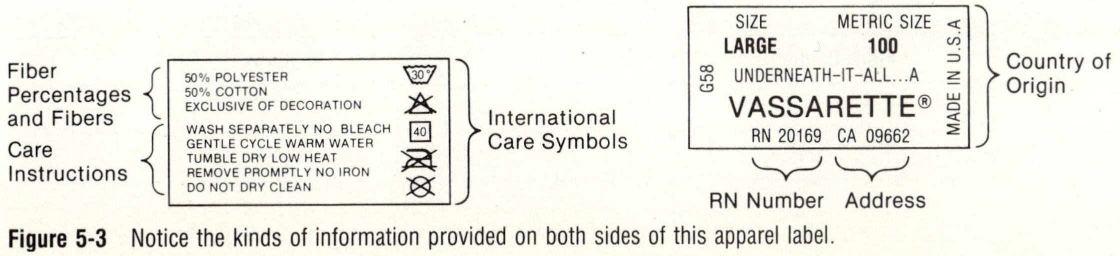

Figure 5-3 Notice the kinds of information provided on both sides of this apparel label.

■ Table 4
Explanation of Care Label Instructions

When Label Reads:	It Means:
Machine-Washable	
Machine wash	Wash, bleach, dry, and press by any customary method including commercial laundering and dry cleaning.
No chlorine bleach	Do not use chlorine bleach. Oxygen bleach may be used.
No bleach	Do not use any type of bleach.
Cold wash, cold rinse	Use cold water from tap or cold washing machine setting.
Warm wash, warm rinse	Use warm water or warm washing machine setting.
Hot wash	Use hot water or hot washing machine setting.
No spin	Remove washload before final machine spin cycle.
Delicate cycle or gentle cycle	Use appropriate machine setting; otherwise, wash by hand.
Durable press cycle, permanent press cycle	Use appropriate machine setting; otherwise, use warm wash, cold rinse, and short spin cycle.
Wash separately	Wash alone or with like colors.
Nonmachine Washing	
Hand-wash	Launder only by hand in lukewarm (hand-comfortable) water. May be bleached. May be dry-cleaned.
Hand-wash only	Same as above, but do not dry-clean.
Hand-wash separately	Hand-wash alone or with like colors.
No bleach	Do not use bleach.
Damp wipe	Surface clean with damp cloth or sponge.

(continued)

Table 4 *(Continued)*

When Label Reads:	It Means:
Home Drying	
Tumble-dry	Dry in tumble dryer at specified setting—high, medium, low, or no heat.
Tumble-dry, remove promptly	Same as above, but in absence of cool-down cycle remove at once when tumbling stops.
Drip dry	Hang wet and allow to dry with hand shaping only.
Line dry	Hang damp and allow to dry.
No wring	Hang dry, drip dry, or dry flat only.
No twist	Handle to prevent wrinkles and distortion.
Dry flat	Lay garment on flat surface.
Block to dry	Maintain original size and shape while drying.
Ironing or Pressing	
Cool iron	Set iron at lowest setting.
Warm iron	Set iron at medium setting.
Hot iron	Set iron at hot setting.
Do not iron	Do not iron or press with heat.
Steam iron	Iron or press with steam.
Iron damp	Dampen garment before ironing.
Miscellaneous	
Dry-clean only	Garment should be dry-cleaned only, including self service.
Professionally dry-clean only	Do not use self-service dry cleaning.
No dry-clean	Use recommended care instructions. No dry-cleaning materials to be used.

Source: Consumer Affairs Committee, American Apparel Manufacturers Association, based on the voluntary Guide of the Textile Industry Advisory Committee for Consumer Interests.

Experience 5-1 ■ Check the Label

Examine 10 different clothing items, such as a coat, dress, blouse, shirt, trousers, and tie. Check the labels that describe the fiber content and care requirements. Some of the items should be all natural fibers, some synthetic, and some blends of both. You may use your own clothing or visit an apparel store to locate these items. Complete the following chart.

Clothing Item	Fiber Content	Care Specifications
1.		
2.		
3.		

(continued)

(Continued)

Clothing Item	Fiber Content	Care Specifications
4.		
5.		
6.		
7.		
8.		
9.		
10.		

■ YARNS

The wear, appearance, and texture of a fabric depend upon the fabric's yarn. A yarn is a continuous thread formed by twisting or spinning strands of fibers together. (See Figure 5-4.) Yarns differ because of their twist, ply, and size.

Twist of the Yarn

Short fibers are twisted into yarn. Filament fibers, which are long and continuous, may also be twisted into yarn. Yarn that is tightly twisted is stronger than yarn that is loosely twisted. Although tightly twisted yarn is more durable, it may also be more scratchy than loosely twisted yarn. Satin is an example of loosely twisted yarn; crepe yarn is an example of tightly twisted yarn.

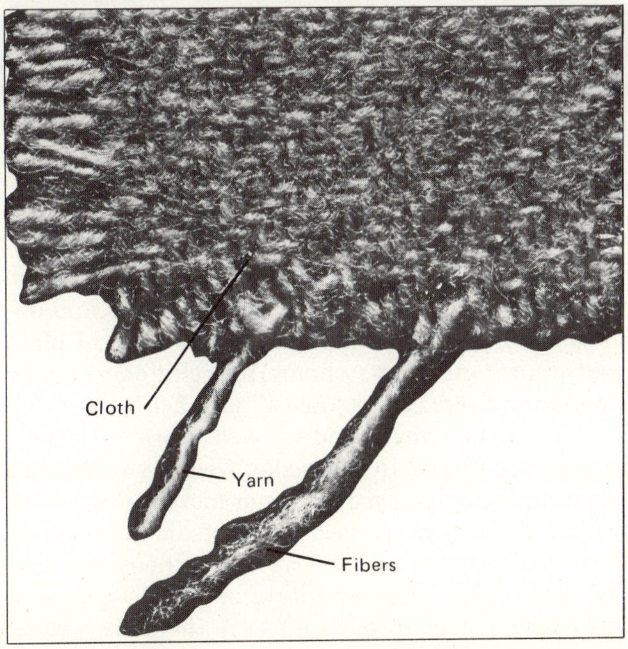

Figure 5-4 Many strands of fibers twisted or spun together form a thread of yarn.

Ply of the Yarn

Ply refers to the number of strands that are twisted together to make the yarn. Single ply describes fibers twisted together to form a single yarn. By twisting together two single yarns, the yarn becomes two ply. A two-ply yarn is stronger than a single yarn of the same thickness and weight. Ply yarns are often used for fabrics that will get hard wear.

Types of Yarn

Yarn can be made from any fiber—natural, man-made, or a blend. How the yarn is constructed and the characteristics of the yarn depend on the fibers used.

Cotton Cotton comes from the cotton plant. Cotton fibers are the long strands attached to the seed inside the boll, or seed pod. When the boll reaches maturity, it bursts open and exposes the cotton, which is picked. The cotton gin removes the seeds from the cotton.

Before cotton fibers are spun or twisted into yarn, they are cleaned and carded. **Carding,** is a process of straightening fibers. All cotton is carded. Some cotton goes through another process called **combing** to remove the short fibers. Combed yarns are finer and stronger than ones that are only carded and are used in higher-quality fabrics.

Linen Flax is a fiber that comes from the flax plant. Linen is the yarn or fabric processed from flax. Like the cotton plant, the flax plant produces both long and short fibers. Yarn that includes both the short and long fibers is called **tow linen.** Yarn that includes only the long fibers is called **line linen.** A tow-linen yarn is similar to a carded cotton; a line-linen yarn may be compared to a carded and combed cotton. Line linen is used for higher-quality linen fabrics.

Wool Wool, which comes from the fleece of animals, is spun into two types of yarn: **worsted** and **woolen.** (See Figure 5-5.) Worsteds are produced by carding and combing wool fibers. Worsteds are firmer, more durable, and take a sharper crease than woolens. Woolens are made from wool yarns containing both long and short fibers. The fibers are carded but not combed. Woolens have a fuzzy surface, are soft, and resist wrinkling. They are warmer than worsteds but do not hold a sharp crease.

Silk Silk comes from the cocoons of silk worms. There are two kinds of silk fibers: reeled and spun.

Figure 5-5 Compare the woolen and worsted fibers illustrated above.

Reeled silk comes from unbroken silkworm cocoons. These long, strong, smooth fibers make the finest grade of silk. **Spun silk,** also known as silk noil or waste silk, is made from the shorter fibers and wastes of broken cocoons. Spun silk is not as smooth and does not have the elasticity, strength, or luster of reeled silk yarn. It has a "cottony" feel.

Man-Made Basically, there are two kinds of man-made fiber yarns: filament yarn and spun yarn. **Filament yarn** is made by twisting two or more strands of continuous filament fibers together. Filament yarn is smoother, more lustrous, and less inclined to pill—form small balls—than spun yarn. **Spun yarn** is made by twisting together shorter lengths of man-made fibers. Spun yarn is usually warmer than filament yarn because air is trapped in the short fibers. It has a fuzzy or fluffy surface. Spun yarn takes longer to dry than filament yarn.

Texturized yarns combine the good features of both filament and spun yarns. Texturized yarns are made from filament yarns, but they look like spun yarns. They have a rougher texture than spun yarns but give greater weight and resist pilling because they have fewer fiber ends.

Experience 5-2 ■ What's the Difference?

1. Describe below the difference between woolens and worsted wool. Locate an example of woolen clothing and worsted wool clothing and bring each to class.

2. Describe below the difference between reeled silk and spun silk. Which is more expensive?

■ FABRIC CONSTRUCTION

In the construction process, yarns are made into fabrics. Yarns may be woven, knitted, knotted, or braided. Weaving and knitting are the two most popular methods. (See Table 5.)

Weaving

Weaving, which is the interlacing of yarns at right angles, is done on a machine called a **loom.** The loom is prepared with lengthwise yarns, called warp yarns. Crosswise yarns, called woof, weft, or filling yarns, are interlaced over and under the warp yarns. (See Figure 5-6.) Woven fabrics are usually firm, strong, and keep their shape. They withstand friction, drape well, and launder well. There are three basic fabric weaves: plain, twill, and satin. All three weaves differ in appearance, even if they are made of the same fibers.

Plain Weave The plain weave is the most basic of all weaves, with one filling yarn going over and under one warp yarn. Examples of fabrics with the plain weave are gingham, chambray, chiffon, challis, tafetta, and seersucker. (See Figure 5-7.)

The basket weave and rib weave are variations of the plain weave. In the basket weave, two or more warp yarns are passed over one or more filling yarns. Oxford cloth, a fabric used for shirts, is an example of the basket weave. In the rib weave, heavier yarns are used in the warp or filling or both. There is a ribbed, corded, or crossbar effect in the plain weave. Broadcloth and poplin are examples of fabrics with the rib weave.

■ **Table 5**
Comparison of Woven and Knitted Fabrics

Wovens	Knits
Tailored, crisp appearance	Soft appearance
Firm, strong fabrics that keep their shape; do not stretch with body movement	Fabrics provide ease of movement because they stretch; fabrics may stretch out of shape or sag
Launder easily but usually require some ironing	Launder easily, often require no ironing
Wrinkle with wear, also wrinkle when packed in a suitcase	Resist wrinkles, also resist wrinkling in a suitcase
Resist friction, wear well	May pill, snag, and stretch out-of-shape
Drape well (hang nicely)	May cling to the body
Can be lightweight yet warm	Can be lightweight yet warm; are warmer than wovens of the same weight

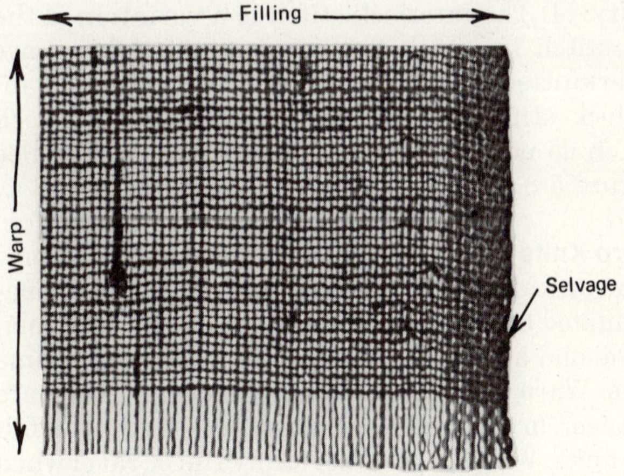

Figure 5-6 Woven fabrics have warp (lengthwise) and filling (crosswise) yarns. Selvage is the finished edge.

Source: American Textile Manufacturers Institute, Inc.

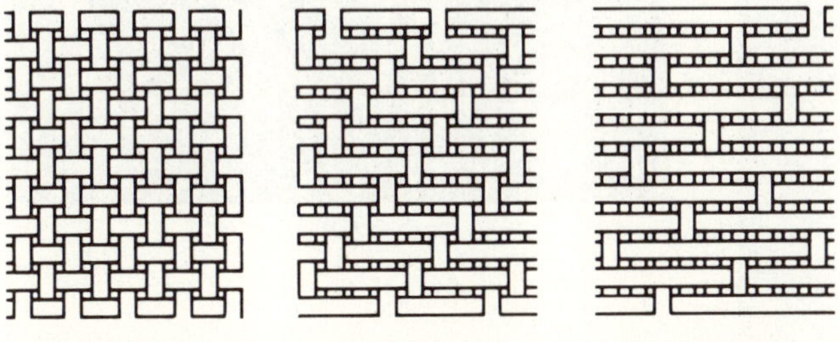

Figure 5-7 Notice how the warp and filling yarns are interlaced in these three popular weaves. Source: American Textile Manufacturers Institute, Inc.

Twill Weave The design of the twill weave fabric has diagonal ridges, called "wales." The twill weave is formed by passing a filling yarn over and under one, two, or three warp yarns, lapping back one warp yarn in each line. The twill weave is very strong and resists wear. Gabardine, flannel, and denim are examples of twill weave fabrics. Herringbone is a variation of the twill weave in which the twill reverses its direction. (See Figure 5-7.)

Satin Weave In the satin weave, either warp or filling yarns float over the surface. The satin weave reflects light and has a lustrous surface. Because satin weaves are easily snagged, they have limited wearing qualities. The satin weave is often used for coat linings and formal wear. Satin and sateen are examples of fabrics constructed with the satin weave. (See Figure 5-7.)

Other Weaves Other variations on the common weaves are as follows:

Pile Formed by interlacing three sets of yarns. Two sets of yarn are used to form the background and a third set (cut or uncut) is used to give a three-dimensional quality. Examples of the pile weave are corduroy, terrycloth, velveteen, velvet, and velour.

Jacquard Created when simple or complex designs are woven into the fabrics. Brocade, damask, and tapestry are examples of jacquard weaves.

Leno Created by twisting warp yarns like a figure eight and passing a filling yarn through the loops made by the twisted warp yarns. The leno weave has an open, lacy appearance. Some fabrics constructed with the leno weave are gauze and marquisette.

Experience 5-3 ■ A Weave...Is a Weave...Is a Weave?

1. Select two to four classmates to visit a local fabric store at a time when the store is not busy. Have one of the students explain to the manager that you are studying fabrics. Ask if you may have at least three different swatches of each of these types: a plain weave, a twill weave, and a satin weave. Try to collect as many swatches as possible. Attach each swatch to a separate sheet of paper. Label each.
2. In class, let all the students carefully examine the labeled fabric samples. On a separate sheet of paper, each student should write a short description of how each fabric feels and looks.

Knitting

Knitting is the looping of yarns together to form a fabric. For apparel use, knit fabrics are more popular than woven fabrics. Knits are known for their give; they are comfortable to wear because they stretch as the body moves. To prevent excess stretching, they should be folded and stored in a drawer rather than on a hanger.

When washed, knits stretch and shrink more than wovens. They are also easier to snag and are subject to pilling. If the knit fabric is made of polyester, nylon, acrylic, or other heat-set fibers, it will usually return to its original shape when washed.

There are two basic categories of knits: filling (or weft) knits and warp knits. Both knits are made on special knitting machines. Each type of knit can be identified by examining the fabric to see in which direction the yarn runs.

Filling or Weft Knits Filling or weft knits involve the use of a single thread. There are four basic stitches used in manufacturing filling knits: (1) the flat stitch has distinct flat vertical lines on the surface and dominant horizontal ribs on the reverse side. Jersey is an example of a flat stitch fabric. (2) The purl stitch produces fabrics that are similar to the reverse of the flat stitch on both sides. It is often used in the construction of bulky sweaters. (3) The rib stitch produces a ribbing effect. It is usually found at the lower edge of sweaters, on sleeves, and at necklines. It is used whenever stretch is desired and gives the fabric excellent elasticity. (4) The interlock stitch is a variation of the rib stitch in which two separate rib fabrics are interknitted. Double knits are produced by the interlock stitch. Fabrics made with the interlock stitch do not sag, stretch, or wrinkle easily. (See Figure 5-8.)

Warp Knits In warp knitting, the loops are formed in a vertical or warpwise direction. Each yarn is manipulated by one specific needle in the knitting machine and all the yarns form the loops at the same time. Warp knits stretch mainly in width. They are stronger, firmer, and more resistant to snags than filling knits. Warp knits do not easily run or ravel when a thread is broken. Tricot knit fabrics, used for lingerie, are an example of a warp knit.

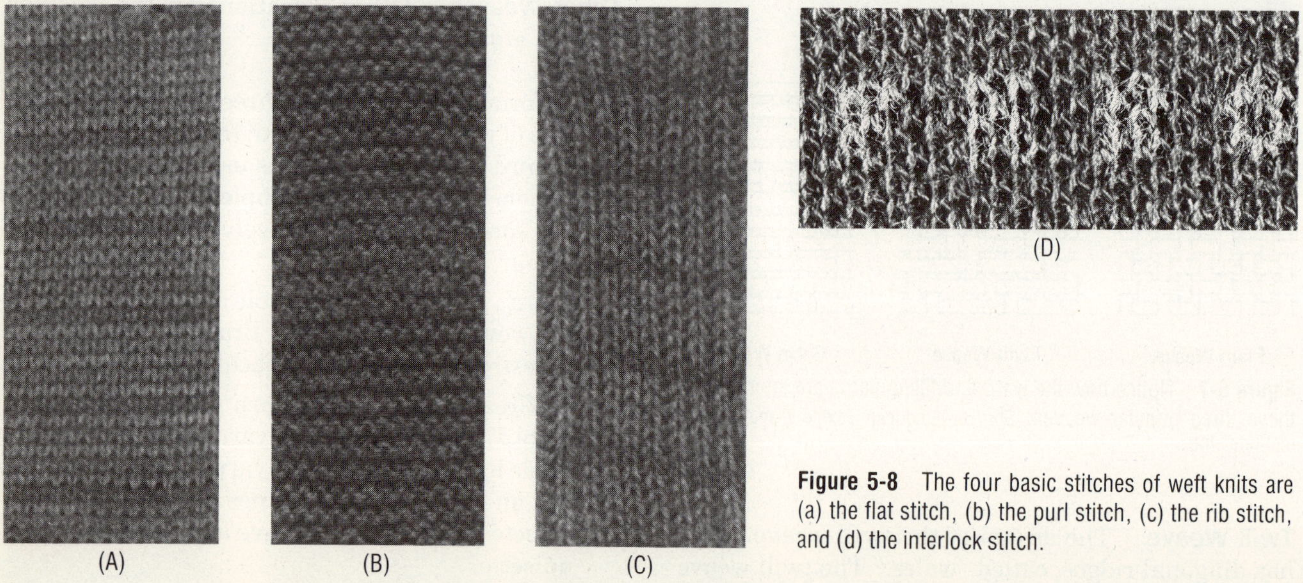

Figure 5-8 The four basic stitches of weft knits are (a) the flat stitch, (b) the purl stitch, (c) the rib stitch, and (d) the interlock stitch.

Name ______________________________ Date ____________________

Experience 5-4 ■ Knitting Versus Weaving: What Are the Benefits?

Bring to class a knit shirt and a woven shirt. Play the role of a shirt salesperson. Have another person role play a customer interested in buying a shirt. Explain to this customer the advantages of buying a knit shirt and the advantages of buying a woven shirt. Be prepared to answer the customer's questions. If possible, use a tape recorder to record the discussion about the knit shirt and the woven shirt.

Other Fabric and Construction Methods

Some fabrics are made directly from fibers rather than yarns. Felting and bonding are some examples of these other fabric construction methods.

Felting Felting is the interlocking of fibers by heat, moisture, and pressure to form a compact fabric. Because the fibers are not securely fastened together, felted fabrics are not as strong as woven or knitted fabrics. However, because they have no threads, felted fabrics do not ravel. Easily shaped, felted fabrics are used to make hats.

Bonding In bonding, fibers are joined together by adhesive. Natural fibers, man-made fibers, or blends may be made into bonded fabrics. Bonded fabrics are often used for interfacings, which add body and shape to collars, lapels, cuffs, pocket flaps, and waistbands.

■ FABRIC FINISHES

A fabric finish is a treatment that changes the fabric's behavior or appearance. Fabric finishes make cloth easier to care for, look better, feel better, and give better service.

Finishes That Improve a Fabric's Appearance

Some finishes that make a fabric more attractive are:

Bleaching Makes a fabric white.

Brushing Raises a fabric's fibers and gives the fabric a fuzzy look.

Calendering Makes a fabric's surface flatter, smoother, and glossier, producing a satiny look.

Embossing Adds a three-dimensional surface design.

Mercerizing Adds shine and luster to the fabric.

Moireing Gives the fabric a watermarked appearance.

Singeing Removes the fabric's fuzz and makes the fabric smoother.

Dyeing, which is the adding of color to a fabric, is another finish that improves a fabric's appearance. Man-made fibers can be dyed before they come out of the spinnerette; this is called **solution dyeing. Stock dyeing** is the dyeing of fibers while they are still loose fibers. **Yarn dyeing** is the dyeing of yarn before it is spun into cloth. **Piece dyeing** is the dyeing of an entire fabric after it is woven or knitted. Piece dyeing is the most used dyeing method today and the most economical.

There are also a number of printing methods that add color or design to a fabric. Many T-shirts and sweatshirts, with words or colorful designs on them, are examples of silk-screening, block, or roller printing.

Customers want the clothing they purchase to be colorfast, which means they want the color to be permanent and resist fading. Some garments are colorfast in soap and water, but may fade when exposed to sunlight, dry-cleaning fluids, perspiration, or certain chemicals. Solution-dyed or stock-dyed fabrics are the most colorfast; piece-dyed fabrics are the least colorfast. Specific colorfast information may be listed on the garment label.

Finishes That Improve a Fabric's Hand

A fabric's **hand,** or feel, can be changed or improved by the following finishes:

Sizing Adds weight, stiffness, and firmness to fabrics. (Sizing may not be permanent; many sizing finishes come out in laundering.)

Weighting Adds a heavy feeling and stiffness to fabrics; silk and linen are often weighted.

Finishes That Make a Fabric More Serviceable

Some finishes that help fabrics provide longer, more comfortable, and safer wear are:

Antistatic Controls or prevents static electricity "cling."

Flame-retardant Makes fabrics resist burning and, if ignited, burn slowly. (Children's sleepwear must pass government flame-retardant regulations.)

Insulating Provides warmth or coolness, depending on the weather.

Mildew-resistant Prevents mildew.

Moth-resistant Makes wool fabrics immune to moths.

Water-repellent Resists water but allows for the passage of air.

Waterproof Allows no water or air to pass through the fabric.

Finishes That Make a Fabric "Easy Care"

Some of the most well-known finishes today make it easier for consumers to care for their clothes:

Crease Resistance Helps such fabrics as linen, cotton, and rayon to resist creasing and lose their wrinkles after hanging.

Permanent Press Allows a fabric to be machine-washed and dried and worn without any ironing. Permanent press fabrics keep their sharp crease and do not wrinkle as easily as other fabrics without this finish. They shrink very little and do not get dirty as quickly as other fabrics. Permanent press fabrics are treated with special resin finishes. These fabrics cannot be washed in very hot water, dried at high temperatures, chlorine bleached, or starched because the resin finish will be lost.

Shrinkage Control Keeps a fabric from losing its shape. There are two types of shrinkage control: **preshrunk** and **Sanforized.** If a garment label says preshrunk, it means the fabric may shrink up to 3 percent of its size. This amount of shrinkage will affect the comfort and fit of clothing. (*Shrunk* and *preshrunk* mean the same thing.) Clothing that carries a Sanforized label means the fabric was shrunk before it was made into a garment. This label is the manufacturer's guarantee that the fabric will not shrink more than 1 percent or one-third of an inch per yard. One percent shrinkage does not noticeably affect the fit of the garment either in the waist or the length.

Some fabric finishes, like embossing, are easily identified. Salespersons and customers may have a difficult time identifying many of the other less obvious fabric finishes, such as stain-repellent or fire-retardant finishes. Quality clothing may mention the fabric finishes on their labels. (See Figure 5-9.) Often, however, garments only have care labels, with no exact description of the fabric finish. Only a few finishes are permanent; most can be destroyed through improper care and laundering. That's why it is important to read the care labels for all fabrics to learn how to care for them properly.

Figure 5-9 What are the special qualities of the fabric finish shown on this label? Courtesy E.I. DuPont de Nemours, Co. Inc.

Experience 5-5 ■ What Should You Say When Customers Ask?

The following are some typical customer questions you will be expected to answer about fabric finishes. Write your answers to these questions on a separate sheet of paper.

1. How much will these jeans shrink? The label says preshrunk.
2. What's the difference between these two raincoats? The beige is water-repellent; the yellow one is waterproof.
3. The label on this silk blouse says it is weighted. What does that mean?
4. This tag says the sizing is permanent. Aren't all sizings permanent?
5. I've heard a lot about flame-retardant sleepwear for children. Does that mean the sleepwear won't burn?
6. This cotton fabric is so shiny that there must be a special finish on it. What do you think that finish could be?
7. How should I wash and dry this cotton/polyester shirt? The label says it's permanent press.
8. This brand of boys' pants cost $5 more than another brand and the only difference I can find is that the more expensive pants are Sanforized. Does that really make a difference?
9. This tag says the garment is colorfast. What does that mean?

■ THE FINISHED FABRIC

As was mentioned earlier in this chapter, most finished fabric has four characteristics: fiber, yarn, fabric construction, and finishes. For example, broadcloth is a finished fabric. It may be made of cotton *fiber,* the *yarn* may be two ply, and it is *woven* in a ribweave. Sometimes, this type of cloth is given a mercerized *finish* to increase the luster.

There are many fabrics in our wardrobes today. The list of common fabrics below includes many fabrics and typical clothing items that these fabrics are used to make. At any one time, not all these fabrics may be popular or fashionable.

Experience 5-6 ■ Spotting the Fashion Fabrics

Look through newspapers, fashion magazines, or clothing catalogs for the past winter and summer seasons. (You can find these newspapers and magazines at a local library.) On a separate sheet of paper, make a list of five winter fabrics and five summer fabrics. What were the three most popular fabrics of each season?

■ A LIST OF COMMON FABRICS

The following list describes fabrics that are commonly used in clothing.[2]

Batiste A fine, thin, plain weave fabric. It generally has a soft texture, but it can have a crisper feel. It is usually made of cotton, but can be made of blends (polyester and cotton), wool, silk, rayon, polyester, and linen. It is used for linings, infant's wear, dresses, blouses, shirts, and lingerie.

Challis A soft, lightweight, plain or perhaps twill weave fabric of wool, cotton, or man-made fibers. It usually has small floral designs. It is used for infants' wear for special occasions, pajamas, dresses, blouses, skirts and shawls for infants and adults.

Chambray A balanced, plain weave cotton fabric with a frosted appearance achieved by the use of colored warps and white filling. It is usually plain colored but may be striped or patterned. Fancy chambray may have small dobby designs (small figured designs) woven in the fabric. It is used for dresses, sportswear, pajamas, and shirting in lighter weights. The heavier weights are used for work clothes.

Chiffon A soft, lightweight, sheer, plain weave silk or man-made fiber (often rayon) fabric. Fine, hard spun (highly twisted) yarns in a balanced weave are used. It is used for formal dresses.

Corduroy A filling cut pile fabric with either narrow or wide wales of pile running the lengthwise direction of the fabric. The background weave may be either plain or twill weave. This was originally a cotton fabric, but man-made fibers are now sometimes used. Corduroy may be solid colored or printed. It is a traditional fall sportswear fabric for men, women, and children.

Covert A medium-weight, closely woven, hard-surfaced twill fabric that has a mottled or mixed color effect in grays, tans, or blues. Cotton, wool, or wool blends are the fibers used. It is used for raincoats, work clothing, and sportswear.

Crepe A large class of fabrics that have crinkled or grained surface effects. Most crepes are made in plain weave with highly twisted yarns used in either warp or filling directions or both. Silk, cotton, worsted, rayon, and man-made fibers may be used. True crepes usually require dry-cleaning to avoid excessive shrinkage.

Denim A strong, yarn-dyed cotton fabric made in a warp-faced twill weave. Various colored yarns may be used for the warps, but white or unbleached yarns are commonly used for the fillings. Heavy denims are used for work clothes, while lighter-weight denims, called jeans, are used for sportswear.

Double knit A knit fabric made with two sets of needles, which gives the knit a firm body and less stretchiness. A common stitch is the double pique stitch, which has a subtle diamond surface effect. Cotton, wool, acrylic, polyester, acetate, and silk have been used in double-knit fabrics. Double knits are used for women's suits, dresses, and sportswear.

Gabardine A tightly woven, durable warp-faced twill fabric that has a steep twill line and a clear finish. It is made of wool, cotton, silk, and man-made staple fibers in solid colors. Gabardine is used for sportswear, suits, uniforms, and raincoats for men, women, children.

Jersey A general term for a plain knitted fabric that does not have a distinct rib. It may be plain, printed, or embroidered. Jersey is used for dresses, coats, men's underwear, T-shirts, children's clothes, women's lingerie, and blouses.

Lace An open-work fabric produced by a network of threads twisted or knotted together. It is made of linen, cotton, nylon, and other fibers. It is used for wedding or other formal dresses, lingerie, scarves, and trimming.

Organdy A fine, sheer, plain weave cotton fabric with a very crisp, clear finish. Most organdy has a permanent crisp finish that is done by a chemical treatment. It is used for women's and girls' summer blouses and dresses.

Satin A smooth, usually lustrous fabric made of filament fibers, such as silk, acetate, or rayon, in a satin weave. It is used for evening wear and blouses.

Seersucker A medium-weight fabric with lengthwise, permanent, crinkle stripes woven into the fabric in a plain weave. It is made of 100 percent cotton or blends of cotton with man-made fibers, such as polyester, rayon, or acetate in stripes, checks, and plaids. Seersucker is used for men's and women's suits and for sportswear.

Serge A clear-finished, even twill with a pronounced diagonal rib on both the right and wrong side. It is usually woolen or worsted but is also made of cotton, silk, man-made fibers, or blends. Serge is used for men's and women's suits and raincoats.

Terry cloth An uncut pile weave fabric with loops on one or both sides of the fabric. It is made of all cotton or blends of cotton with man-made fibers, such as rayon or stretch nylon, and is used for beach wear, baby clothes, and robes.

For more information about fabrics, see Chapter 7, page 99.

Chapter Note

1. "High-tech Workout Wear: How to Choose the Right Clothes at the Right Price," *Glamour,* February 1987, p. 298.

2. *Fibers and Fabrics,* EIMC, 1976, Austin, Texas.

CHAPTER 6

Women's Clothing

In 1976, Liz Claiborne started a small sportswear company with her husband, using $50,000 of their own money and another $200,000 borrowed from family and friends. Today, Liz Claiborne Inc. is one of the biggest names in the apparel business. In fact, in January 1988, *Forbes* ranked the company first in profitability and sales growth in the apparel group.[1]

Liz Claiborne created her design empire by producing clothing that appeals to career working women. Jerome Chazen, Claiborne's executive vice president for marketing, defines her target market as "the executive and professional career woman who is a little more updated in her taste level, as opposed to a very traditional customer who might wear structured suits and blouses with ties."[2] This customer group, which has increased throughout the 1980s, is willing to pay for Liz Claiborne clothes, which sell in the "better" price range.

Before you may be considered a professional in women's clothing, you must be familiar with the many women's apparel manufacturers, including Liz Claiborne, Evan Picone, Counterparts, and Chaus. You need to be aware of the different styles, colors, and textures offered by manufacturers each season. Customers will also expect your advice on choosing clothing that is a good fit, compensates for their figure problems, and coordinates with the other items in their wardrobes.

■ WOMEN'S CLOTHING STYLES

Each of the major classifications of women's clothing consists of many styles, or specific looks. For example, some of the many dress styles are the shirtwaist, sheath, sundress, and sweater dress.

You will be recognized as being knowledgeable about women's clothing when you refer to a particular clothing item by its correct style name. By using specific style terms, you will be able to accurately describe and identify your store's assortment of women's apparel to customers and co-workers.

Coat Styles

The women's coat assortment may include reefers, Chesterfields, balmacaans, wraparounds, trench coats, and parkas. Depending on the fabric used, these coats can be worn in winter or spring and fall. The coat fabric also determines whether or not the coat has a dressy or casual look. (See Figure 6-1.)

Dress Styles

Dress styles follow three basic silhouettes, or shapes: the bouffant (bell-shaped), the bustle (back fullness), and the straight or tubular. Each of these silhouettes goes in and out of fashion. In the 1980s, all three of the shapes were popular. For example, the fit and flare dress is a variation of the bell silhouette. The wedge, with its accentuated shoulders, is the inverse of the bell silhouette. The peplum suit, with its defined waist and fullness at the hips and back, resembles the bustle. The sweater dress is the straight silhouette. Other popular dress styles are the shirtwaist, coatdress, slip dress, wrap dress, dropped-waist dress, and halter dress. (See Figure 6-2.)

Tops: Blouses, Shirts, Sweaters, and Jackets

Current blouse and shirt styles are the bow blouse, classic man-tailored long-sleeve shirt, tuxedo-front shirt, camp shirt, camisole, polo shirt, and T-shirt.

Classic sweater styles are the cardigan and pullover with crew neck, V neck, and turtleneck.

The Chanel-style jacket, fitted jacket, blazer, denim jean jacket, safari jacket, and shirt jacket are jacket styles that customers often request.

Figure 6-1 Familiarize yourself with the basic women's coat styles shown in this illustration.
© Daisy de Puthod.

Many salespeople and customers describe women's tops by referring to various details of the garment. For example, a customer may ask for a particular blouse by saying it has a jewel neckline, dolman sleeves, and keyhole back. The illustrations in Figure 6-3 will help you describe a garment (not every style name is included). Try to use these style terms often. Remember, customers and co-workers respect salespeople who can describe and explain clothing styles.

Bottoms and Swimwear

There are many styles of women's bottoms, including skirts, pants, and shorts that clothing salespeople should know.

Popular skirt styles are the sheath or slim skirt; A-line; dirndl, with soft pleats; wrap; pressed pleat; and circular skirt.

Classic straight-leg pants are the most popular. Cropped pants are worn above the ankle. Stirrup pants have straps that go around the bottom of the foot. Tapered or straight-leg pants were popular during most of the 1980s, although a pleated wide-leg pant became more fashionable toward the end of the decade. Flared pants were popular in the 1970s. Jeans and pedal pushers are two casual pant styles that are still fashionable today. (See Figure 6-4.)

Examples of different styles of shorts are the running short, walking short, Bermuda short, and pull-on (elastic waist) short. (See Figure 6-4.)

Swimwear styles are classified as one piece or two piece. Variations of the one-piece suit are the maillot (pronounced "my oh"), bandeau, blouson, swimdress, and sheath. The bikini is a popular two-piece suit. (See Figure 6-5.)

Hosiery, Intimate Apparel, and Sleepwear

Hosiery provides a coordinated look with women's clothing. Today, hosiery is available in many differ-

Figure 6-2 Learn to recognize these popular dress styles. © Daisy de Puthod.

ent textures, including pinstripes, dots, and floral designs. It is available in dark, neutral, and pastel colors that can be matched to a woman's clothing.

The intimate apparel area sells bras, panties, girdles, slips, unitards, nightgowns, pajamas, robes, and lounging clothes. These garments vary in size, fabric, and style. For example, full slips are fitted according to bust measurement and length. Half-slips are sized petite or extra small, small, medium, large, and extra large according to waist measurement, and petite, short, average, and long according to length.

Experience 6-1 ■ Name That Style

1. Give a detailed description of one of your female classmate's clothing. Name the style of her dress, shirt, sweater, jacket, skirt, or pants. Also name the neckline or collar and sleeve style. On a separate sheet of paper, sketch and label each clothing item.
2. Look through a fashion magazine, newspaper, or clothing catalog. Cut out five different styles of the following classifications: coats, dresses, shirts, sweaters, jackets, skirts, pants, shorts, and swimwear. Using a separate sheet of paper for each classification, attach the pictures to the paper. Write the correct style names for each picture on another paper titled "answer key."

   Trade your pictures with another classmate who is also completing this exercise. Then write the correct style names beneath the pictures your classmate gives you.

   Return the pictures to your classmate when you have both completed naming the styles. Check your answer key to see how many he or she was able to identify correctly.

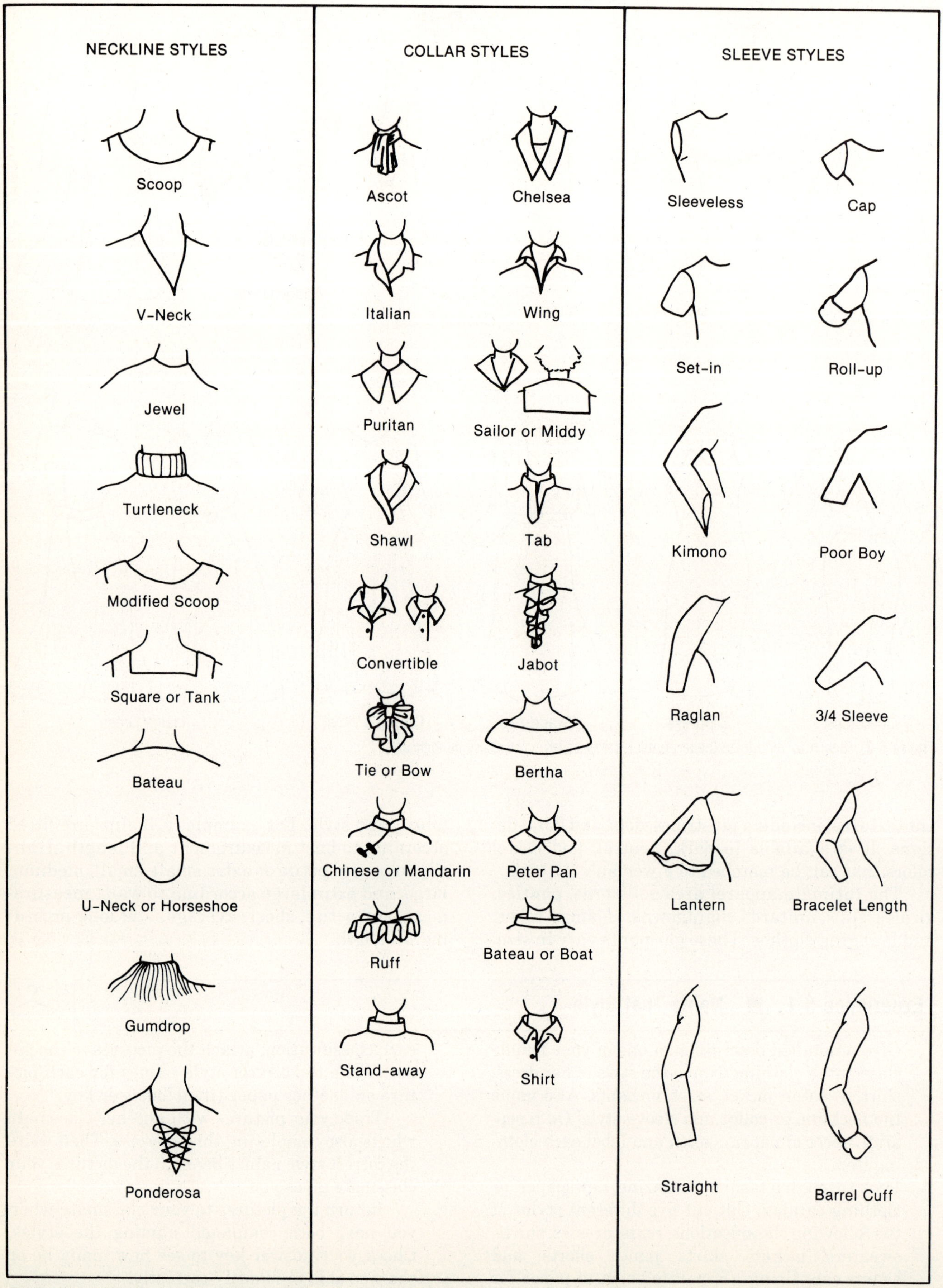

Figure 6-3 You can describe women's clothing if you know these details.

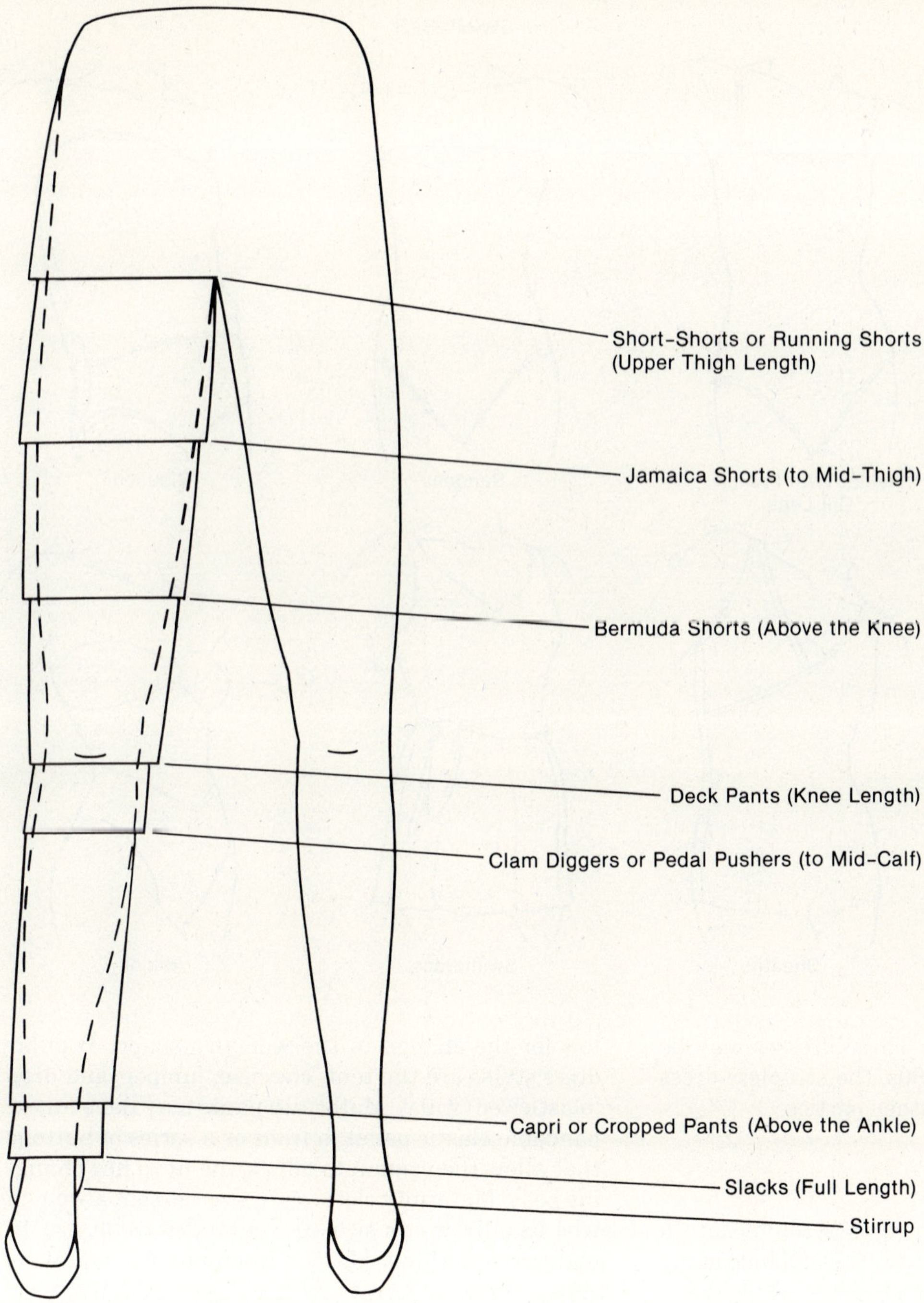

Figure 6-4 Pants are identified by their length.

Bridal Attire and Formal Wear

Another important classification of women's clothing is bridal and formal wear. Some stores provide bridal consultants to help coordinate the entire bridal outfit. These specialized salespeople help a bride and her attendants choose coordinated dresses and headpieces.

Bridal gowns are sold differently from other apparel. Instead of buying a complete size range of a certain dress, a store will stock a sample dress in each style, usually size 10. Women look at the samples and try on the dresses they like best. When they decide which dress they want, the dress is special ordered in the size needed.

Traditional bridal gowns are white floor-length dresses with a veil and train. They may be styled to have a fitted waist, empire or raised waist, or no waist, such as the A-line. Because bridal gowns are special dresses for a very special day, there are many style variations so that each dress looks unique or different. Luxurious fabrics—satin, taffeta, silk, brocade, organdy, and lace—are used.

Evening gowns and party dresses may be sold at a specialty shop or in the "special occasion" depart-

Figure 6-5 Some of the many styles of women's swimwear are shown here.

ment of a department store. These dresses may be short or long. In the late 1980s, the strapless dress was popular for proms and other parties.

Maternity Clothes

Maternity clothes are designed to accommodate a pregnant woman's growing body. They include many loose-fitting tops, dresses, and coats, expandable pants, and underwear (such as panties, hosiery, slips, and bras). Dresses are designed to fit a fuller bust. They are longer in the front than in the back to allow for the change in the woman's shape. Popular dress styles are the tent, chemise, jumper, and drop (elasticized) waist. Maternity pants may have an expandable elastic panel in front or a series of buttons that allow the woman to adjust the fit to her changing body. Maternity clothes are sized so that a woman who usually wears size 10 would also wear size 10 maternity clothing. Many women buy good-quality, classic-styled maternity clothes. They often save their maternity clothes in case they become pregnant again. Or they give their clothes to other pregnant women.

Experience 6-2 ■ Special Dresses for Special Days

Visit a store that sells bridal gowns or formal wear for women. Choose three different dresses that you especially like. On a separate sheet of paper, describe in detail the style of each dress and its fabric. Next to each description, make a simple sketch of the dress. Choose accessories (jewelry, headpiece, shoes, and so on) that would turn this dress into a coordinated ensemble. Describe these accessories on your paper. If you are unable to visit a store, locate a copy of a bridal magazine and choose dresses and accessories from it.

WOMEN'S CLOTHING SIZES

Women's clothing is available in many size groupings—junior, misses, petite, tall, women's, and women's petite sizes. You will soon find out that not all manufacturers produce clothing in the complete size range. For example, even though the complete misses size range goes from size 2 through 20, a manufacturer may offer clothing in sizes 4 through 16 only. Also, stores do not usually purchase the complete size range of a particular style. One store may include sizes 4 through 12 in their assortment; another may stock sizes 6 through 16 of the same clothing. Both variations are because manufacturers and stores adjust their inventories to meet the needs of their specific target markets.

The same size from different manufacturers may fit differently. For example, a size 8 straight skirt from Ellen Tracy may be looser than a similarly styled size 8 skirt from Jones New York. Because the fit may differ, customers should try on clothing before making a purchase decision.

Choosing the Right Size

When helping women choose clothing, you should know the "most important" measurement. If a garment fits the customer's "most important" measurement, it will be comfortable, attractive, and fit well. For example, for pants or skirts, the most important measurement is the hip measurement. If a customer has 36-inch hips, size 7 pants should fit her (see Table 1). For shirts, sweaters, jackets, or vests, the most important measurement is the bust measurement. The bust, waist, and hip measurements are all important when fitting a dress, suit, or coat.

The best way to help your customer get the right fit is to suggest she try on the clothing item. After examining the fit, it is easy to decide if she needs a larger or smaller size.

Junior Sizes

Many stores, such as 5-7-9 Shop, Susie's Casuals, and Foxmoor Casuals, sell sportswear, dresses, and coats for junior-sized customers. Junior sizes range from size 1 to 15. They are designed for women with a smaller frame and a shorter shoulder-to-waist measurement than women who wear misses sizes. (See Table 1.) Junior-sized clothing is often very trendy, that is, the styles, colors, and patterns reflect the very latest fashion look and change often. The styles featured in this size range appeal mostly to teenagers and women in their 20s.

Misses Sizes

Misses sizes are the most popular size grouping for women. They range from sizes 2 to 20 and fit women who have an average frame and figure. (See Table 2.)

Misses and Junior Petite and Tall Sizes

Petite clothing is for women of average proportions who are 5 feet 3 inches and under. The clothing is sized so that it is shorter in total length, in the torso (from shoulder to waist), and in the sleeve. Details, such as pockets, collars, and lapels, are also sized proportionately smaller. (See Figure 6-6.)

Some petite clothing manufacturers size their clothing to fit women of the same measurements as misses or juniors. For example, two women with the same bust, waist, and hip measurements (35-26-37) would wear the same size, size 10. The woman who is 5 feet 1 inch

Table 1
Junior Sizes

Junior sizes: for women with smaller frames and slightly shorter torso proportions than misses sizes.
Petite junior sizes: for women with smaller frames and slightly shorter torso proportions than misses sizes who are 5 feet 3 inches or under in height.

Size	3	5	7	9	11	13	15
If bust is (in inches)	31½–32	32½–33	33½–34	34½–35	35½–36½	37–38	38½–39
If waist is (in inches)	22½–23	23½–24	24½–25	25½–26	26½–27½	28–29	29½–30½
If hips are (in inches)	33½–34	34½–35	35½–36	36½–37	37½–38	39–40	40½–41½

Source: Sears Roebuck and Co.

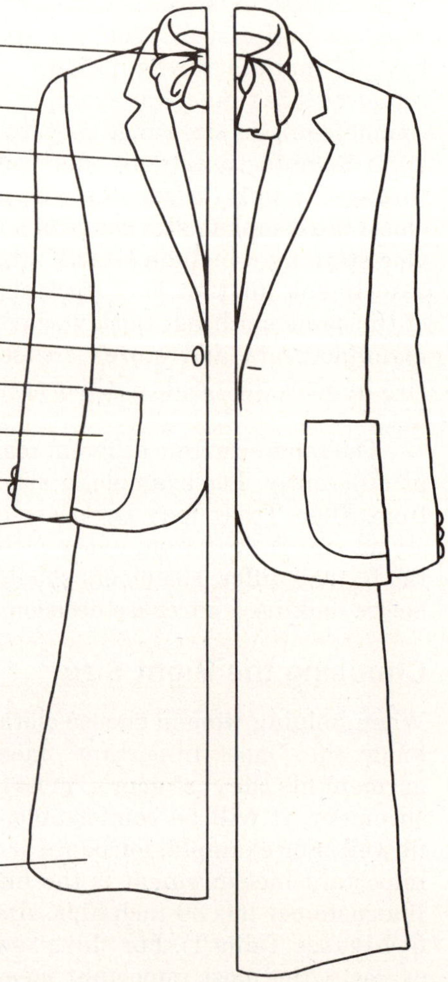

Figure 6-6 Petite sizes are not only shorter than regular misses sizes, they are scaled down in other ways to fit and flatter the proportions of the petite figure. Based on material from Leslie Fay.

would wear a size 10 petite. A woman who is 5 feet 6 inches would wear a size 10 misses. Other manufacturers design their petite clothing one size larger. For example, if the customer's measurements indicate that she should wear a size 10 misses, and if she is 5 feet 2 inches she would need a size 12 petite.

Women's clothing manufacturers also produce tall sizes for women over 5 feet 7 inches tall. These

■ **Table 2**
Misses Sizes

Misses sizes: for women with average frame and average proportions.
Petite misses sizes: for women with average frames and average proportions who are under 5 feet 3 inches.

Size	6	8	10	12	14	16	18	20
If bust is (in inches)	32½–33	33½–34	34½–35	35½–36½	37–38	38½–39½	40–41½	42–43½
If waist is (in inches)	23½–24	24½–25	25½–26	26½–27½	28–29	29½–30½	31–32½	33–34½
If hips are (in inches)	34½–35	35½–36	36½–37	37½–38½	39–40	40½–41½	42–43½	44–45½

Source: Sears Roebuck and Co.

sizes are longer in total length, torso, and sleeves. The details are sized proportionately larger than misses. The tag on the clothing indicates if it is a misses or junior, petite or tall size.

Women's Sizes

Women's sizes are for women between 5 feet 4 inches and 5 feet 7 inches who have medium to heavy builds. This size range is designed for women with full hips, busts, and arms. It includes sizes 14W through 32W. (See Table 3.) Few stores stock an extensive assortment of women's sizes, with the exception of specialty stores such as Lane Bryant and Woman's World Shops. Your knowledge of this size range will help you advise the larger woman how to get a better fit and look more attractive.

Women's Petite Sizes

Women's petite, or WP, sizes are for shorter women with medium to heavy frames. The clothing, which ranges in size from 14WP to 32WP, is cut shorter from shoulder to waist than women's sizes. (See Table 3.)

Small, Medium, Large, and Extra Large Sizes

Sometimes sweaters, tops, bottoms, leisurewear, and workout clothing are sized according to small, medium, large, and extra large rather than numbered sizes. In junior sizes, small (S) is sizes 5 to 7, medium (M) is sizes 9 to 11, and large is sizes 13 to 15. In misses sizes, small is 4 to 6, medium is 8 to 10, large is 12 to 14, and extra large is 16 to 18.

■ Table 3
Women's Sizes

Women's sizes: for women who are 5 feet 4½ inches to 5 feet 7 inches in stocking feet who have fuller, more mature figures, with longer torsos and arm proportions than women's petite sizes.
Women's petite sizes: for women who are 5 feet 4 inches and under in stocking feet with medium to heavy frames, and shorter torsos and arm proportions than women's sizes.

If your bust is	37½–39	39½–41	41½–43	43½–45	45½–47	47½–49	49½–51	51½–53	53 ½–55	55½–57
If your waist is (in inches)	29½–31	31½–33	33½–35	35½–37	37½–39	39½–41	41½–43	43½–45	45½–47	47½–49
If your hips are (in inches)	39½–41	41½–43	43½–45	45½–47	47½–49	49½–51	51½–53	53½–55	55½–57	57½–59
Women's size	14W	16W	18W	20W	22W	24W	26W	28W	30W	32W
Women's petite size	14WP	16WP	18WP	20WP	22WP	24WP	26WP	28WP	30WP	32WP

Source: Sears Roebuck and Co.

Experience 6-3 ■ Size Them Up

1. Estimate the height and proportions of three female classmates. Record your estimates in the following chart. Then ask each for her actual measurements. If they are not sure, use a measuring tape and yardstick to determine the accurate measurements of your classmates.

	Classmate 1		Classmate 2		Classmate 3	
Measurements	**Estimate**	**Actual**	**Estimate**	**Actual**	**Estimate**	**Actual**
Height						
Bust						
Waist						
Hips						

(continued)

2. Refer to the size charts in Tables 1 to 4. Based on your three classmates' actual measurements, decide which size pants, blouse, jacket, and dress each wears. Record the sizes in the following chart. Check with your classmates to see if the sizes determined from the size chart are the same sizes that they last purchased in each of the items listed in the chart below.

Item	Classmate 1		Classmate 2		Classmate 3	
	Size from Chart	Size from Classmate	Size from Chart	Size from Classmate	Size from Chart	Size from Classmate
Pants						
Blouse						
Jacket						
Dress						

■ FITTING WOMEN'S CLOTHING

The way clothing fits is very important. Customers want clothing that is comfortable when sitting, standing, and moving around. They also want attractive clothing. Clothing that fits well will be comfortable and attractive.

Customers will often ask your advice on whether or not a particular item fits properly. Sometimes a customer is not used to seeing a particular style and has a difficult time recognizing proper fit. For example, if a customer is not used to the "over-sized" look or big shoulder pads, she may think a dress doesn't fit.

When working with customers, check the fit of the clothing they like. In general, look for too much fullness or unwanted wrinkles. These are signs that a garment does not fit properly. For example, check the pants for horizontal wrinkles from the crotch to the hips. These wrinkles indicate that the fabric is pulling because the pants are too small at the hips or thighs. A coat, dress, or blouse collar should fit close to the neck at the back and sides. The sleeves should hang straight and, if the blouse has set-in sleeves, the seams should be at the shoulder bone. Darts, which add fullness or taper the fit of a garment at the bust, shoulders, waist, elbows, or back, should make the garment fit better and look more attractive. Lengths go up and down depending on fashion. Check the hem on skirts and dresses to be sure it is straight.

After you have identified a problem with fit, find the customer a larger or smaller size to correct the problem. For example, if a size 10 is too large in the jacket, suggest the customer try a size 8 or perhaps a size 10 petite. It is better to show the customer a similar style in the correct size than to significantly alter a poorly fitting garment. Major alterations are expensive and may not produce the desired fit.

You will be better able to help customers with fitting problems when they wear the same kind of undergarments and shoes as they will normally wear with the clothing. The undergarments and shoes help you get the real picture of how a garment drapes on the customer's body and if the length is correct.

When customers wear the wrong shoes or undergarments, point out how the clothing will look with all the proper accessories. If you work in a store that sells intimate apparel and accessories, bring the customer the additional items and put together the total look. For example, bring the customer a strapless bra and attractive necklace to wear with a strapless party dress. Customers will get an idea of the finished look and thank you for your extra effort in helping them. And this extra touch will pay off in bigger sales for you.

■ ANALYZING CUSTOMER FIGURE TYPES

A good figure is a matter of proportion. Each part of the body is seen in relation to the other parts as well as to the whole body. Pleasing female proportions follow these guidelines:

- About two-fifths of the body height is above the waist and three-fifths of the body height is below the waist.
- The body can be cut in quarters at the bustline, hips, and knees (see Figure 6-7).
- The shoulders are slightly wider than the hips and the waist is narrow enough to complement both shoulders and hips.

Few customers have a perfect figure and fewer still are satisfied with the way they look. They want to be taller, thinner, bigger one place, or smaller another. Sometimes they want to camouflage large hips

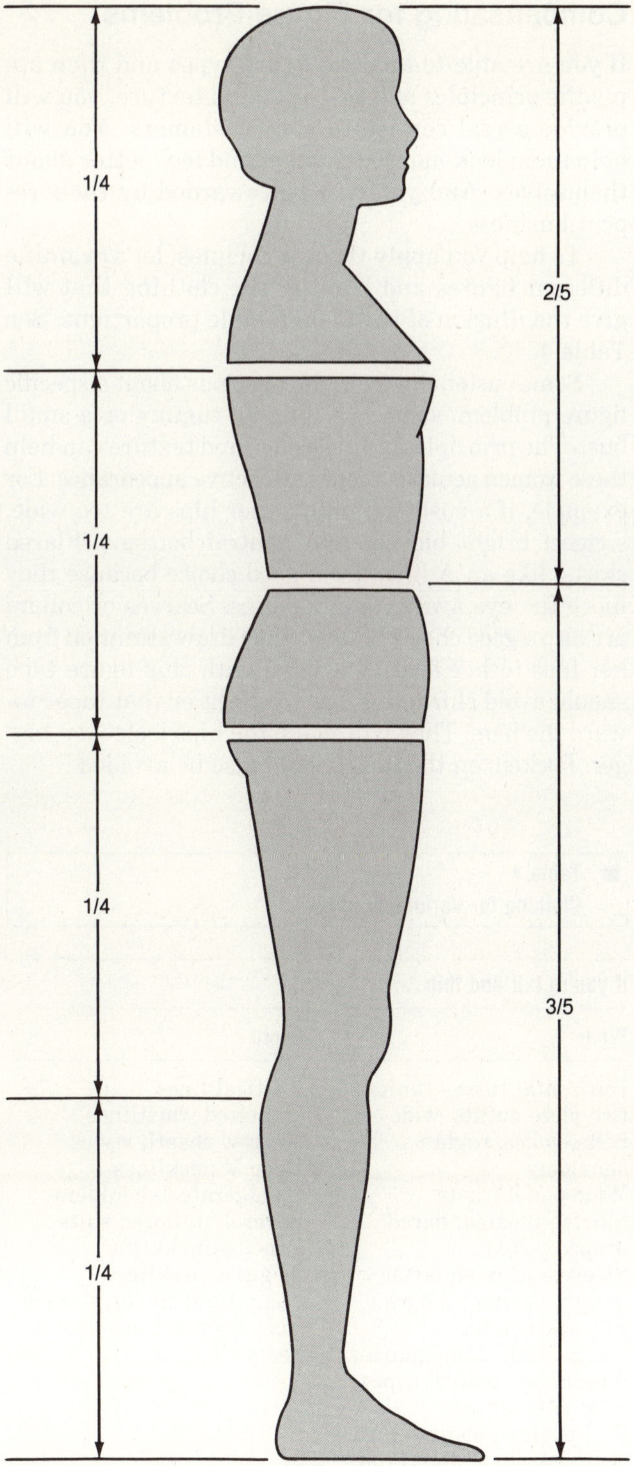

Figure 6-7 The ideal female figure is proportioned according to the guidelines shown above. *Source: Dress*, Bennett Pub., 1975.

or a bulging tummy. As a salesperson, you can help a customer achieve the look she wants by analyzing her figure and choosing styles or clothing lines, colors, and textures that best suit her unique figure characteristics.

Well-chosen clothing calls attention to the customer's best features and away from her worst. Because of the large selection of clothing available today, women can experiment with clothing lines, colors, and textures to find their most attractive looks.

Clothing Lines

The lines of a garment can fool your eyes. For example, vertical lines, which carry your eyes upward, add height to a figure. Horizontal lines, which carry your eyes around the body, add weight and shorten a woman's figure. Curved lines add softness and fullness.

When helping customers select clothing, remember the following:

- Garments with a single center vertical line add height and slim the figure.
- An asymmetrical or off-center vertical line adds height and cuts the weight of the figure.
- A series of vertical parallel lines, such as stripes or pleats, force the eyes to travel around the figure. These multivertical lines add height and width to the figure. The closer the parallel lines are, the less wide the figure appears to be.
- A vertical line cut by a horizontal line at the hips cuts the figure's height. It is a good choice for a tall person.
- An arrow line, like a halter dress or raglan sleeves, adds height and narrows the shoulders. It is a good choice for a short, broad-shouldered woman.
- Diagonal lines add interest and eye appeal. The more vertical a diagonal line is, the thinner the figure will look. The more horizontal a diagonal line is, the broader the figure will look.
- Horizontal lines make any figure appear shorter and heavier. The lower the horizontal line, the shorter the woman looks. A low horizontal line, produced by wearing a long tunic or three-quarter length coat, may make a short woman look dumpy and a tall woman look more attractive. A short woman can wear a high horizontal line, such as an empire waist.
- Big prints will make a figure appear larger. For example, a large geometric or floral print blouse will call attention to the figure above a woman's waist.
- A garment with ruffles or large sleeves adds to the apparent body weight. Although some fullness can make a thin person look more attractive, too much fullness will dwarf the thin person.
- Narrow belts do not attract as much attention as wide belts. Belts in the same fabric as a garment,

called self belts, are good for women with large waists and broad hips because they seem to blend into the garment.

- Deep lapels, vertical darts, and long V necklines are slenderizing because they add height. Broad V necklines widen the figure.

Clothing Colors and Textures

Color is very important in helping a customer achieve the look she wants. Colors that can be described as warm or bright will attract attention to the parts of the body they cover. Cool- or neutral-colored clothing will not attract attention to the figure. Very dark colors, such as navy blue or black, also tend to make the figure appear slimmer. For example, a taupe (neutral) skirt with a sunshine yellow (warm) blouse will attract the eye above the waist. This would be a good color combination for a woman with large hips and small bust because it would visually balance the figure.

A shape will appear longer and slimmer if it is clothed in one color rather than multiple or contrasting colors. For example, a pale peach dress with matching-color hosiery and shoes will help a short woman look taller than if she wore a white blouse with navy skirt, neutral hose and white shoes.

Use bright colors to call attention to a woman's best features. For example, a hot pink collar on a plum-colored dress will attract attention to the face and away from a bulging tummy. Contrasting-color yokes, pockets, and belts can make a woman look shorter because contrasting colors interrupt the vertical line of the figure.

You may work with customers who refer to themselves as a fall, winter, spring, or summer person. These customers are referring to the color analysis system that was made popular in the book *Color Me Beautiful.* Each season represents a certain group of colors in various tints or shades that enhance particular skin, eye, and hair combinations. Many customers who believe in this color-coding system shop with their own color palette, which includes chips of the exact clothing colors they should wear. Customers who mention their "season" have very specific clothing needs. They usually want to see only those garments that are available in their particular colors.

In addition to color, the clothing's texture can also create an illusion that a woman is larger or smaller than she really is. Shiny or glossy fabrics, such as satin or sequins, reflect light and make a figure look larger. Dull textures, such as flannel or jersey, absorb the light and make a figure look smaller. Smooth or sheer fabrics, such as poplin or voile, also make a figure appear smaller. Coarse or heavily textured fabrics, such as mohair or corduroy, give the illusion of fullness.

Compensating for Figure Problems

If you are able to analyze figure types and then apply the principles of line, color, and texture, you will provide a real service to your customers. You will help them look more attractive and feel better about themselves. And you will be rewarded by their repeat business.

To help you apply these principles, let's examine different figures and identify the clothing that will give the illusion of the ideal female proportions. See Table 4.

Some customers feel self-conscious about a specific figure problem, such as a bulging tummy or a small bust. The principles of line, color, and texture can help these women achieve a more attractive appearance. For example, if a customer thinks her hips are too wide, suggest bright blouses and neutral bottoms. Flared skirts, like an A-line, are a good choice because they move the eye away from her hips. Scarves or collars are also a good choice because they draw attention from her hips to her face. A woman with this figure type should avoid skirts that are too tight or that taper toward the hem. They will make the hips look even bigger. Pockets on the hips should also be avoided.

■ Table 4
Clothing for Various Figures

If you're tall and thin...

Wear	Avoid
Horizontal lines—tunics, two-piece outfits, wide belts, yokes, raglan shoulders	Vertical lines
Wide-panel fronts	Unbelted waistlines
Skirts—pleated, flared, draped yoked	Narrow sheath styles
Sleeves—three-quarter length, bracelet length, full; wide cuffs	Deep V necklines
Coats—full, three-quarter length, and belted; capes	Exaggerated shoulders
Tent silhouettes	Severely tailored suits
Full bodices, bloused tops	Pencil-slim skirts
Round necks, large collars	Angular necklines
Large accessories—big handbags and pieces of jewelry	Long, tight-fitting sleeves
Shoes with medium heels	Very high or very low heels
Pants—any length and cut	
Contrasting or bold colors	
Plaids, tweeds; large or bold prints	
Crisp, bulky, or shiny fabrics	

(continued)

Table 4 *(Continued)*

If you're tall and heavy...

Wear	Avoid
Vertical lines	Horizontal lines
Asymmetrical lines	Yokes, if bust is full
Diagonal lines	Princess lines
T-, Y-, and arrow lines	Wide-panel fronts
A-line or center-pleated skirts	Multiverticals
V necklines	Wide belts
Dropped waist	Full or pleated skirts or very straight skirts
Pointed collars	Round necklines
Self belts (less than 1½ inches wide)	Tent silhouettes
Moderate-length jackets	Bloused bodices
Full-length coats	Sleeveless dresses, cap or tight sleeves
Set-in sleeves	Boleros or short jackets
Shoes with medium heels	Jerseys, chiffon, satin, stiff materials
Subdued colors	Plaids or large prints
Pants—straight and simple in solid colors	Tight-fitting styles

If you're short and thin...

Wear	Avoid
Horizontal lines	Deep V necklines
Modified vertical lines—T-, arrow, and Y-lines	Exaggerated lines of any kind
Narrow-panel and wide-panel fronts, princess lines	Very wide shoulders, capes
Multivertical lines	Wide belts
Pleated, full, flared, or draped skirts	Long jackets
Semifitted sheaths	Tunics
Bolero or short jackets	Tent silhouettes
	Pencil-slim skirts
	Tight-fitting sleeves
	Heavy trimmings
	Large handbags and accessories

(continued)

Table 4 *(Continued)*

If you're short and thin...

Wear	Avoid
Short V necklines, round collars and necklines	Chunky jewelry
Bloused bodices	Clinging fabrics
Empire waistlines	
Delicate trimmings	
Bulky, shiny, or stiff fabrics	
Bold colors	
Small prints	
Pants—any cut	

If you're short and heavy...

Wear	Avoid
Vertical lines	Horizontal lines
Asymmetrical lines	Wide-panel fronts
Diagonal lines	Princess lines
Y-lines	Multiverticals
V necklines	Wide belts
Set-in sleeves	Boleros and bulky tunics
Narrow, half- or self belts or beltless styles	Extremely long jackets
A-line skirts	Draped skirts
Center-pleated skirts	Tent silhouettes
Jackets not much below hipbone	Pencil-slim skirts
Subdued colors	Full skirts
Shoes with medium heels	Capes and three quarter length coats
Medium-sized handbags	Blouson tops
Pants—straight cut in solid colors	Plaids and bold prints
	Bright colors
	Clinging or stiff fabrics, bulky or crisp textures, shiny finishes

Source: Helen Whitcomb and Laura S. Cochran, *The Modern Ms.*, New York: McGraw-Hill Book Co., 1975.

Experience 6-4 ■ Solving Figure Problems

As a clothing salesperson, you will have to be able to analyze customers' figures and know which clothing items look best on certain figure types.

If you are female, follow the steps below to analyze your own figure. If you are male, ask a female friend or relative to help you complete this exercise. Examine her figure type and suggest appropriate clothing that will compensate for specific figure problems. Use a separate sheet of paper to record your answers.

1. Look in a full-length mirror. Write your best two figure assets and two figure liabilities.
2. Have someone measure your height with a yardstick. Remove your shoes and stand against a wall or door. Is two-fifths of your body height above your waist? If not, are you short or long-waisted?
3. Can your body height be cut in quarters at your bustline, hips, and knees?
4. Based on the above information, what are your most attractive clothing styles? Where should you wear warm or bright colors? Cool, dark or neutral colors? Shiny textures? Big patterns? Dull textures? Special details?
5. Look through fashion magazines and catalogs and cut out three items that would look most attractive on you.

■ COORDINATING WOMEN'S CLOTHING

A complete, coordinated outfit includes many different clothing items—intimate apparel, shoes and accessories, as well as any combination of coat, suit, dress, skirt, pant, blouse, sweater, or jacket. All items should add up to one "put together" look. Customers often ask and want suggestions on what to wear and "what goes with what." Salespeople who suggest coordinated, attractive outfits by combining different clothing items and accessories earn the reward of higher sales, more commission, and the appreciation of their store management.

For example, a customer who is buying an expensive wool gabardine suit will welcome a salesperson's suggestion of buying two blouses: a white tuxedo-front oxford-cloth shirt, and a bright yellow silk T-shirt. The two blouses provide different looks and increase the versatility and value of the expensive suit. The right pin, scarf, or belt can turn a rather boring outfit into one that commands attention. These related suggestions will increase the customer's satisfaction with her purchase as well as increase the sale for the salesperson.

Clothing coordination is largely a matter of personal taste. Watch your customers' reaction to the styles, colors, textures, and accessories you put together. Your recommendations should be based on their response and may need to be altered if their feedback is unfavorable. Customers who wear clothing that is either too fancy or too casual for an occasion will be considered poorly dressed by their peers. If the customer purchases inappropriate clothing based on your suggestions, it is doubtful that she will shop with you again. On the other hand, when you help customers purchase the "right" clothing, they may tell friends about the service you provided. Your quality service could bring you additional customers.

Experience 6-5 ■ Coordinate!

Choose a particular occasion or event that you or a female friend will be attending. Put together a coordinated outfit by deciding on the type of clothing that would be appropriate to wear. Visit a clothing store or use a catalog and choose the clothing items and accessories that could be worn. Explain how the coordinated look was achieved. Complete the following chart by describing each clothing item in detail. (It is not necessary to include all these items if they do not apply.)

Occasion: ______________________________

Type of occasion (circle one):

formal dressy business casual sporty

Clothing Item

Coat ______________________________

Hat ______________________________

Dress ______________________________

Skirt ______________________________

Pants ______________________________

Blouse ______________________________

Sweater ______________________________

Jacket ______________________________

Jewelry ______________________________

Scarf ______________________________

Belt ______________________________

Hosiery ______________________________

Intimate apparel ______________________________

Shoes ______________________________

Handbag ______________________________

Chapter Notes

1. *Forbes,* January 1988, p. 74.

2. Rayna Skalnick, "Liz the Wiz," *Sales and Marketing Management,* September 9, 1985, p. 50.

CHAPTER 7

Menswear

The polo shirt is a clothing classic. American men usually own several of these knitted, soft collar, short-sleeve shirts with the extra long tail. No one knows who designed the polo shirt, but it was first noted in 1893 when it was worn by polo players in Buenos Aires. In 1920, Lewis Lace, an Argentinian polo star, started selling these shirts with the logo of a polo player on a horse. In 1933, French tennis star, René Lacoste, produced a polo shirt with a crocodile logo. (Lacoste was known as the "crocodile" for his snappy play on the tennis court.) A company called Izod began selling the Izod Lacoste polo shirt in the United States in the 1950s. In 1972, American designer Ralph Lauren introduced his own version of the polo shirt with polo player and pony. Polo/Ralph Lauren (the name of Lauren's company) says it sells about 4 million of these shirts every year.[1]

Lauren's polo shirt is a trendy, fashion item for today's contemporary male. It is just one of the many product lines available in upscale department or specialty stores.

This chapter will prepare you to sell men's tailored clothing including suits, sportcoats, dress slacks, and overcoats. It also discusses men's furnishings, such as shirts, ties, hoisery, and pajamas and robes, and men's casual wear and active sportswear.

■ SUITS, SPORTCOATS, AND DRESS SLACKS

A man's suit consists of a jacket and trousers that are specifically designed to be worn together. A suit offers a man one look because the jacket is seldom worn without the trousers. Men who remove their jackets once they get to the office often find that their trousers wear out faster than their suit jacket.

There are also three-piece suits which consist of a jacket, trousers, and a vest. A vested suit is versatile; it can be worn with or without the vest. The vest presents a more formal look for the man who likes to remove his suit jacket. (See Figure 7-1.)

Many men wear tailored sportcoats and dress slacks that have not been matched as a suit. The color, pattern, or texture of a sportcoat is different from that of the slacks. A blazer is a popular style of sportcoat. The traditional blazer is blue or camel. It has metal buttons on the front closings and cuffs. Blazers are very versatile because they can be worn over matching or contrasting slacks.

Dress slacks are available in a wide range of colors and patterns. Also versatile, they can be worn with a blazer or sportcoat or with a shirt and/or sweater without a coat.

Sizes

The size suit and sportcoat a man wears is determined by his height and chest measurement. Men between 5 feet 3 inches and 5 feet 6 inches wear a short. Men between 5 feet 7 inches and 5 feet 11 inches wear a regular. Men between 5 feet 11 inches and 6 feet 3 inches wear a long. There are also extra long suits and sportcoats for men over 6 feet 3 inches.

Many men's clothing stores stock suits and sportcoats in even sizes ranging from 36 to 46 short, regular, long, and extra long. Some stores include odd sizes (37, 39, 41, and 43) and larger sizes (48, 50, and 52). Some stores also carry stouts or portly sizes, to fit the short, heavyset customer. (See Figure 7-2.) The suit or coat size is usually written on a tag attached to the left coat sleeve.

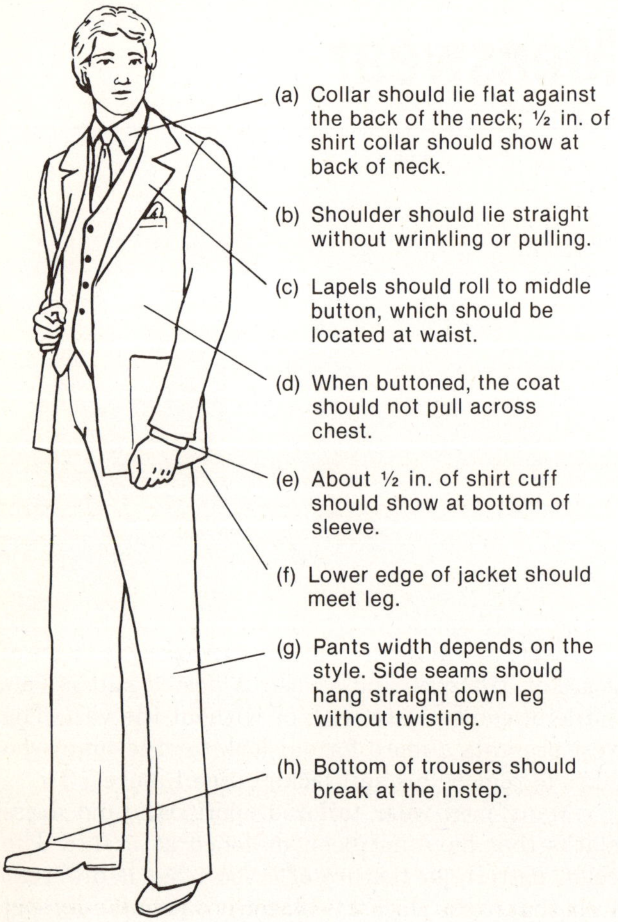

Figure 7-1 Some pointers for fitting a man's three-piece suit are shown above.

To measure a customer for a suit or coat, wrap a measuring tape around his chest, under his arms, and over his shoulder blades. Hold the tape snugly but not too tightly. This will give you the customer's chest measurement. Then estimate the customer's height or ask him. Depending on the customer's height and chest measurement, you can then choose the appropriate size. For example, a man who has a chest measurement of 39½ inches and is 5 feet 10 inches would wear a size 40 regular.

Another way to determine suit or coat size is to estimate the size needed and try that size coat on the customer. If the coat is too large, locate a smaller size; if it is too small, choose a larger size. After some experience and practice, a salesperson finds it easy to estimate the correct size. In fact, professional suit salespeople can accurately determine a customer's size just by looking at him.

Dress slacks are sized in inches and according to the waist and inseam measurements. The inseam is the measurement from the crotch to the bottom of the pants. Slacks that are sized 30–33 will fit a person with a 30-inch waist and 33-inch inseam measurement. The size is usually on a tag attached to the waist.

More expensive dress slacks do not have a finished bottom. After a man has decided to buy the slacks, the salesperson measures the slacks for correct length, marking exactly where the slacks will be hemmed or cuffed.

Although most customers know what size slacks they wear, a men's clothing salesperson may need to measure a customer's waist to determine the size needed. To measure the waist, wrap the measuring tape around the waistline, just above the hip bones. To measure the inseam, place the tape measure below the zipper and measure the length according to the customer's preference.

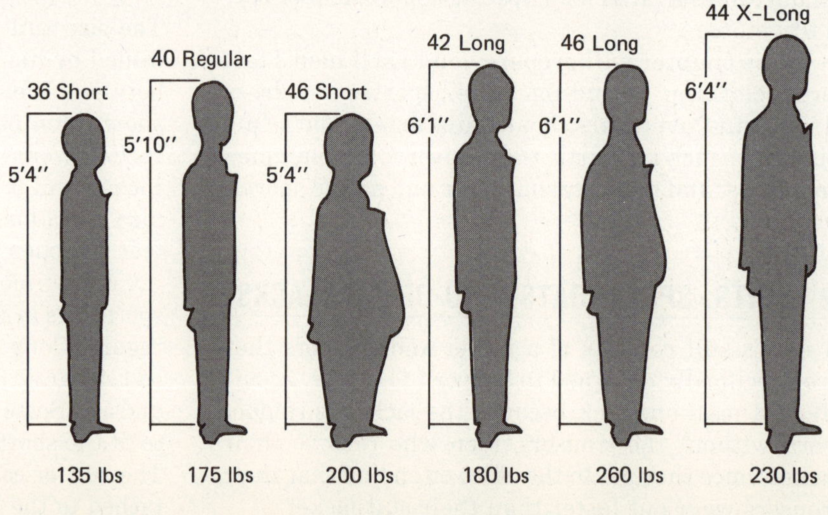

Figure 7-2 A man's suit size is related to his height and weight.

Experience 7-1 ■ Size Them Up

1. Estimate the height, chest, and waist measurements of three male classmates. Record your estimates in the following chart. After you have made your estimates, use a measuring tape and yardstick to determine their accurate measurements. Record the actual measurements below.

	Classmate 1		Classmate 2		Classmate 3	
Measurements	**Estimate**	**Actual**	**Estimate**	**Actual**	**Estimate**	**Actual**
Height						
Chest						
Waist						

2. Based on the actual measurements of your classmates, decide what size suit, sportcoat, and dress slacks each would wear. Record the sizes below. Check with your classmates to see if the sizes you estimated are the sizes they last purchased in each item.

	Classmate 1		Classmate 2		Classmate 3	
Item	**Estimate**	**Actual**	**Estimate**	**Actual**	**Estimate**	**Actual**
Suit						
Sportcoat						
Dress slacks						

Styles

Suit styles are often referred to as models. There are three basic suit models: the traditional, the contemporary, and the European. (See Figure 7-3.)

The traditional model, also referred to as the American silhouette, comes from the famous Brooks Brothers "sack" suit of the 1950s. It is the most conservative, body-concealing, boxy-looking of the suit styles. It is cut straight and full with **natural shoulders** (soft shoulders with a minimum of padding); low, roomy armholes; medium-width notched lapels; and a shallow center vent in the jacket. (The center vent allows the American male, who is often larger in the hips than the European, more freedom of movement.) The traditional or classic model is usually single-breasted with two or three buttons. It has the standard 6-inch **drop,** that is, the difference between the chest measurement and the waist measurement is 6 inches. (A man who wears a size 40 regular traditional suit has a 40-inch chest and 34-inch waist.) The trousers have a plain front (without pleats) and regular **rise** (the distance from the crotch to the waist—a regular rise is about 11 inches), with straight, loose legs and belt loops. Some men prefer to wear their trouser legs cuffed.

The contemporary model is a more fashionable version of the traditional model. It can be quickly identified by its shaped, nipped-in waist. Its shoulders may be natural or slightly padded. The details of the contemporary model vary according to current fashion. For example, the lapels may be notched or peaked. This model may have center or side vents. The pockets may be hacking flaps or patch pockets. The trousers may have a plain or pleated front, usually with a shorter rise than the traditional model. They may fit the legs and buttocks loosely or snugly. The contemporary model may be available in a 7-inch drop, to fit the athletic build. In summary, this model is more fitted and fashionable than the traditional suit but not as fitted or fashionable as the European model. It appeals to the wide majority of American males.

The European model is known for its body-hugging, fitted, high-fashion look. This style fits snugly around a man's chest, with heavily padded or peaked shoulders; high, tight-fitting armholes, and defined waist. It is as often double-breasted as single-breasted. The lapels are peaked. The back of the

© Duval © Hartmarx © Hartmarx

Figure 7-3 Note the distinguishing features of the traditional model, the contemporary model, and the European model of men's suits.

jacket may be ventless or with high side vents. (The ventless jacket fits snugly around the hips and waist and is balanced by the broad shoulders.) The trousers are pleated and cuffless; 7-inch drops are not uncommon, designed to fit the man with smaller waist and hips. The European model appeals to fashion-conscious males with slender waists and narrow hips. Its details change the most frequently of the three basic suit models.

Sportcoats are styled similarly to suit jackets. Dress slacks are styled basically the same as trousers. (See Figure 7-4.) Some slacks have a continuous elasticized waistband with no belt loops, called a continental waistband. Slacks usually have inset, scoop, or quarter front pockets, and welt or flap back pockets. (See Figure 7-5.) Although a basic style often stays fashionable for a long time, designers may make slight detail changes.

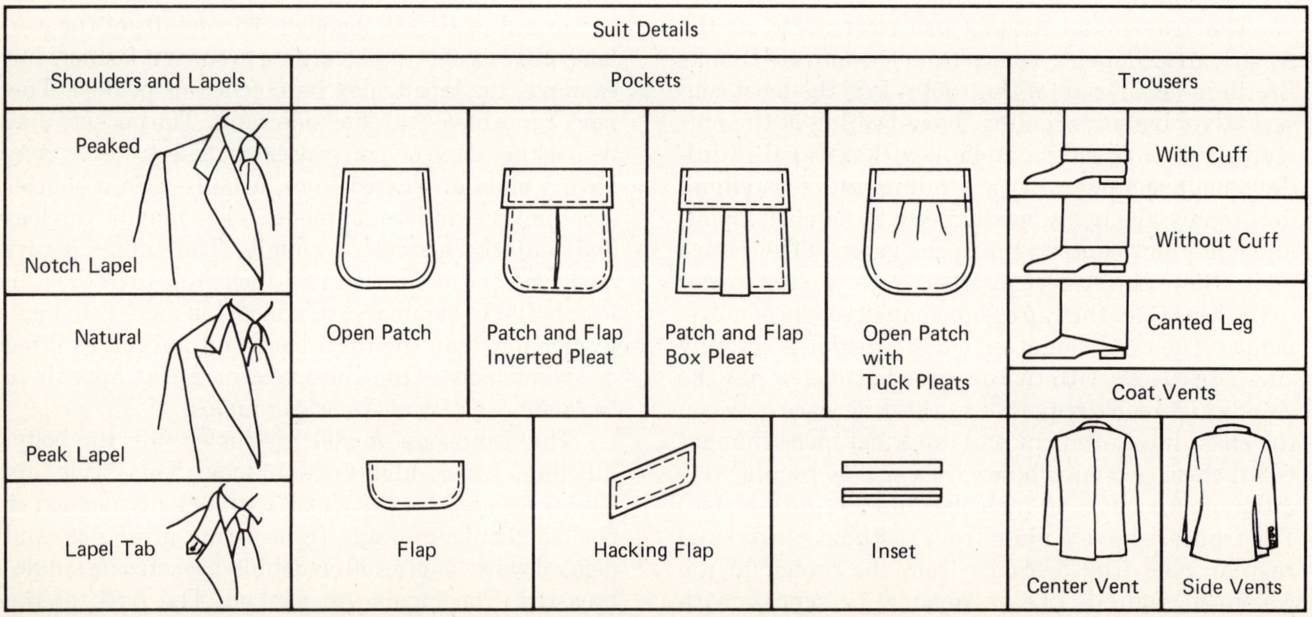

Figure 7-4 Men's suits can be varied with these shoulder and lapel, pocket, vent, and trouser details.

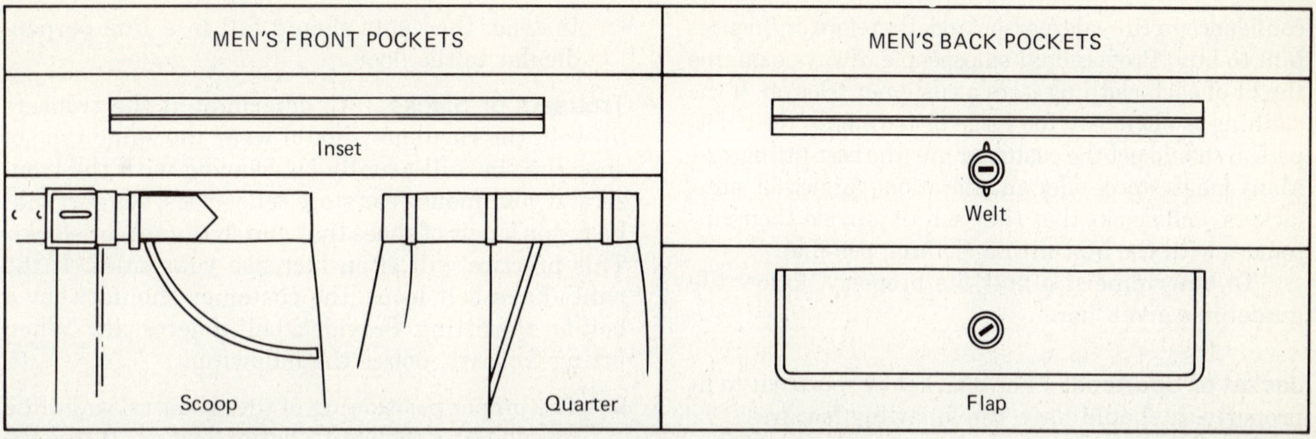

Figure 7-5 Men's pant styles can vary according to the use of the pocket details illustrated here.

Experience 7-2 ■ Style Details: Can You Describe Them?

In a men's fashion magazine, such as *Gentlemen's Quarterly, M,* or *Esquire,* or in your newspaper, locate two illustrated advertisements showing suits with different style features. If possible, bring the pictures to class with you.

1. Examine the style features of each suit. On a separate sheet of paper, write a detailed description of the style features.
2. Without showing your classmate the advertisements, read descriptions of both suits. Based on your descriptions, ask the classmate to sketch both suits on a separate sheet of paper.
3. Compare the classmate's sketches to the advertisements. What additional information should you have provided your classmate to get a better likeness of both suits? Write your answer on a separate sheet of paper.

Fabrics

Suits, sportcoats, and dress slacks are made from different weight fabrics. A suit may be made from regular-weight, lightweight, or year-round fabrics. Regular weight is designed for fall and winter wear. Fabrics include cashmere, cavalry twill, cheviot, flannel, and tweed. Lightweight fabrics, which are worn in warm weather, include chino, poplin, sharkskin, and tropical worsteds. Examples of year-round fabrics are silk and worsted, gabardine and whipcord. The following gives a brief description of several fabrics.

Cashmere Soft, luxurious fabric with a slightly fuzzy surface. Wrinkles moderately.

Cavalry twill Closely woven worsted, with obvious slanted lines in texture. Easy to care for and press.

Cheviot Crisp but loosely woven worsted, ideal for herringbone and other patterns. Wrinkles readily but is easy to press.

Chino Smooth, shiny surface, twill weave, made of cotton or cotton/polyester blend.

Flannel Soft-feeling fabric with slight nap, either wool or wool blend.

Gabardine Closely woven worsted or wool/polyester blend with fine diagonal ribbing. Wrinkle resistant, easy to press.

Poplin Smooth, closely woven, lustrous fabric, usually in cotton or a cotton/polyester blend.

Sharkskin Finely spun and closely woven worsted, fine and light in weight.

Silk and worsted Lightweight, closely woven, lustrous, and very luxurious. Wrinkles less easily than pure silk.

Tweed Coarsely woven wool with a fuzzy feel. Comes in twill, herringbone, and basket weave in a wide variety of colors and patterns.

Tropical worsteds Crisp, lightweight fabric for hot-weather suits.

Whipcord Twill weave worsted with diagonally ribbed surface. Wrinkles only slightly.

Checking the Fit

Good fit is important to men. They want clothing to add, not detract, from their appearance. Menswear salespeople must be able to recognize good fit and point it out to their customers. This increases the customer's confidence in the salesperson and, therefore, influences him to buy. Professional salespeople always examine the fit of each clothing item a customer tries on. If the clothing is obviously too large or too small, the salesperson should get the customer into the best-fitting size. Many men's stores offer an alterations service for suits, jackets, and slacks that they sell to provide their customers with the best-fitting clothing possible.

To determine if a suit fits properly, follow the guidelines given here.

Jacket or Sportcoat For a jacket or sportcoat to fit properly, it should have the following features:

- The garment should hang smoothly with no wrinkles or bulges at the neck and no pulling across the shoulders. If the shoulders are too tight, try the next larger size; if they are loose, try the next smaller size.
- The collar should lie flat and smooth against the back of the neck; about ½ inch of the shirt collar should be seen above the coat collar in back.
- The shoulder line of the coat should form a straight line from the collar to the tip of the shoulder; there should be no wrinkles or bulges.
- The lapels should lie flat and smooth.
- The armholes should be deep enough for easy movement but not look baggy.
- When buttoned, the coat should not pull across the chest. (Button the top button of a two-button coat and the middle button of a three-button coat.) If it is a double-breasted coat, all buttons should be buttoned, including the interior guard button.
- The sleeve should end at the wrist, allowing about ½ inch of the shirt cuff to show.
- The coat should be long enough to cover the curvature of the buttocks. The vents should never pull apart allowing the seat of the pants to show. Instead, the vents should fall in a line perpendicular to the floor.

Trousers or Slacks To determine if the trousers fit well, the customer should wear the same type of shoe that he will usually be wearing with the trousers. If your menswear store sells shoes, suggest that he try on a pair of shoes that match the suit or slacks. This practice will often increase your sales. If the pants have belt loops, the customer should wear a belt for the fitting. Provide a belt if necessary. When fitting trousers, notice the following:

- The proper positioning of the trousers' waistline should be on the man's natural waist. If this position is uncomfortable to him, ask the customer to place the waistband where it feels the most comfortable.
- The seat and crotch of the trousers should not look baggy or wrinkled. The trousers should look smooth and feel comfortable when the customer is sitting or standing.
- The length of the trousers is a matter of customer preference. Many men like a break in their trousers, that is, a slight break in the crease over the instep. Uncuffed trousers are ½ inch shorter in the front than in the back. Cuffed trousers are parallel to the floor and, depending on the man's preference, may or may not have a break.

Vest When a customer buys a suit with a vest, the jacket should be fitted with the vest on. Because a vest should fit close to the body, it is important that the customer wear the same weight shirt he will wear under the vest. Notice the following when fitting a vest:

- The vest should fit with no pulling or sagging.
- The armholes should be open enough for comfort but should not sag.
- The back should have no wrinkles and should not ride up when the customer moves his arms.

Experience 7-3 ■ Which Model Fits Best?

For Male Students
Visit a men's clothing store. Locate the three basic suit models in your size. Try on each model. On a separate sheet of paper, list the differences in style and fit. Describe why each model is a good fit, poor fit, or needs alteration.

For Female Students
With a male friend, visit a men's clothing store. Locate the three basic suit models in his size. Have him try on each model. Complete the assignment as described to the left.

Alterations

Even though men's suits, sportcoats, and dress slacks are available in many sizes, some men need their clothing shortened, lengthened, taken in, or let out. These changes are called **alterations.** At many finer stores alterations are made free of charge or for a small fee. Some men's clothing stores or departments have tailors who mark the suit jacket and trousers for alterations. Besides marking clothing, the tailors also make the required changes. At other men's stores or departments, menswear salespeople mark the alterations. After the clothing is marked, the salespeople send the clothing to the tailor, who makes the alterations.

Most stores require their tailors or salespeople to complete an alteration ticket when any clothing needs to be altered. Information about the customer, the clothing item, and alterations needed are written on this ticket. Usually, the ticket is perforated so the customer can use the bottom part as a claim check when picking up the merchandise. (See Figure 7-6.)

Alteration Symbols The tailors or salespeople who mark alterations use certain symbols. For example, a row of short, unconnected lines means the garment should be shortened or taken in. A row of unconnected, slashed lines means the garment needs to be lengthened or let out. When marking alterations, the salesperson uses marking chalk. Even though this chalk comes out when the garment is steam pressed, the chalk should be used sparingly. On light-colored clothing, pins are used to mark the alterations, because chalk lines do not show up on light colors.

Trouser Alterations Before making any alteration, check the fit of the trousers, noting the points mentioned on page 96.

The most common trouser alterations are shortening or lengthening the leg and taking in or letting out the waist. Some stores use pins to hold the leg at certain lengths temporarily. When the customer has decided on the correct length, the salesperson marks the trousers with marking chalk. Only one leg is marked for alterations. If the customer has one hip lower than the other or one leg shorter than the other, both legs should be marked.

If the trousers have unfinished bottoms or finished bottoms that are too long, fold the excess material and determine the correct length. Using the chalk, mark the desired length by using a horizontal dotted line. Some stores make the mark on the inseam, some at the heel, and some mark the trousers at the front crease. (See Figure 7-7.) If the trousers are to be cuffed, be sure to include that information on the alteration ticket. Because the majority of dress slacks have unfinished bottoms, the measurement for lengthening pant legs is uncommon.

To alter the waist, ask the customer to position the trousers at his waist. The customer should not be wearing a belt. If the waist is too loose, pinch the trousers at the waist along the backseam to take up the excess material. Taper the excess material downward in a wedge shape toward the crotch. Be sure the amount of material on both sides of the backseam is equal. Mark the sides of the wedge with a dotted line on both sides of the backseam. The width of this wedge tells the tailor how much material should be taken in. (See Figure 7-8.)

Letting out the waist calls for more experience and skill than taking it in. The best way to determine how much to let out is to have the customer open the waistband to the point where the pants feel comfortable. Then measure the width of the distance the trousers lack. Mark this distance along both sides of the backseam with slashed lines. (See Figure 7-8.) Another way is to measure the customer's waist. Then take the difference between this measurement and the waist size on the trousers. The difference is the number of inches the waist should be let out. If the waist is taken in or let out more than 2 inches, the pockets and seams will be out of position. In this case, the customer should wear a different size.

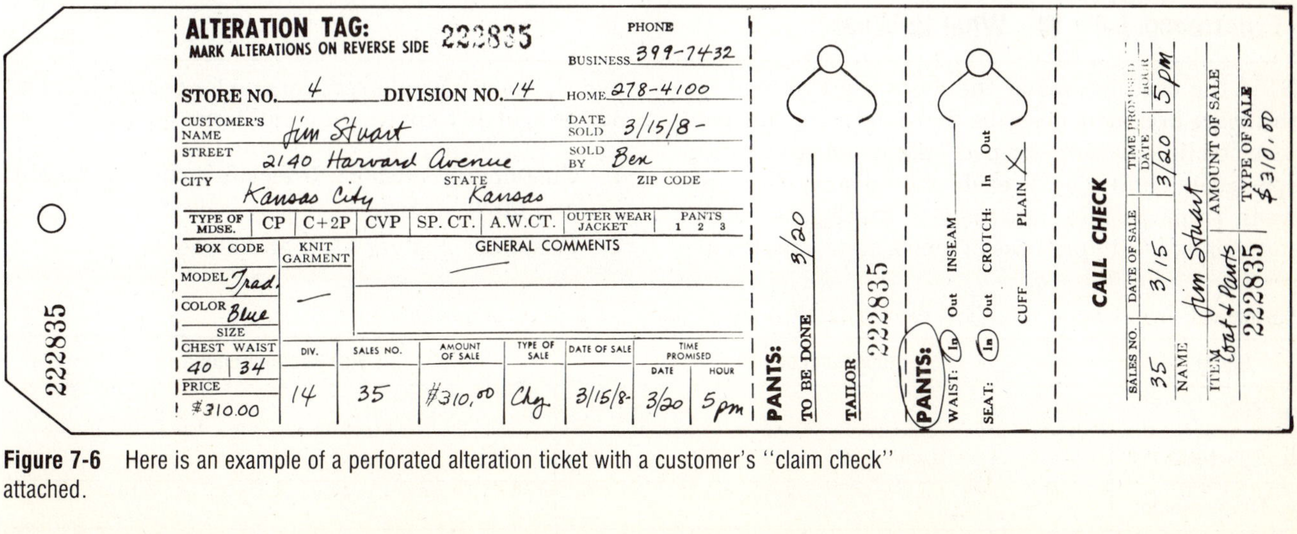
ALTERATION TAG: 222835
MARK ALTERATIONS ON REVERSE SIDE
PHONE
BUSINESS 399-7432
STORE NO. 4 DIVISION NO. 14 HOME 278-4100
CUSTOMER'S NAME Jim Stuart
DATE SOLD 3/15/8-
STREET 2140 Harvard Avenue
SOLD BY Ben
CITY Kansas City STATE Kansas ZIP CODE
TYPE OF MDSE. CP | C+2P | CVP | SP. CT. | A.W.CT. | OUTER WEAR JACKET | PANTS 1 2 3
BOX CODE | KNIT GARMENT | GENERAL COMMENTS
MODEL Trad.
COLOR Blue
SIZE CHEST 40 WAIST 34
PRICE $310.00
DIV. 14 | SALES NO. 35 | AMOUNT OF SALE $310.00 | TYPE OF SALE Chg. | DATE OF SALE 3/15/8- | TIME PROMISED DATE 3/20 HOUR 5pm
222835
PANTS: TO BE DONE 3/20 TAILOR 222835
PANTS: WAIST: In Out INSEAM SEAT: In Out CROTCH: In Out CUFF___ PLAIN X
CALL CHECK
SALES NO. 35 | DATE OF SALE 3/15 | TIME PROMISED DATE 3/20 HOUR 5pm
NAME Jim Stuart
ITEM Coat + Pants 222835
AMOUNT OF SALE TYPE OF SALE $310.00

Figure 7-6 Here is an example of a perforated alteration ticket with a customer's "claim check" attached.

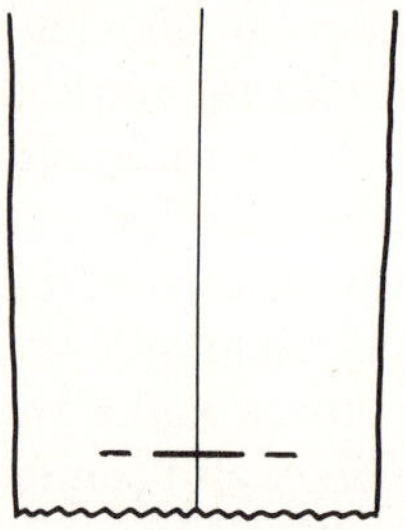

Figure 7-7 Here is the mark used for shortening men's pant legs.

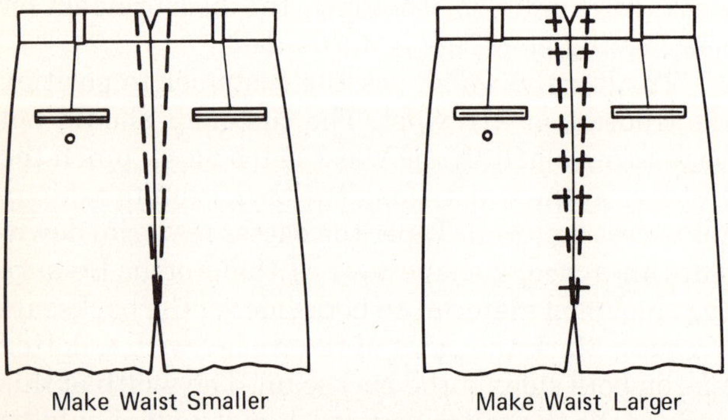

Figure 7-8 Here are the marks used for altering the waists of men's pants.

Other alterations, such as letting out or taking in the crotch and leg width, may also improve the fit of the trousers. Because these alterations require skill and experience, their measurements should be made by a tailor.

Jacket or Coat Alterations The most common jacket or coat alterations are lengthening or shortening the sleeves. The same procedure as lengthening or shortening trouser legs is used for these sleeve alterations. Other jacket alterations are lowering the collar, letting out the armholes, and taking wrinkles from the shoulder. These alterations are complicated and require skill in marking.

■ ANALYZING CUSTOMER BUILDS

A man of average build will look good in almost any style, pattern, and color of suit, slacks, or coat. A man who is tall, short, thin, or heavy may not be able to wear all styles and colors. Menswear salespeople should help customers choose clothing that make them look their best.

A short, slender man will look best in clothing that makes him appear taller. A pinstripe or solid color suit will look good on him. Slacks that are tapered at the bottom will help him look taller. Jackets with lapels that form a deep V will also add height. Sportcoats and slacks in contrasting colors, as well as slacks with cuffs, will cut the man's height. A short, stout man will look taller and leaner in a pinstripe, subtle patterned, or solid color suit with tapered legs. He should wear a single-breasted rather than a double-breasted jacket. And he should avoid checks, plaids, contrasting colors, cuffs, and patch pockets.

A tall, slender man should wear clothes that make him appear broader. A suit with a horizontal pattern or a sportcoat with wide lapels and patch pockets gives the impression of more weight. Jackets that are full at the waist and padded or raised at the shoulders will also add weight. Double-breasted jackets are a good choice for the tall, slender man. A person with this body build can easily wear cuffs and contrasting colors because of his height. However, he should avoid garments that make him appear taller and thinner, such as jackets with deep lapels and pinstripes or vertical patterns.

A tall man with a heavyset build should wear clothing that makes him look better proportioned, such as a pinstripe or chalk stripe suit. Plaids, checks, horizontal patterns, and contrasting colors should be avoided.

Experience 7-4 ■ What to Wear

Read the descriptions of the following two men. In the space provided, describe the style, color, and pattern (solid, tweed, striped, plaid) of the suits, sportcoats, and slacks each should and shouldn't wear. In magazines, newspapers, catalogs, or clothing stores, locate pictures of men's suits, sportcoats, and slacks you would recommend for these men. Cut out or sketch your recommendations for each.

1. Customer 1 is short, 5 feet 5 inches in height, and fairly heavy.
2. Customer 2 is very tall and slender.

Customer 1

Clothing	Recommended	Not Recommended
a. Suit		
b. Sportcoat		
c. Dress slacks		

Name ______________________ Date ______________________

Customer 2

Clothing	Recommended	Not Recommended
a. Suit		
b. Sportcoat		
c. Dress slacks		

Special Reminders for Selling Suits

Generally, men have a fairly specific idea of what they want to buy. If they are not browsing, they should be approached promptly. Since suits, sportcoats, and dress slacks are often expensive, the menswear salesperson must be professional and use appropriate selling skills.

Introduce Yourself Many menswear stores encourage salespeople to approach customers, introduce themselves, and shake the customer's hand. For example, "Good morning. My name is Timothy Henricks. Welcome to our store." Customers often respond to this approach by telling the salesperson their name.

Acknowledge the Woman Many men ask their wife or a female friend to go shopping with them. If the male customer is accompanied by a woman, it's very important to pay attention to her. After greeting the man and woman, find out what merchandise they are interested in and get the woman comfortably seated next to where the man will be trying on clothing. Since most men will rely on the woman's opinions more than on the salesperson's, it is very important to get her as actively involved in the sale as possible.

What's Appropriate? Men buy suits for many reasons. Men who wear suits to work will probably have between four and seven suits in their wardrobe. Men who do not wear suits regularly will probably own only one or two. For a man who does not wear many suits, a dark-colored suit will be the most versatile. Men who own many suits will probably want a different color or style than they already own, but one that is appropriate for the stated occasion. Ask questions to pinpoint the customer's needs. The customer's build will give clues to what sizes and styles will look best on him. If the customer is wearing a suit, sportcoat, or dress slacks, the salesperson should observe them. This gives the salesperson a clue to the quality of clothing the customer is used to wearing.

Trying on Merchandise When selling a sportcoat or suit, the salesperson should offer to help the customer remove his coat so that he can try on one of the store's coats. Trying on the coat increases the customer's desire to buy.

Help the customer off with his coat and carefully hang it up or drape it over a chair. It's important to treat the customer's clothing with respect. When selling a coat or suit, proceed as follows:

Putting on the Coat When helping a customer into a coat or jacket, make sure the customer's shirt sleeves are rolled down. Hold the coat by the collar, lift it high on the customer's neck, and drop it forward to the collar line. Starting at the neck, smooth the coat over the customer's shoulder.

At the Mirror If the coat fits well, take the customer to a three-way mirror. Position the customer in front of the center glass so that he can get an overall view of the front of the coat and how the sleeves fit. Position the customer in front of the left mirror so that he will see the reflection of his side by turning his head and looking in the center mirror. He will see the reflection of his back in the right mirror.

If the customer is with a woman, take him over to her for her approval, since it is usually a must for completion of the sale.

To get an idea of the overall look of the suit, suggest that the customer also try on the matching trousers. Select dress slacks for customers who are trying on sportcoats and encourage them to try on the slacks with the coat. Show the customer to the dressing room. When he comes out, take him to the three-way mirror and point out the benefits of the suit.

Talk About Comfort and Appearance Customers are interested in comfort and appearance. While in front of the mirror, ask the customer to move around to see how the coat or suit feels. Point out why the coat or suit looks good on him. Talk about the color. Make your sales demonstration stronger by actually pointing out specific fit features, such as the way the lapels and collar lie flat and smooth. If the garment needs alterations, say so and explain how the suit will look after it is altered.

Talk About Quality Before he will make a purchase, the customer must be convinced that he is getting value for his money. Point out the quality features, such as fabric, styling, and workmanship, that make the garment worth its price. Some quality workmanship features of a well-made suit are:

- Patterns, plaids, and stripes that match up at the seams and armholes.
- Wide seams, which make alterations easy.
- Reinforced points of stress, such as the crotch.
- Good grade, nonchip buttons.
- Well-bound, reinforced buttonholes, showing no fraying edges.
- Interfacings (extra padding) which help garments keep their shape and hang better.
- A collar with a flat flannel lining on the underside.
- A lining which helps prevent wrinkles or stretching of the fabric and protects the outer part of the coat or suit from perspiration and wear.

Experience 7-5 ■ Mirror Maneuvers

Play the part of a men's clothing salesperson. Have a male classmate play the part of a customer who is looking for a sportcoat. Visit a local store that has a three-way mirror. (If you have a three-way mirror in your classroom, you can complete the experience at school. Have the classmate who is playing the part of the customer bring a sportcoat from home.)

1. Choose a sportcoat that would complement the customer's appearance.
2. If the customer is wearing a coat, demonstrate how you would help him take off his coat. Then demonstrate how you would help him try on the sportcoat.
3. If the coat fits well, position the customer in front of the three-way mirror and proceed to point out the fit of the jacket.

■ DRESS SHIRTS

Dress shirts, which are designed to be worn with a tie, are sized by the man's neck and sleeve measurement, in inches. They come in a variety of sizes, fabrics, and colors.

Sizes

Most men's stores stock shirts in neck sizes 14½, 15, 15½, 16, 16½, 17, and 17½. Long-sleeve shirts are stocked in exact (32 inches to 36 inches) or average sleeve length (such as 32/33). An example of an exact sleeve length shirt is 15½-34. Shirts with average sleeve length sizing are designed to fit two arm lengths, such as 15½-inch neck and 32/33 sleeve.

To determine a man's dress shirt size, place the measuring tape around the base of the neck. The measurement should be taken to the nearest half inch. If the neck size falls between sizes, show the customer the larger size shirt. To determine the sleeve length, have the customer hold his arm at shoulder level and bend the elbow at a 45-degree angle. Measure from the back of the customer's neck, at the spine, to the wrist joint.

Styles

Dress shirts are available in a variety of collar styles, cuff styles, and cuts. (See Figure 7-9.) There are three basic collar styles: button-down, spread, and plain collar. The button-down collar is very versatile and can be worn by most facial shapes. The spread collar flatters a thin, narrow face. The plain collar is most flattering to a broad, round face.

Some important "shirt" words for menswear salespeople to know are:

Collar stays The plastic or metal strips inside the collar that help keep the points of the shirt collar straight and keep the collar down on the shirt front. Stays can be removable or permanent.

Points The distance from the neckband to the collar's tips.

Placket The opening in a garment that allows for ease in dressing at the throat or cuffs.

Shirttail Bottom The part of the shirt below the waist, short at the seams and longer in the front and back.

Spread The distance between the collar tips.

Yoke The fitted or shaped piece at the top of a shirt over the shoulders.

Fabrics and Patterns

The most common dress shirt fabrics are oxford cloth and broadcloth. Oxford cloth is a soft but sturdy basket weave made of cotton or cotton/polyester blend. Broadcloth is a plain, tightly woven fabric with fine crosswise ribs. It is the dressier of the two fabrics.

In addition to solid-color dress shirts, there are many patterned shirts available. Some popular shirt patterns are the tape stripe, tattersall, end-on-end, and tone-on-tone.

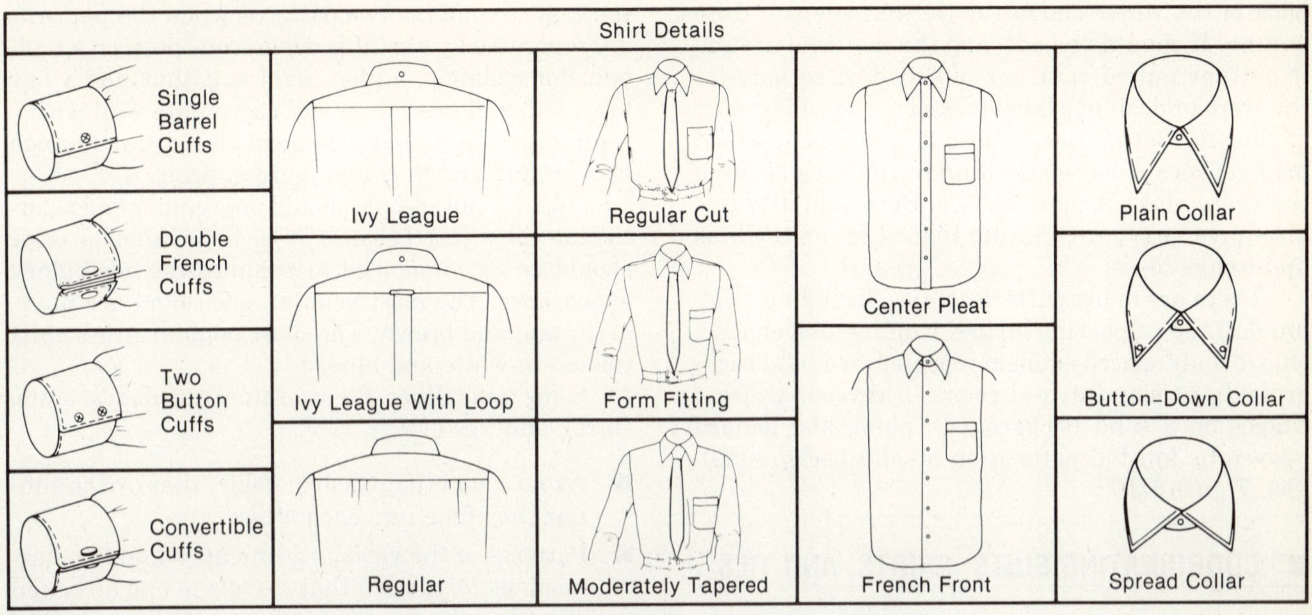

Figure 7-9 Men's dress shirts come in a variety of styles as these cuffs, cut, and collar details show.

Checking the Fit

Most men know their shirt size; however, as a man gains or loses weight, his size may change. It's a good idea to measure a regular customer about once a year to determine the proper size. A salesperson should check the following areas to determine if a shirt is a good fit:

- The waist should be smooth, with no bagginess or bunching of material. It should be loose enough so the material does not pull at the buttons when the customer moves.
- The shirt should be long enough so it does not pull out of the pants when the customer bends.
- The collar should not wrinkle. If it does, it probably is too tight. Button-down collars should have a soft-roll. When a man who seldom wears a dress shirt and tie complains that the correct-sized collar is too tight, show him a shirt that is a half size larger in the neck. Explain that the larger next size may be more comfortable.
- The armholes should be open enough for easy movement, but the sleeves should not look baggy.
- The shirt sleeves should stop just below the wrist bone, about ½ inch below the jacket.
- Buttoned cuffs should fit close to the wrist and still allow movement. Many average sleeve length shirts have two buttons to permit adjustment.

■ TIES

Two popular tie styles are the four-in-hand, or necktie, and the bow tie. The four-in-hand is cut on the bias and tied in a slip knot so that the long flared ends overlap vertically. It is worn for business and many social occasions. The bow tie, which is short and tied in a bow knot, is most often worn for formal occasions. It also passes in and out of fashion for streetwear. A variation of the bow tie is the string tie, or bolo, sometimes worn as part of the country western look.

Four-in-hand ties are available in different lengths. The standard length is 54 to 56 inches; extra long ties for tall men are 60 inches. When tied properly, the tip of the tie should cover the belt buckle. Tie width varies with fashion. Tie width usually corresponds to the width of the jacket lapels. Four-in-hand ties may be knotted in a Windsor knot, half-Windsor, or four-in-hand. The Windsor, a wide knot, is best worn with wide-spread collar shirts, the four-in-hand knot with the plain collar or button-down collar.

Ties are made from silk, polyester, wool, wool challis, and various blends of these fibers. Silk ties are known for their rich colors and ease in knotting.

Most ties are wovens, but a few are knit. Woven neckties have pointed ends; knit neckties have a square end.

When examining a tie for workmanship look for the bar tack, a short, horizontal stitch on the back of the wider end of the tie that keeps it from gaping. It should be neat, and the material around it neither pulled taut nor bunched. Also look for the store or designer label tacked to the widest part of the tie. Better ties will be slip-stitched, which means there is loose stitching in the back seam of the tie to make it more elastic. Better-quality ties also have heavy interlining that is cut on the bias and hangs loose.

There are many patterns of ties, including solid, pin-dot, rep (repeating stripes running diagonally), club (evenly spaced emblem repeated on a solid background), paisley (several colors in curved, abstract shapes on a solid background), plaid, and foulard (woven or knitted pattern on a solid background). (See Figure 7-10.)

■ COORDINATING SUITS, SHIRTS, AND TIES

Because there are so many colors, textures, and styles of men's clothing, a menswear salesperson must be able to coordinate the customer's wardrobe. The easiest clothing coordination is achieved by combining a solid color suit and shirt with a patterned tie, such as a paisley or rep that picks up the color of the suit.

Many men wear a combination of patterns (checks, plaids, or stripes). A salesperson needs talent to combine multiple patterns so they don't clash. When combining patterns, identify the boldest pattern first. Usually it is the tie. This bold pattern becomes the focal point. The other patterned clothing should be less bold and not clash with the bold item. It's easy to combine two patterns when the patterns are separated by a solid and when one pattern stands out. For example, a glen plaid suit that has a fine blue, red, and green line will coordinate well with a solid white shirt and a foulard tie with a maroon background and blue and maroon print.

Apparel salespeople should not recommend a particular color just because it is in fashion; a color should be recommended to enhance the customer's appearance. The most popular suit colors are navy, gray, tan, and brown. The most popular dress shirt colors are white and blue.

Some guidelines for coordinating men's suits, shirts, and ties are:

- Avoid colors that clash or colors that are so similar they fade into each other.
- Patterns of the various garments should be harmonious. A pattern that travels in one direction should not be placed next to a pattern in another direction. For example, a pinstripe suit with a vertical pattern will clash with a plaid shirt that has an obvious horizontal pattern.
- Similar style clothing should be worn together. For example, clothing that has a conservative style, such as a traditional model suit, looks mismatched when worn with a Western-style shirt.
- Help the customer visualize how the suit, shirt, and tie look together. Slip the shirt inside the jacket front and, using the salesperson's knot (as shown in Figure 7-11), place the tie on top of the shirt, near the collar.

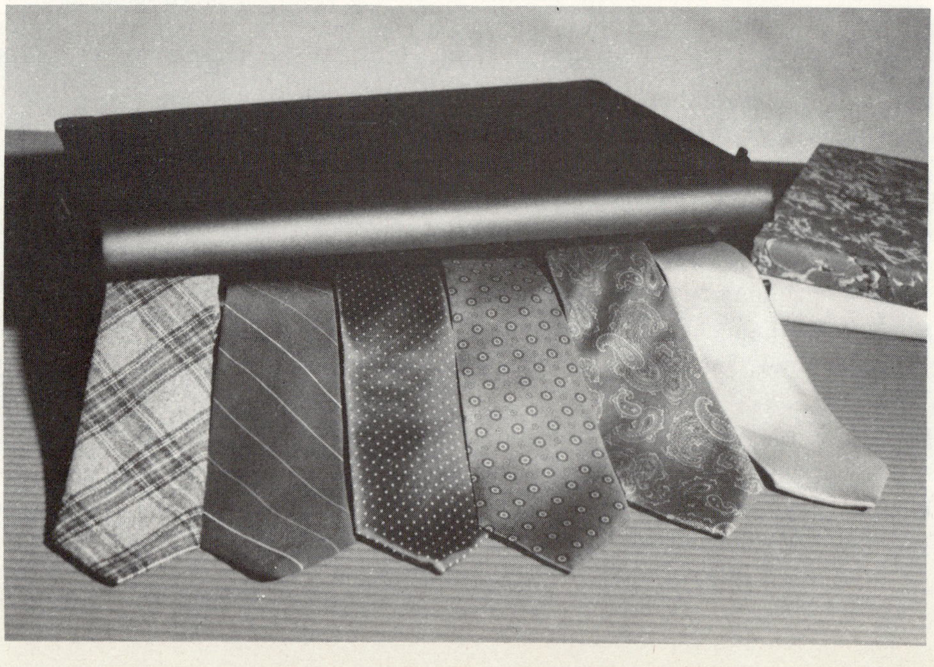

Figure 7-10 Note these popular tie patterns, from left to right: plaid, rep, pin-dot, foulard, paisley, and solid. © Jules Allen Photography.

Experience 7-6 ■ Suits + Shirts + Ties

In this exercise, you will coordinate suits, shirts, and ties. To complete the exercise properly, you will need a camera. If you do not have a camera, use pictures from magazines or catalogs.

Visit a men's clothing store. Explain to the manager that you are studying men's apparel and would like permission to coordinate three men's suit, shirt, and tie ensembles. Choose a solid colored suit, a striped suit, and a plaid suit. Next choose a shirt and tie to match each suit. You may choose patterned or solid shirts and ties as long as each combination looks attractive. Place the shirt inside the suit coat. Using the salesperson's knot, place the tie on top of the shirt. When the suit, shirt, and tie are arranged, take a picture of your three clothing combinations.

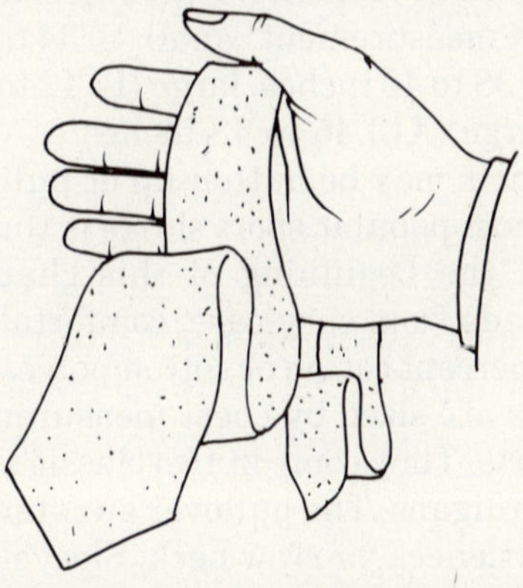

1. Drape the tie, folded in half, over the palm of your hand. The narrow end should extend half way down your hand.

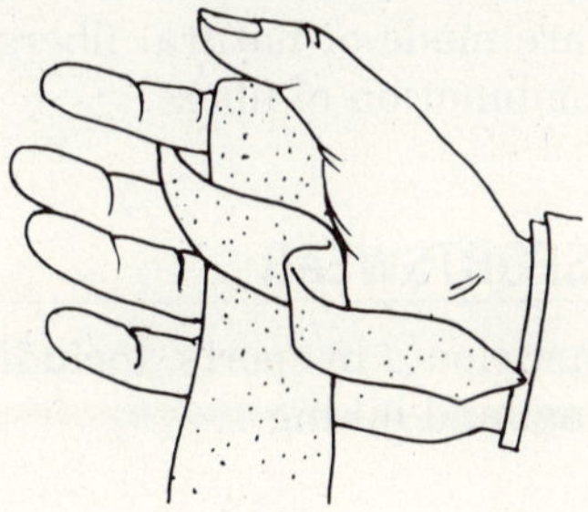

2. Draw the narrow end between your first and second fingers, over the wide end.

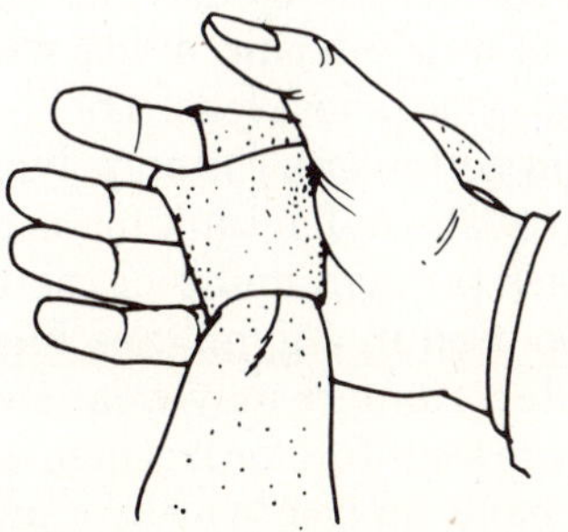

3. Bring the narrow end around behind the thumb. Hold the knot in place with your thumb. Form a dimple in the tie right under the knot.

Figure 7-11 Follow the steps shown here to make a salesperson's knot.

■ OUTERWEAR

There are four basic types of men's **outerwear:** overcoats, topcoats, three-quarter–length coats, and all-weather coats. Overcoats are full-length heavy coats worn in cold weather. They are made of fabrics like cashmere, camel's hair, or tweed. Topcoats are full-length coats worn in the spring and fall; they are made of lighter weight fabrics. Both overcoats and topcoats are designed to be worn over a suit. Their actual length is determined by fashion.

Three-quarter–length coats are shorter and more casual. All-weather coats usually have a water-repellent or water-resistant finish. They will protect the wearer from mild rain. They usually have a zip-out lining that makes them warm enough to be worn in cold weather. Without the lining, they can be worn in chilly spring and fall weather. An example of an all-weather coat is a trench coat. (See Figure 7-12.)

Men's outerwear is sized the same as men's suits, by chest sizes (36 to 46) and height (short, regular, long, extra long). Thus, for example, the label for a coat may read 38L, meaning the coat should be worn by a person 5 feet 11 inches to 6 feet 2 inches who has a 38-inch chest.

■ FORMAL WEAR

Men need formal wear for weddings, special evening social events, and semiformal functions, such as proms. When an occasion is referred to as being "black tie," it means men are expected to wear tuxedos. Many men's stores sell or rent men's formal wear, including tuxedos and dinner jackets in a range of colors and styles.

There is a wide variation of formal wear available to men. The classic tuxedo or "tux" is a finely tailored, black suit. The jacket, which may be single breasted or double breasted, has satin or faille (a ribbed silk or rayon fabric) lapels. The trousers have a stripe of the same material running along the outer seam from the waist to the bottom hemline. The tuxedo is worn with a pleated front shirt with a wing collar or conventional collar; a vest, cummerbund, or suspenders; and a bow tie or an ascot; with patent leather oxfords or spats. The bow tie and cummerbund are often the same color and pattern.

Contemporary tuxedos are colorful variations of the traditional black tuxedo, such as white, gray, or pale blue. Grooms often choose light color tuxedos for summer and dark colors for fall. Two men's formal wear styles you should be familiar with are shown in Figure 7-13.

■ CASUAL WEAR

Casual wear—slacks, shorts, jeans, sport shirts, and sweaters—is a major part of a man's wardrobe. The popularity of casual wear has grown because of its comfort.

Pants and Shorts

Casual pants and shorts are popular menswear items, worn for both leisure and active sports. When worn for active sports, they must allow room for freedom of movement. Casual pants are styled like dress slacks; however, they are often in more casual fabrics, such as corduroy or cotton/polyester blends. Casual pants are sized by waist and inseam measurement. Men's shorts vary in length from just above the knee, as in walking shorts, to high on the thigh, as in tennis shorts. Shorts are sized by waist measurement only.

Jeans

People of all ages often have many pairs of jeans in their wardrobes. The classic jean has five pockets with rivets. It is made of 14-ounce cotton in indigo blue. Jeans are available in two leg styles: straight leg and boot cut, flared bottoms to fit over boots. Some jeans are styled with a button fly; others have a zipper.

Today, jeans are available in different styles, colors, and fabrics. Fashion jeans, such as Guess or Union Bay, may have designs on their pockets. Jean colors range from solids, like blue, gray, black, and white, to patterns, such as stripes and prints. One hundred percent cotton jeans are often prewashed or stonewashed so that they feel "broken-in" from the first wearing. Many cotton jeans are treated with a finish that controls shrinkage because 100 percent cotton can be expected to shrink 1 inch in the waist and 2 inches in the length. Jeans made from a cotton/polyester blend do not shrink as much.

Jeans are sized by waist and inseam measurement. Because women often buy men's jeans, a salesperson must be able to convert her hip size into waist size. (For example, a woman with 35-inch hips would wear a 29-inch waist.)

There are also different fits to consider. A trim fit is worn slightly below the waist and snugly fits the seat and thighs. A regular fit is worn at the waist and is slightly fuller in the seat and thighs.

Sport Shirts and Sweaters

Sport shirts with short or long sleeves are designed to be worn without a tie. Most woven sport shirts are sized by a man's neck measurement into four categories: small (S) 14 to 14½ inches, medium (M) 15 to 15½ inches, large (L) 16 to 16½ inches, and extra large (XL) 17 to 17½ inches. Knit shirts are sized by a man's chest measurement: small (S) 34 to 36 inches, medium (M) 38 to 40 inches, large (L) 42 to 44 inches, and extra large (XL) 46 to 48 inches.

Sport shirts may be button-up or pullover style. One of the most popular sport shirts is the polo shirt discussed at the beginning of this chapter. Sport shirts are made from easy-care, comfortable fabrics, such as 100 percent cotton or cotton/polyester blends.

Sweaters are sized by chest measurements, just like knit shirts. They come in two classic styles, pullovers and cardigans. The pullover sweater may have a V neck, turtleneck, or crew neck. Sleeveless V neck sweaters are called sweater vests. Cardigan sweaters button down the front. They may or may not have a collar, and they may or may not be belted.

Sweaters are available in a variety of colors and patterns. They are chosen to match or contrast with pants. They are made of natural fibers, man-made fibers, or a combination of fibers.

■ ACTIVE SPORTSWEAR

Many men participate in sports, including running, golf, swimming, and biking.

Running Clothes

Warm weather running clothes consist of shorts and a T-shirt or tank top, usually in bright colors so the wearer can be seen easily by drivers. The shorts are usually made of quick-drying nylon with an elasticized waist and wide, short legs. The T-shirt is usually 100 percent cotton or a cotton/polyester blend to absorb moisture. The nylon tank top is perforated to allow air to pass through and cool the runner. This clothing is also used in a gym. (See Figure 7-14.)

In the winter, runners may wear sweat suits, or "sweats." They are cut full for freedom of movement. The tops may have zip fronts or be a pullover style. The waist, wrist, and leg openings are elasticized to prevent air from chilling the runner. Sweats are made from fleece-backed sweatshirt material that absorbs moisture. Warm-up suits of acrylic, nylon, or

Figure 7-12 Men's coats are available in a variety of styles. Learn to recognize these basic styles.

Figure 7-13 Learn the elements that go to make up these two popular looks in men's formal wear.

velour are also popular with nonrunners, who wear them for leisure.

Swimwear

Men's swimwear is most often made of quick-drying 100 percent nylon and spandex. The common styles are racing bikinis, briefs, and boxer shorts. Swimwear colors are bright and often patterned. Swimwear is sized by waist measurement. (See Figure 7-14.)

Golf Clothes

The short-sleeve polo shirt, brightly colored golf slacks, and zip-front, hip-length jacket are common attire for golfers. The pants are usually very bright, like Kelly green or yellow, or pastel, like pale blue. Most golf pants have a "shirt-gripping" elastic waistband that keeps the shirt inside the pants.

Bicycle Racing Clothes

Bicycle racers and serious riders wear special clothing, including crewneck shirts with pockets to hold supplies (food, gloves, sunglasses) and padded shorts, for comfort, that end just above the knee (for ease of movement). The top and shorts are made of stretch knit and worn snug to the body to reduce wind resistance. The shorts have become a fashion item that is sometimes worn with a T-shirt or tank top. (See Figure 7-14.)

■ MEN'S FURNISHINGS

The following men's furnishings are sold in most department stores and men's specialty shops.

Hosiery

Most men's hosiery is one size that stretches to fit all sizes. These are called stretch socks. Nonstretch hosiery sizes run in half sizes from 9½ to 13. Nonstretch sock sizes do not correspond to shoe sizes.

The most popular hosiery colors for suits and dress slacks are black, navy, brown, gray, green, and olive. Usually, the sock color matches or blends with the man's slacks. Dress socks, which are available in mid-calf and over-the-calf lengths, have become a fashion accessory. They no longer come in solid colors only. Many men wear argyle, paisley, or bird's-eye pattern socks. Athletic socks are usually white with colorful horizontal stripes at the top.

© 1988 Jon Feingersh/The Stock Market.

© Adidas

Figure 7-14 Active sportswear features details that make the wearer comfortable for participation in particular sports.

© Supplied by PGA Tour Division of Ram Golf Corp.

© Bruce Herman/Speedo.

© Catalina.

Underwear

Undershirts and undershorts make up a man's underwear wardrobe. Customers look for fit, comfort, absorbency, and easy care when buying their undershirts and shorts.

Undershirts come in two styles—T-shirts and athletic shirts. T-shirts have short sleeves and either a V neck or a crew neck. Athletic shirts have a scoop neck and no sleeves. Undershirts are sized by chest measurement: small (30 to 34 inches), medium (36 to 40 inches), large (42 to 44 inches), and extra large (46 to 52 inches). They are available in a range of fashion colors, but white is most popular.

Two styles of undershorts are briefs and boxer shorts. Briefs are knit and have an elasticized waistband and legs. Boxer shorts are woven and have an elasticized waist and hemmed legs. Undershorts are sized by waist measurement, 28 to 44, in even inches. They are available in many colors and different patterns.

Pajamas and Robes

Sleepwear, such as pajamas and robes, is available in a wide range of colors and patterns. Popular sleepwear fabrics are flannel and broadcloth. Pajamas are sized by chest measurement: small (34 to 36 inches), medium (38 to 40 inches), large (42 to 44 inches), and extra large (46 to 48 inches). Pajamas are also sized by letters A, B, C, and D. Size A pajamas are small; B are medium; C, large; and D, extra large. In robes, usually one size fits all.

Two popular pajama styles are the pullover with a V neck or crew neck and the coat style, which buttons up the front. For hot weather, some men wear short pajamas with pants above the knee. Others prefer nightshirts.

Men's robes are usually the wraparound style. Often, robes are made of terry cloth. More luxurious robes are made of velour or "silky" fabrics. For warmth, men wear flannel robes.

Experience 7-7 ■ Mix and Match a Wardrobe

1. Create a casual wear wardrobe for a customer who is going on a three-day vacation to San Diego, California. The temperature there is usually mild, in the upper 60s. This man plans on playing some golf, jogging, and sightseeing in southern California. The customer wants to take as few clothing items with him as possible. He has just purchased a 100 percent worsted wool navy blue blazer to take on this trip.

   Look through magazines or newspapers and locate the casual wear clothing and furnishings this customer should take with him. Cut out or sketch these clothing items and attach them to a separate sheet of paper. On that same paper, explain why you chose the clothing items. Discuss the texture, colors, and versatility of each casual wear item.
2. If you are male, look through your wardrobe and identify all the casual wear. If you are female, analyze a male friend's casual wardrobe. On a separate sheet of paper, describe these items by style, color, texture. Determine how many attractive combinations you can create by mixing and matching these casual wear items. Write all the combinations on paper.

Chapter Notes

1. "A Popular Shirt Tale," *Time,* September 1, 1986, p. 60.

CHAPTER 8

Children's Wear

Kids' clothing has changed over the years. The traditional look of pinafores and overalls has given way to fashionable, trendy clothing. Stirrup pants with novelty sweaters and sweatshirts can be seen in nursery schools.

Girls' clothing takes its cue from the junior and misses market. A little girl wants to look like her big sister or favorite TV star. Boys' clothing is often a scaled down version of styles found in the young men's, men's sportswear, and tailored departments, just like what dad or big brother wears.

Expensive designer-label clothing for children can be found in specialty shops such as Nieman-Marcus. Many parents who have the money are even willing to buy their 3- or 4-year-old dry-clean-only items, like a trench coat for $120 or an $80 wool sweater.

Most young girls and boys make their own clothing decisions. Although mothers may decide how their children will dress for the first few years, peer pressure and the child's own determination most often win out. In fact, as soon as kids begin watching television, they become more independent in their clothing choices. They see what other kids on TV wear and start to notice their friends' clothing.

As a group, kids are very fashion conscious. They ask for specific clothing made by specific manufacturers, such as the McKids line sold exclusively by Sears Roebuck and Co. When you work with children's clothing, be prepared for a demanding young customer.

■ CHILDREN'S WEAR SIZES

Children's wear is classified into various size groups: infants, toddlers, children or preschool, girls, boys, young junior, and student. These size groups reflect the growth stages of children. Although there is some overlap between size groups, and some manufacturers may use different sizing standards, children's sizes are usually based on height and weight.

Children's sizes sometimes relate to the age of the child. Many 4-year-olds do wear a size 4, for example. However, there are many variations in the body build of children. A 6-year-old may not wear a size 6 because children of the same age often vary in height and weight.

When a customer asks, "What size do I need for my nephew? He's 6," you do not have enough information to determine the correct size. Also, "He's big for his age" will not help you, because you will not know how big. Try to find out the child's height and weight and then refer to the size charts for the merchandise your store sells. Many manufacturers print sizing guides for their clothing on merchandise boxes or bags. Let's examine infants' and children's size groups.

Infants' Sizes

Infants' clothing is for babies from birth until the time they begin to walk. Two methods are used to size infants' clothing. Both methods are based on a baby's height and weight. (See Table 1.) One method uses the following sizes: 3 months, 6 months, 12 months, 18 months, and 24 months. The other uses these sizes: newborn, small, medium, large, and extra large.

Toddlers' Sizes

A toddler is a young child who is learning to walk. Clothes for the toddler are shorter in the legs and

■ **Table 1**
Infants' Sizes

Size	Height (in inches)	Approx. Weight (in pounds)	Size	Height (in inches)	Approx. Weight (in pounds)
3 mo.	24	13	Newborn (NB)	To 24	To 14
6 mo.	26½	18	Small (S)	24½–28	15–19
12 mo.	29	22	Medium (M)	28½–32	20–25
18 mo.	31½	26	Large (L)	32½–36	26–31
24 mo.	34	29	Extra Large (XL)	36½–38	32–36

Note: Infant size charts are often listed in the merchandise tag or on the package. Height is the most important measurement to determine the correct size.

wider at the waist than clothing for older children. Toddler sizes, which range from T1 to T4 are made to fit a child who wears diapers or training pants or hasn't lost baby roundness. (The T stands for toddler.) The sizes are based on height and weight. (See Table 2.)

Children's Sizes

Children's sizes, or preschool sizes, are mostly worn by boys and girls between the ages of 3 and 6. These children have "round" figures and have not developed waistlines. Many children's wear manufacturers produce sizes 4, 5, 6, and 6x for girls and sizes 4, 5, 6, and 7 for boys. (See Table 3.) There is some overlap between toddlers' and children's sizes. For example, a T4 has the same height and weight measurements as a children's size 4.

Girls' Sizes

Girls' sizes are for school-aged girls, usually 6 to 13 years of age, who have slimmer, longer figures than those in the younger groups. The U.S. Department of Commerce, National Bureau of Standards, has proposed 18 sizes for girls' clothing. Each size is based on height and girth (broadness). The height and weight, plus bust, waist, and hip measurements, for girls' clothing are shown in Table 4.

The numbers in each size correspond with the height of the girl in her stocking feet. Weight and body measurements determine whether she wears a slim, regular, or chubby size. For example, a girl who is 55 inches wears a size 10. If she weighs 69 pounds, she will wear a 10 slim.

■ **Table 2**
Toddlers' Sizes

Size	Height (in inches)	Approx. Weight (in pounds)
T1	31	25
T2	34	29
T3	37	34
T4	40	38

Note: Height is the most important measurement to determine the correct size for toddlers.

■ **Table 3**
Children's Sizes

Size	Height (in inches)	Approx. Weight (in pounds)
4	40	38
5	43	44
6	46	49
6x (girls)	48	54
7 (boys)	48	54

Boys' Sizes

Most stores stock boys sizes in even numbers from 8 through 20. Odd-number sizes are used less frequently. Some boyswear manufacturers also produce clothing in sizes 6 and 22 through 26. Boys' clothing is classified into three size groups: regular (R), slim (S), and husky (H). (See Table 5.) The height of the boy in his stocking feet is the primary guide to size. Chest, waist, and hip measurements determine whether the boy wears a slim, regular, or husky. For

Name ________________________ Date ____________________

■ Table 4
Girls' Sizes

1. Find the girl's height on the top line of the chart.
2. Look through the groups of measurements labeled Group A, B, or C.
3. Find the group that best matches the girl's measurements.

Height (in inches)	49–51½	52–53½	54–55½	56–58	58½–60½	61–63
Regular						
Group A	7R	8R	10R	12R	14R	16R
Bust (in inches)	25½–26½	26½–27½	28–29	29–30½	31–32	32½–33½
Waist (in inches)	22½–23	23–23½	24–24½	25–25½	26–26½	27–27½
Hip (in inches)	27½–28	28–29	29½–30½	31–32½	33–34½	35–36½
Weight (in pounds)	60	66	74	84	96	112
Slim						
Group B	7S	8S	10S	12S	14S	16S
Bust (in inches)	24–25	25–26	26½–27½	28–29	29½–30½	31–32
Waist (in inches)	20½–21	21–21½	22–22½	23–23½	24–24½	25–25½
Hip (in inches)	25½–26	26–27	27½–28½	29–30½	31–32½	33–34½
Weight (in pounds)	53	57	67	77	89	103
Chubby						
Group C	7½	8½	10½	12½	14½	16½
Bust (in inches)	28–29	29–30	30½–31½	32–33	33½–34½	35–36
Waist (in inches)	26–26½	26½–27	27½–28	28½–29	29½–30	30½–31
Hip (in inches)	30½–31	31–32	32½–33½	34–35½	36–37½	38–39½
Weight (in pounds)	71	79	89	102	116	132

Source: Voluntary Product Standard PS 54-72, *Body Measurements for the Sizing of Girls' Apparel.* U.S. Department of Commerce, National Bureau of Standards.

example, all boys 4 feet 2 inches tall wear a size 8. A boy 4 feet 2 inches tall with a 26-inch waist and 29-inch hips would wear 8H (husky) pants.

Young Junior

Young junior sizes, for girls whose figures are beginning to develop, range from 6 through 14. These sizes are determined mainly by height and bust, waist, and hip measurements. (See Table 6.) The young junior customer ranges in age from 9 to 15 years, depending on her size and style preferences. The young junior customer may be either a young girl growing out of girls 7 to 14 or a junior customer. These clothes have a more adult styling than girls' sizes. When girls outgrow young junior sizes, they move into the junior sizes described in Chapter 6.

Student Sizes

Student sizes are for young men, usually between the ages of 12 and 16, whose bodies are maturing. Often there is a separate young men's, student, or prep department for these customers. Student sizes are determined in the same way as men's clothing: pant sizes by waist and inseam, shirts and jackets by chest measurement. (See Table 7.)

Experience 8-1 ■ What's the Size?

Using Tables 4 and 5, determine the correct size clothing for the following boys and girls. In the space provided, write the size that will best fit each child.

1. Boy, age 8, height 53 inches, weight 66 pounds. ____
2. Girl, age 9, height 57½ inches, weight 100 pounds. ____

3. Girl, age 7, height 51 inches, weight 55 pounds. ____

4. Boy, age 11, height 60 inches, weight 109 pounds. ____

5. Boy, age 14, height 60 inches, weight 94 pounds. ____

6. Girl, age 13, height 62½ inches, weight 102 pounds. ____

■ Table 5
Boys' Sizes

Slim (S)					
Size	**8**	**10**	**12**	**14**	
Height (in inches)	49–50½	51–54½	55–58½	59–61½	
Chest (in inches)	25–25½	26–27	27½–29	29½–30½	
Waist (in inches)	21½–22	22½–23	23½–24	24½ –25	
Hip (in inches)	25–25½	26–27½	28–29½	30–31	
Approx. weight (in pounds)	53	65	77	90	
Size	**16**	**18**	**20**	**22**	**24**
Height (in inches)	62–64½	65–66½	67–68½	69–70½	71–72½
Chest (in inches)	31–32	32½–33½	34–35	35½–36½	37–38
Waist (in inches)	25½–26	26½–27	27½–28	28½–29	29½–30
Hip (in inches)	31½–33	33½–34½	35–36	36½ 37½	38–39
Approx. weight (in pounds)	104	115	126	138	152
Regular (R)					
Size	**8**	**10**	**12**	**14**	
Height (in inches)	49–50½	51–54½	55–58½	59–61½	
Chest (in inches)	26½–27	27½–28½	29–30	30½–32	
Waist (in inches)	23½–24	24½–25	25½–26	26½–27	
Hip (in inches)	26½–27	27½–28½	29–30½	31–32½	
Approx. weight (in pounds)	59	73	87	100	
Size	**16**	**18**	**20**	**22**	**24**
Height (in inches)	62–64½	65–66½	67–68½	69–70½	71–72½
Chest (in inches)	32½–33½	34–35	35½–36½	37–38	38½–39½
Waist (in inches)	27½–28	28–29	29½–30	30½–31	31½–32
Hip (in inches)	33–34½	35–36	36½–37½	38–38½	39–40
Approx. weight (in pounds)	115	126	138	152	168
Husky (H)					
Size	**8**	**10**	**12**	**14**	
Height (in inches)	49–50½	51–54½	55–58½	59–61½	
Chest (in inches)	27½–28	28½–29½	30–31½	32–33½	
Waist (in inches)	25½–26	26½–27	27½–28	28½–29½	
Hip (in inches)	28–29	29½–30½	31–32½	33–35	
Approx. weight (in pounds)	67	81	95	112	

(continued)

Table 5 *(Continued)*

Husky					
Size	**16**	**18**	**20**	**22**	**24**
Height (in inches)	62–64½	65–66½	67–68½	69–70½	71–72½
Chest (in inches)	34–35½	36–37	37½–38½	39–40	40½–41½
Waist (in inches)	30–30½	30½–31½	32–32½	33–33½	34–34½
Hip (in inches)	35½–36½	37–38	38½–39½	40–40½	41–42
Approx. weight (in pounds)	130	143	156	170	186

Note: This chart represents a range of measurements for each size for each key measurement. It is included only to identify boys' body types and sizes, and is not a requirement of the standard.

Source: Adapted from Voluntary Product Standard PS 36-70. *Body Measurements for Sizing Boys' Apparel.* U.S. Department of Commerce, National Bureau of Standards.

■ Table 6
Young Junior Sizes

Size	Height (in inches)	Weight (in pounds)	Bust (in inches)	Waist (in inches)	Hips (in inches)
6	57½	69	27	22½	28
8	58½	79	28½	23½	30
10	59½	89	30	24½	32
12	60½	99	31½	25½	34
14	61½	109	33	26½	36

■ Table 7
Student Sizes

Pants						
Waist (in inches)	25	26	27	28	29	30
Inseam (in inches)	27–32	27–34	27–34	28–36	28–36	29–36

Shirts						
Size	14	16	18	20	22	24
Chest (in inches)	35¼	37¼	39¼	41¼	41¾	42¼

■ MEASURING CHILDREN

The best way to fit children's clothing is to know their height and weight. To help customers choose the correct size of children's clothing, many stores provide a measuring stick attached to the wall. Salespeople may ask the child to stand next to the measuring stick in stocking feet to measure his or her height. Although some stores have scales, most of them rely on their salespeople to estimate each child's weight.

When determining clothing sizes for children, other measurements besides height and weight are important. For example, the chest measurement is important to determine coat, jacket, sweater, shirt, or blouse sizes. Height, waist, and seat measurements are important for determining the size slacks, undershorts, and swimwear a boy wears. A girl's height, waist, and hip measurements are needed to determine sizes for her pants and skirts. Salespeople should use a tape measure to determine these measurements. To measure a child's waist, hips, and chest or bust, follow the procedure described in the chapters on women's and men's apparel. (See Figure 8-1.)

How to Measure for Girls' and Teens' Apparel

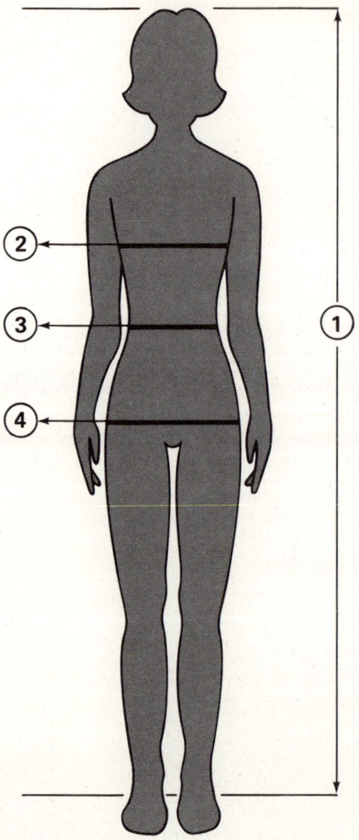

(1) **Height:** Have girl stand against the wall without shoes. Make a mark level with the top of her head. Measure from this point down to the floor.

(2) **Chest or Bust:** Measure around girl's body at fullest part of her chest or bust. Hold tape firm but not tight. Keep the tape straight across her back. Be sure girl is standing naturally.

(3) **Waist:** Measure around the smallest part of girl's natural waistline. Hold the tape firm but not tight.

(4) **Hips:** Have girl stand naturally with her feet together. Measure in a straight line around body at fullest part of hips. Hold the tape firm but not tight.

Figure 8-1 When measuring girls and teenagers, follow the guidelines shown here. Source: Sears, Roebuck and Company.

Experience 8-2 ■ Size Them Up

1. Choose three children. (If possible, one should be a baby, another about 10 years old, and the third somewhere in between.) Estimate the height and weight of each child. Record your estimates in the space provided. After you have made your estimates, use a yardstick or measuring tape to determine each child's height. Weigh each child on a scale. For children over 5 years old, also estimate and then measure their chest, waist, and hips. Record the measurements in the space provided.

2. Based on the actual measurements you took, decide what size coat, slacks, and shirts or blouse each child should wear. Refer to the size charts in Tables 1 through 7. Record the sizes in the space provided. Check with the children or their parents to see if the sizes you determined from Tables 1 through 7 are the same sizes they last bought in each of the items.

Measurements	Child 1: Age ____		Child 2: Age ____		Child 3: Age ____	
	Estimate	Actual	Estimate	Actual	Estimate	Actual
Height						
Weight						
Chest						
Waist						
Hips						

Item	Child 1		Child 2		Child 3	
	Size From Table	Actual Last Size	Size From Table	Actual Last Size	Size From Table	Actual Last Size
Coat						
Slacks						
Shirt or blouse						

■ FITTING CHILDREN'S CLOTHING

When fitting children's clothing, it is very important to consider comfort. Each garment should allow the child to move around, sit, stoop, stand, run, and jump. Clothing that is too tight or too large restricts a child's movement. Uncomfortable, ill-fitting clothing may cause a child to be irritable or to have an accident.

When possible, the child should be taken shopping to try on clothing. The child gets the best fit, and parents do not have to return poorly fitting clothing to the store. Customers who shop alone for children's clothing will minimize returns if they know their children's height, weight, and other measurements.

When the child tries on clothing, the salesperson should check for proper fit. Note how the "Checklist for Proper Fit," prepared by Sears, Roebuck and Co., will help you identify the proper fit for children's clothing. If you can answer yes to each question on the checklist, the garment fits the child well.

If you sell infants' clothing, you will often help customers who are buying gifts for new or expected babies. Because this clothing is usually for future use, help them select items that will be appropriate to the season when the baby will be large enough to wear them.

When you are helping customers buy clothing, remind them that it is important to buy the right size. Clothing that is too large may be worn out by the time the child actually grows into it, or the child may never grow into that exact size. It is a safety hazard because it may trip the child. Also, garments that are too large will not be comfortable or look attractive.

Checklist for Proper Fit

By answering the following questions, one can judge whether a garment fits a child properly.

YES NO

- ☐ ☐ Can the child move freely? Ask the child to sit, bend, stoop, reach, stretch.
- ☐ ☐ Does garment fit smoothly, neither bagging nor binding?
- ☐ ☐ Do clothes hang from the shoulders rather than the neck?
- ☐ ☐ Do necklines and collars fit well without binding or rubbing and yet go over the head easily?
- ☐ ☐ Are armholes and sleeves roomy enough for action?
- ☐ ☐ Are separate tops long enough to stay tucked in as child moves and plays?
- ☐ ☐ Are pant legs and dresses short enough to prevent the child from stepping on hems and tripping as he or she moves?
- ☐ ☐ Are coats, jackets, snowsuits, large enough to fit comfortably over other clothing?

Experience 8-3 ■ How Does It Fit?

Take a small child between the ages of 3 and 8 shopping at a clothing store. Choose three different clothing items the child could wear. (Try to estimate accurately the size the child needs.) Help the child try on all three items and analyze the fit of each. Use the "Checklist for Proper Fit" to help you. On a separate sheet of paper, describe the clothing items the child tried on and the good and poor fitting features of each.

■ CHILDREN'S WEAR STYLES

When you work in children's wear, you will need to be familiar with the many styles of children's clothing sold. You also need to know about the activities of children at different ages so that you can suggest appropriate clothing. Because school-aged children's clothing is similar to men's and women's clothing, the emphasis here is on infants', toddler's, and preschoolers' apparel.

Infants' Clothing

If you work in the infants' department, you will often be asked to help choose a layette. A **layette** is a combination of articles a newborn baby needs: diapers, sleepwear, playtime wear, bedding, and items for feeding and bathing. Sometimes bottles, scales, and infant carriers are included. (See Figure 8-2.)

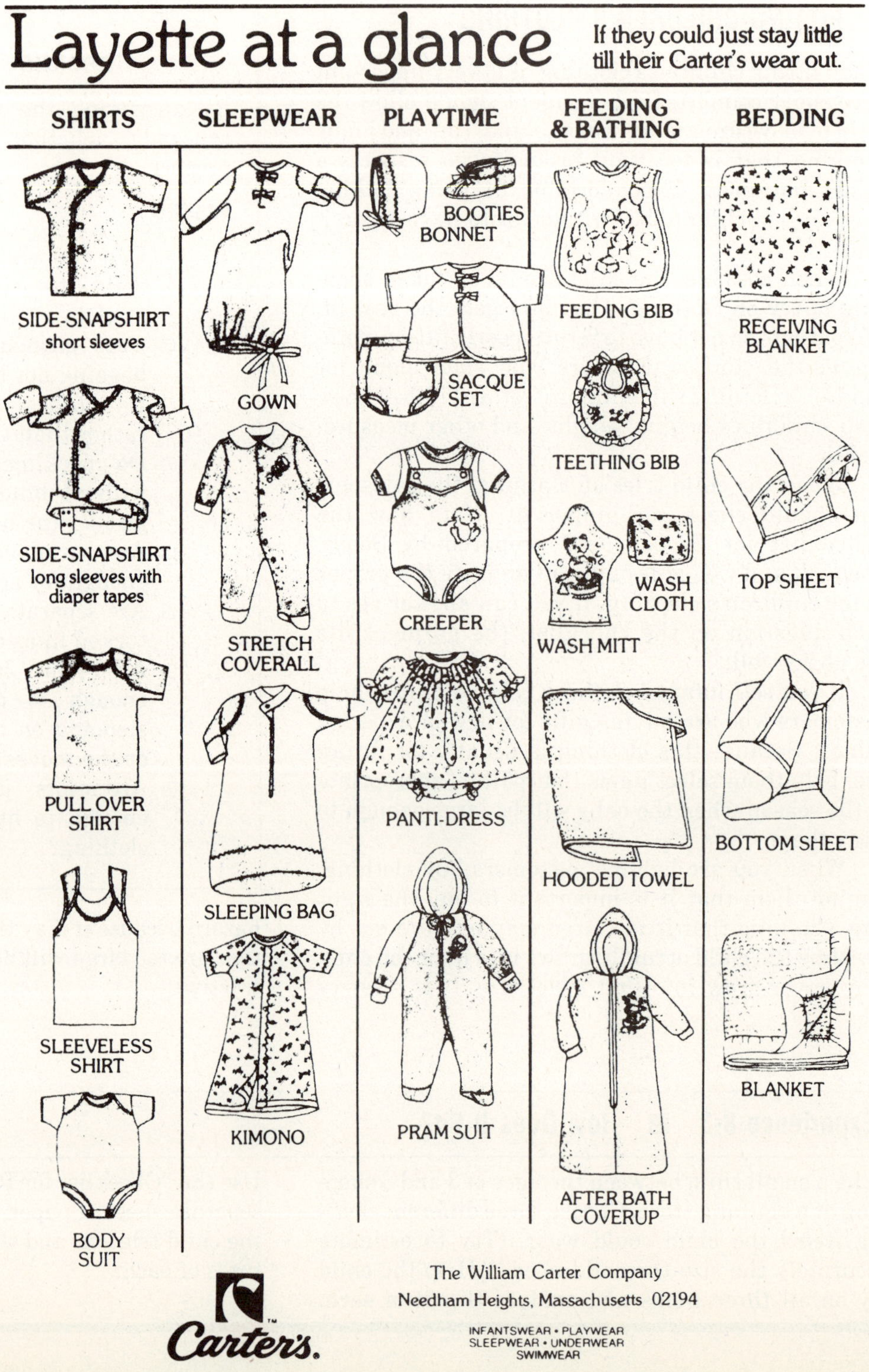

Figure 8-2 The clothing and supplies needed for a newborn baby are included in a layette.
© Carter's.

Although most parents prefer disposable diapers or a diaper service, some purchase diapers and waterproof pants for their babies. Cotton diapers are also used to clean up after baby's feedings.

Infant shirts are designed for warmth. Two popular styles are slip-on and snap. Snap shirts are convenient because they open down the front and do not have to be slipped over the baby's head. Both types of shirts are worn under other clothing or alone.

There are many sleepwear styles for babies. **Sleepers,** also called stretch coveralls, are one-piece suits that cover the baby's feet. They may have snaps down the front, crotch, and legs or a zipper down the front for easy dressing and changing. **Sleeping gowns** have set-in sleeves with cuffs that can be turned back to cover the baby's hands and a drawstring bottom that covers the baby's legs and feet. **Kimonos** are the length of sleeping gowns, have long sleeves, and close at the neck with two snaps or ties; they are open around the feet. **Two-piece sleepers,** worn by older babies, include a pullover top and bottoms that snap together at the waist and may have snaps down the crotch and legs.

Because newborn babies sleep much of the time, they are often dressed in sleepwear. These sleepwear styles keep babies warm even when they kick off the covers. For dress up, baby girls wear dresses and baby boys wear suits.

During the winter, infants wear a bunting bag or bootie suit when out of doors. The **bunting bag,** shaped like a bag, covers the baby's hands, legs, feet, and head. The **bootie suit,** or pram suit, has separate legs. (See Figure 8-3.)

Infants beyond the newborn stage wear coveralls, overalls, pants, and shirts for casual occasions or for play. Coveralls have long pants, long sleeves, and a high neck. Overalls have long pants and shoulder straps. In cooler weather, babies wear a knit shirt under their overalls for warmth. (See Figure 8-4.) Snaps along the inseam make it possible to change diapers without completely undressing the infant.

Baby's clothing styles appeal to the purchaser, not the wearer. Parents like clothing that is washable and requires no ironing. They want clothing that is styled for baby's comfort and easy movement.

Toddlers' Clothing

Toddlers are just beginning to walk. They need clothing styled for maximum freedom of movement. Clothing with baggy legs or big cuffs can impair movement or cause accidents. Also toddlers need clothing that is easy to take off, because this period is when most children are toilet trained. Because toddlers don't have a definite waist, their clothing should have elasticized waists or adjustable straps.

Toddlers' clothing is similar to infants' clothing. Boys and girls usually wear sleepers with gripper feet to bed for warmth. They may wear training pants, rather than diapers, and undershirts for warmth. Toddlers often wear pants and pullover shirts or overalls for play. Overalls often have suspenders or straps that button on the waist. Usually straps and suspenders are adjustable and can be changed as the toddler grows. They hold up the toddler's pants.

Girls' play clothes often have puffed or ruffled sleeves, round collars, and flowery designs. Boys' play

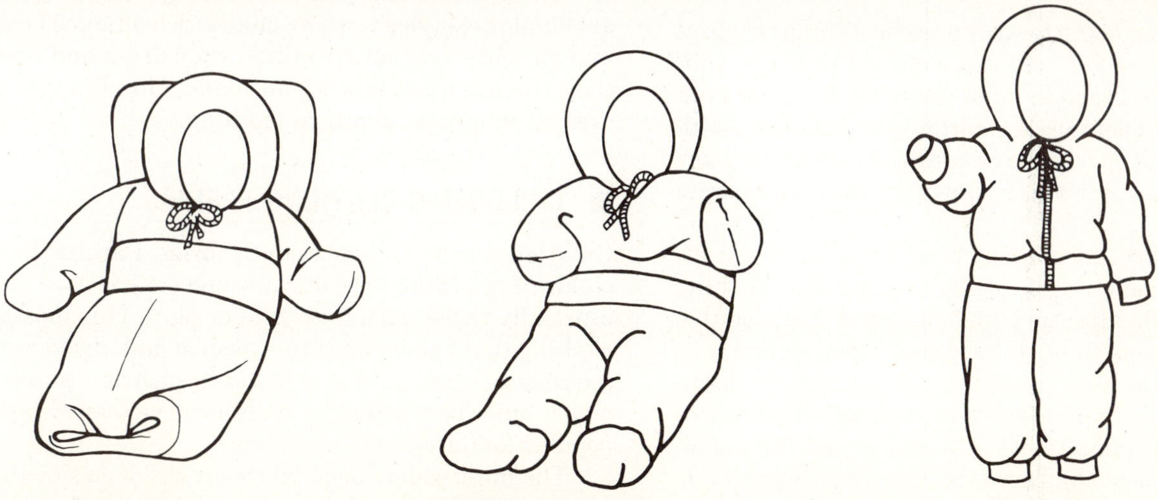

Figure 8-3 Become familiar with the three basic styles of infants' and toddlers' outerwear pictured above.

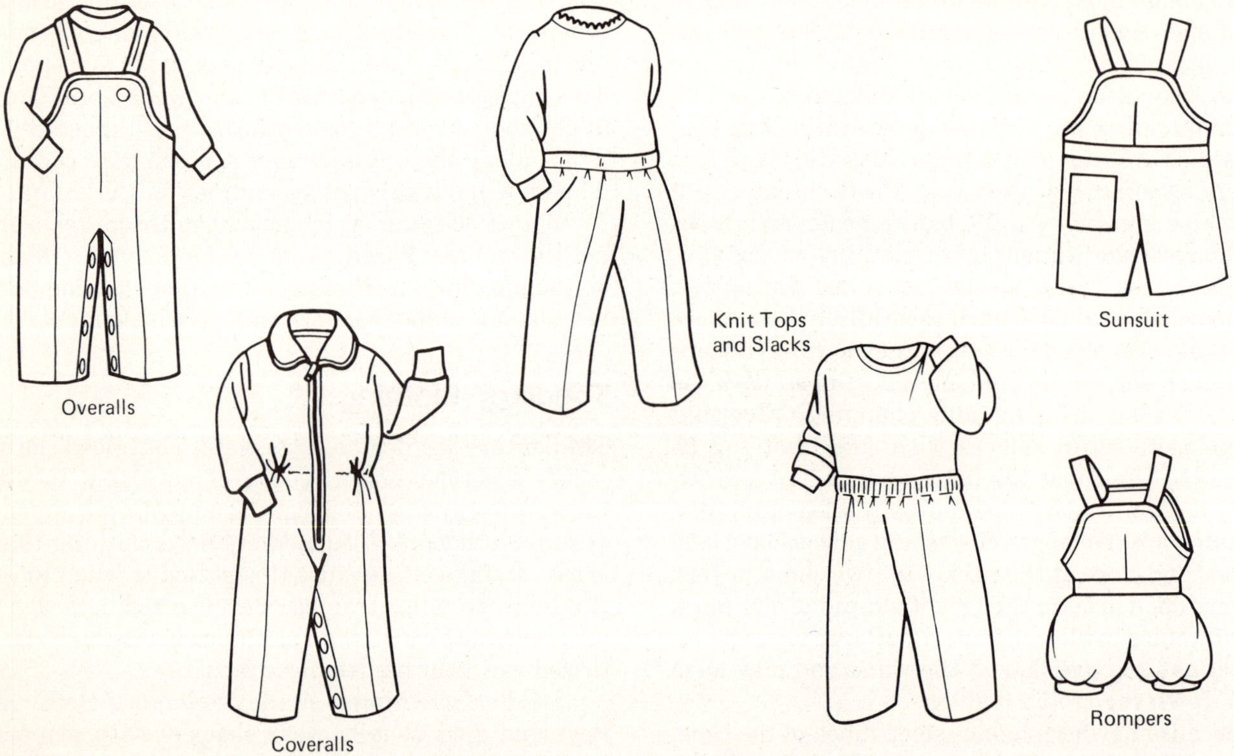

Figure 8-4 Learn to recognize the different garments available for infants' and toddlers' play wear.

clothes usually have squared collars, tailored sleeves, and flat trimmings. For dress occasions, little girls wear short dresses, smocks, a jumper and blouse, or a dress with a pinafore. (A **pinafore** is a sleeveless garment that looks like an apron.) Boys usually wear long or short pants with a shirt and jacket. Both boys' and girls' clothing often includes **appliqués,** special designs that are sewn or ironed on the garment.

Preschoolers' Clothing

Most children begin to wear preschool children's sizes when they are 2½ or 3 years old. At this age, children begin to learn to dress themselves. Large armholes, front opening, big buttons, and clothing with the front and back alike or so different that it is easy to tell them apart make it easy for children to dress themselves.

Around this age, many children begin to pay attention to their clothing. They realize that clothing brings them attention and approval from adults. They often ask to be dressed like other children.

Preschool girls usually wear pajamas or nightgowns rather than sleepers for bedtime. Boys at this age prefer pajamas. Both also wear robes. For underwear, preschool children wear sleeveless shirts or T-shirts, with short sleeves. Girls wear banded- or ruffled-leg panties, sometimes trimmed in lace. They wear slips under their dresses. Boys wear knitted briefs or woven undershorts and undershirts or T-shirts.

Playtime wear usually consists of pants, overalls, and pullover shirts. For dress occasions, girls wear dresses; boys wear Eton suits or pants, shirts, and jackets.

School-Aged Children's Clothing

School-aged children wear clothes for school, play, and dressy occasions. The clothing styles they wear are similar to styles worn by adults. School-aged boys and girls are very aware of how they dress and how their friends dress. It is not uncommon to see a group dressed in almost identical clothing.

■ CHILDREN'S CLOTHING FABRICS

Children like soft, comfortable fabrics. Parents like children's clothing to be durable and easy to care for, especially those worn for school or play. This means clothing that can be machine-washed and -dried and requires no ironing. Delicate fabrics, such as organdy, which must be washed by hand or dry-cleaned, are popular for "dress up" occasions.

The most widely used fabric for children's clothing is cotton. Because it is cool, soft, lightweight, and absorbent, it is comfortable. Cotton is also washable and needs no ironing if it is knitted or permanent press. Much infant clothing is made from combed cotton because it is very soft and does not irritate a baby's skin.

Children's clothing is also made from blends of man-made and natural fibers; cotton and polyester is the most common blend. This blend combines the long-wearing, easy-care features of man-made fibers with the coolness, absorbency, and feel of the natural fibers.

Many children's clothing items are also made from wool. Wool is warm and can insulate the child from cold, damp weather. It is often used for children's outerwear, such as sweaters and coats, and is long-wearing and always fashionable.

All children's sleepwear manufactured and sold in the United States must be made of fabrics that can pass current government flammability standards. (See Figure 8-5.) Today, clothing stores sell flame-resistant and flame-retardant sleepwear. The words are used interchangeably and mean the same thing. They do not mean the garment is fireproof. Rather they mean that the fabric, if accidentally ignited, will burn very slowly, and when removed from the flame, will be self-extinguishing or easy to extinguish.

Advise the customer to follow the care instructions

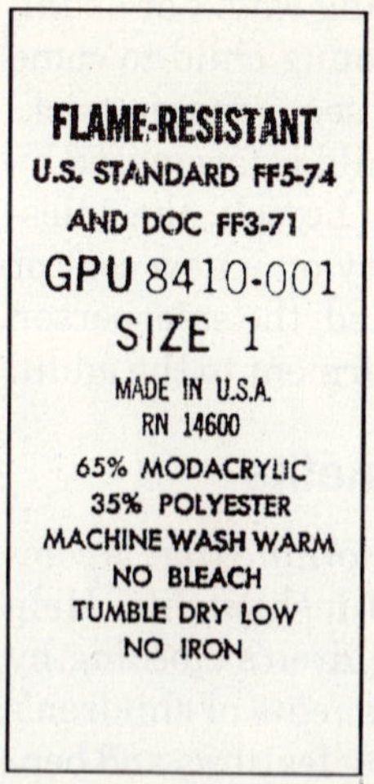

Figure 8-5 Here's a typical example of a label from a garment made of a flame-resistant fabric. Courtesy Nazareth Mills, Inc. A Division of Kayser-Roth Corp.

attached to each clothing item. Any soap product, including bar soap, powder, or liquid soap, should not be used when washing a flame-retardant garment. This is because an afterwash residue can build up on the fabric and become flammable. In addition, this soap buildup can cause stiffness in the fabric, yellowing, and often, a rancid odor. Instead, use a phosphate powdered detergent or a heavy-duty liquid detergent.

Experience 8-4 ■ Design a Wardrobe

Using the children's clothing featured in a catalog, plan a wardrobe for a boy and one for a girl of any age. Choose five different playtime outfits, two dress outfits, and appropriate outerwear for cold winter months. Also choose the underwear and sleepwear needed for three days. Keep the total cost of each wardrobe below $400.

Cut out your choices and attach them to a separate sheet of paper or sketch the clothing you choose for the boy and girl. Next to each item, write the manufacturer, style name, fabric content, and price.

Show your wardrobe choices to another classmate who also completed this exercise. Discuss the reasons why you selected those clothing items. (When choosing the items, consider their versatility, practicality, and attractiveness.)

■ SPECIAL REMINDER FOR SELLING CHILDREN'S CLOTHING

In selling children's clothing, you will often work with two shoppers: the parent, who pays for the merchandise, and the child, who is a strong influence on the parent's decision of what to buy. When selling to a parent and child, try to determine who is the decision maker and gear your sales approach to that person.

Involving the Child in the Sale

Many young children dislike shopping trips even when the purpose of the trip is to buy them new clothes. They don't like waiting around or looking at so many items. They get bored listening to salespeople talk about styles, fit, or construction details in language they don't understand. When the salesperson gears the total sales presentation to the "grown up," young shoppers probably wish they were out playing.

It takes a skillful salesperson to make shopping fun for both parents and children. Successful children's wear salespeople have to work with children. They need to understand that children try to "show off" for strangers. They must realize that children who are looking for attention will do almost anything to get it.

Successful children's wear salespeople show a real interest in children, which makes both children and adults feel good. These salespeople pay attention to their young customers and talk to them in words they understand about things that interest them. Often they crouch down so that they are on an eye level with a child. They do everything possible to make the child feel as though he or she has a part in the sale.

The children's wear salesperson must present the

merchandise in an attention-getting way. For example, the salesperson may ask a young child to come over to the three-way mirror to see a magic trick. While at the mirror, where the child is fascinated by seeing three images of himself or herself, the salesperson can be helping the child try on a new coat or sweater. While the child is occupied, the salesperson can point out the benefits of the garment to the adult.

Pointing Out Important Benefits

Children's wear customers want clothing that is comfortable, durable, a good fit, and fashionable. Help customers decide to purchase children's clothing by pointing out the advantages or benefits of children's wear. Try to point out the following features and benefits:

Comfort Some styles and fabrics are more comfortable than others. For example, clothing that hangs from the shoulders is usually more comfortable than clothing that hangs from the waist. Nonbinding clothing with simple designs and minimum trim is comfortable. Stretch fabrics and knits, as well as soft, lightweight fabrics, are comfortable. Clothing that fits the child well is also comfortable.

Ease in Self-Dressing Parents appreciate clothing that children can put on and take off by themselves. For example, front buttons or zippers can be opened and closed by a child.

Room for Growth Because children outgrow their clothing quickly, parents look for clothing that has room for growth. These features allow a child to wear clothing for a long period of time. Some of these features are:

- Raglan sleeves, which have no definite shoulder line
- Adjustable shoulder straps
- Deep hems
- Deep cuffs
- A set-in band across the front of a dress rather than a seamed waistline
- Undefined waistline
- Stretch fabrics
- Two rows of buttons or snaps for attaching tops and bottoms

Easy Care Because children get their clothing dirty, items that can be machine-washed and -dried and requires no ironing will appeal to customers. Soil-resistant or soil-release finishes will keep the clothing clean longer and cut down on washings. Care labels attached to children's wear provide information on how to take care of the garment.

Clothing Coordination Part of growing up involves learning how to coordinate clothes. When young children begin to dress themselves, they often combine garments that do not go together. They may wear clothing that clashes, or they may combine a dressy blouse with casual pants. To help children learn how to coordinate clothing, many manufacturers produce clothing that has the same design or color pattern on the slacks and shirt. This helps children learn "what goes with what" at a young age.

Experience 8-5 ■ Choosing the "Best" Clothes

Using a catalog that includes children's clothing, complete the following:

1. Choose five different clothing items that would be comfortable for a girl age 9 months. Cut out and attach these pictures to a separate sheet of paper. Write the reasons you identified these clothing items as comfortable on the paper.
2. Choose five different clothing items that would be easy for a girl of three to put on and take off. Cut out and attach these pictures to a separate sheet of paper. Write the reasons you identified these clothing items as "easy dressing" on the paper.
3. Choose five different clothing items that have "room for growth" for a boy aged seven. Cut out and attach these pictures to a separate sheet of paper. Write the reasons you identified these clothing items as having "room for growth" on the paper.
4. Choose three different outfits that would be easy for a boy of three to coordinate. Cut out and attach these pictures to a separate sheet of paper. Write the reasons you identified these outfits as "easy to coordinate" on the paper.

Chapter Note

1. Sears Roebuck and Co.

CHAPTER 9

Shoes

Reebok, the manufacturer of popular athletic shoes, began selling shoes in the United States in 1979 to serious athletes and joggers. At that time, aerobics was just catching on. This company, which started in England in the 1890s, noticed that people were doing aerobic exercises in shoes made for other activities, resulting in user discomfort and risk of injury.

Reebok talked to aerobic instructors and exercisers. For over two years, the company designed and tested a shoe called the Freestyle. This was the first shoe of garment-quality leather designed for comfort and fit, the first shoe designed for aerobics, and the first athletic shoe designed for women. Today, Reebok sells a complete line of aerobic shoes for light, moderate, and heavy aerobic exercisers.

Product information about athletic shoes is one of the many topics included in this chapter. This merchandise information is very important because it enables salespeople to increase their sales and earning capacity. It creates enthusiasm for the product and adds to a salesperson's self-confidence. And when you talk knowledgeably about the shoes you sell, you build the customer's confidence in you, the merchandise, and the store.

■ PARTS OF A SHOE

"It seems to rub at the topline," "it doesn't fit my instep," and "the vamp is cut too low" are examples of customer statements often heard by shoe sales professionals. Co-workers and manufacturer reps also use a specific vocabulary when discussing new shoe styles. Therefore, you need to know and use the correct names for the parts of a shoe and parts of a foot.

Shoes are made to fit the various parts of the foot: toes, ball, arch, heel, instep, waist, and ball joint. (See Figure 9-1.) Basically, a shoe consists of three parts: sole, heel, and upper. The **sole** is the bottom of the shoe, the part of the shoe that comes in contact with the ground. It is also called the outsole or outersole. There is also the **insole.** This is the sole to which the upper part of the shoe and the outsole are attached. The **shank** is a rigid, semi-rigid, or flexible piece that may be inserted between the insole and outsole. A shank runs between the heel and the ball of the foot. It provides support for the arch and helps maintain the shape of the shoe. (See Figure 9-2.)

The **heel** is the part of the shoe at the back of the foot. The **heel lift** is attached to the heel. It is a protective rubber or plastic cap on the bottom of the heel. The front of the heel is called the **heel breast.**

The **upper** is the top part of the shoe. The **vamp** is the front part of the upper. It stretches across the ball of the foot and is attached to the sole. The **quarter** is the back part of the upper. It fits around the heel and sides of the foot. The edge of the upper is called the **topline.** The edge or topline of the vamp, in front of the instep, is called the **throat.** Throat

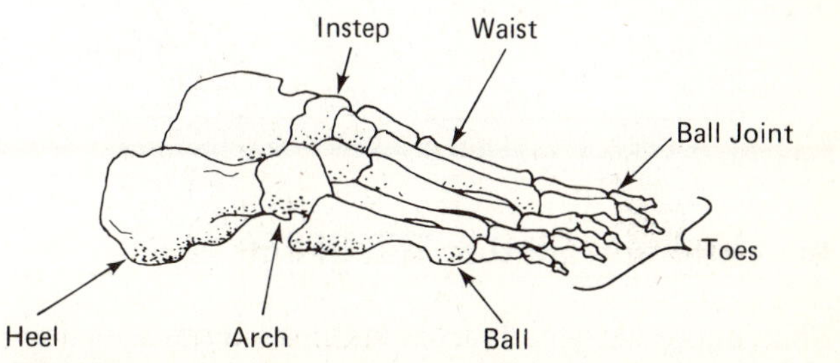

Figure 9-1 You should learn the various parts of the foot as illustrated here. Courtesy National Shoe Retailer's Association.

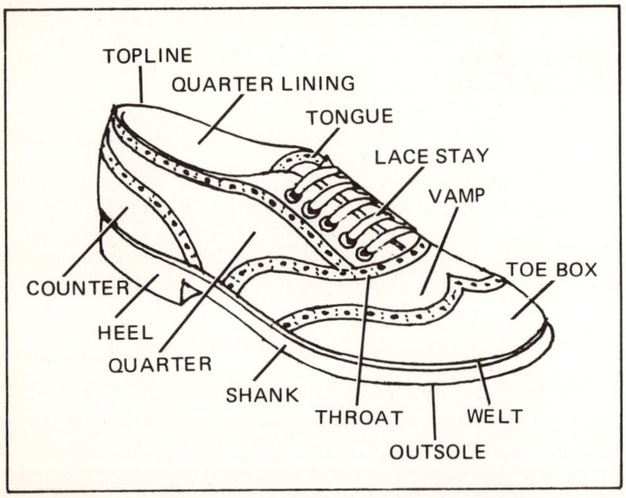

Figure 9-2 By studying this diagram, you can learn the parts of a shoe. These parts are described in more detail beginning on page 125.

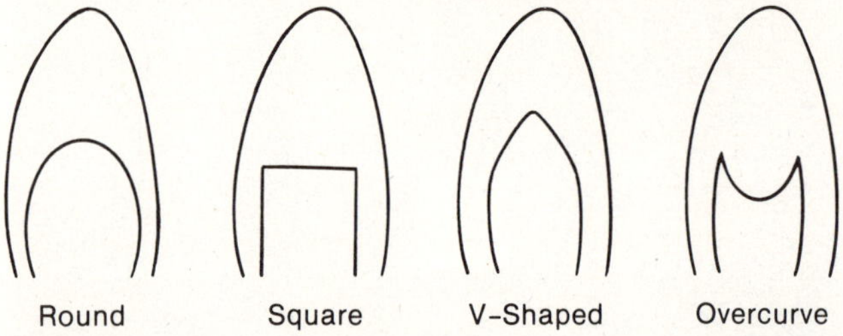

Figure 9-3 Some of the throat lines used on pumps are shown here.

lines can be round, square, V shaped, or scalloped. (See Figure 9-3.) The **tongue** is a strip of leather or other material that is attached inside the throat of a laced shoe. The tongue comes up to the top of the shoe under the lacing. The **toe box** is the tip of the shoe at the toe. The shape of the toe (as well as the heel) changes with fashion trends. At the back of the shoe, between the outside and the lining of the quarter is the **counter.** It is a stiff piece of material that makes the shoe fit snugly at the heel. It also strengthens the back part of the shoe. The **linings** are those parts of the shoe that add stiffness; they help the shoe keep its shape and make it more comfortable to wear.

Experience 9-1 ■ Name Those Parts

Identify the lettered parts of the shoe in the space provided.

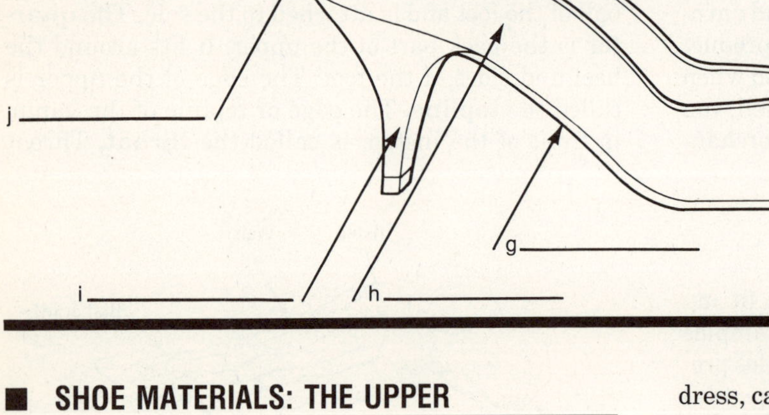

■ SHOE MATERIALS: THE UPPER

Shoe uppers are made from leathers, man-made materials, or fabric. For example, men's oxfords may be made with cowhide. Vinyl uppers are featured in dress, casual, sport shoes, and boots for men, women, and children. Linen and satin uppers are preferred for women's shoes that will be dyed another color. (See Table 1.)

■ **Table 1**
Comparison of Leather, Synthetic Materials, and Woven Fabrics

	Leather (animal skins and hides)	Synthetic Materials (combination of organic and inorganic materials)	Woven Fabrics (natural and synthetic fibers)
Porosity	Will absorb moisture, keeps foot cool	Usually nonabsorbent, nonbreathable unless porous (poromeric) material	Usually will absorb moisture, will keep foot cool
Durability	Extremely durable, tear resistant; long wear life with proper care	Almost indestructible with normal wear; long wear life	Easily damaged, short wear life
Flexibility	Flexible and elastic (i.e., goes back to original shape after wearing); leather becomes supple with use	Not as flexible; material may crack with wear	Varies in flexibility depending on fabric
Care	Care required (polish, creams, etc.)	Little care required (damp cloth)	Care required depends on fabric; some fabric uppers are easily cleaned
Appearance	Luxurious texture and feel; capable of extensive detailing	Smooth, mostly shiny surface; can be made to look like leather	Varies with construction and fabric; range from satin to canvas
Price	Expensive	Moderate to expensive	Moderate

Leather

Leather comes from the hides and skins of cows and calves (cowhide and calfskin), pigs (pigskin), goats and kids (goatskin and kidskin), sheep (sheepskin), and horses (cordovan). The raw skins are converted into finished shoe leather by a process known as tanning. Leather, used for the shoe upper and sole, has the following characteristics:

1. *Leather is flexible.* Leather stretches to fit the shape of the foot and returns to its original shape so the shoe does not lose its style. This flexibility makes for a very comfortable fit.
2. *Leather is porous.* Pores allow air to pass through the upper of the shoe, keeping the foot cool, and permit moisture to evaporate, keeping the foot dry.
3. *Leather is renewable with polish or creams.* Properly cared for leather shoes can provide customers with years of attractive wear.

Types of Leather As you read product literature about leather shoes, you will come across the following words to describe various types of leather:

Side Leather Side leather refer to all cowhides. A cowhide is so large that it is cut down the center, along the spine, into two parts. Cowhide is a durable, pliable shoe leather.

Top Grain Leather The hair side of an animal's skin is called the grain side. The side next to the flesh of an animal is called the flesh side. The grain side is more durable, firmer, and smoother than the flesh side. If the grain side of the leather has not been altered, the leather is called **top grain.** Top grain leather is the finest quality leather. It is the most attractive and wears the longest. Only top grain leather may be called "genuine leather."

Split Leather Heavy, thick hides are split into layers. The under layers, which have no grain surface, are called splits. Split leathers may have artificial grains stamped on them, making it difficult to tell if a shoe is made from top grain leather or split leather. However, split leather is less expensive, has a rougher, coarser appearance, and does not wear as well as top grain leather.

Facts About Leather Because so many synthetic materials look like leather, it is difficult to tell if a shoe is made from leather or some other material. The best way to determine the material in a shoe is by reading the label on the sole of the shoe or inside the heel. If the sole reads "genuine leather," the entire shoe is made of top grain leather, except for the heel, inside stiffenings, and ornamentation. Many shoes are labeled on the sole, *vero cuoio,* which is Italian for genuine leather and indicates the shoe upper and sole were made of leather.

■ **Table 2**
Materials Used in Shoe Soles

Quality	Crepe	Leather	Neolite	Wood
Durability	Very durable, but will darken	Not as durable as crepe, will wear out	Most durable	Not as durable as crepe, leather, or neolite
Comfort	Most comfortable, is flexible, and offers give	Comfortable, flexible, and offers give; lets foot breathe	Does not give, not very comfortable, hot	Not as comfortable as leather, does give somewhat
Price	More expensive than leather	More expensive than neolite	Least expensive	Costs about the same as leather

Synthetic Materials

Vinyl is the most popular synthetic or man-made material for shoes. A vinyl upper can be made to resemble any type of leather. The texture of vinyl can range from shiny patent leather to suede. The label inside the heel or on the sole of a vinyl shoe will say "man-made." Originally, shoes made with vinyl had a reputation for being very hot because vinyls could not "breathe" like leather. However, a synthetic material called **poromeric** was created. Like leather, poromeric material can breathe. It is much more comfortable to wear than nonbreathable vinyl and much more durable and easy to care for than leather.

Woven Fabrics

The upper material of adult's or children's shoes may be made of woven fabrics. The decision regarding which fabrics to use as shoe uppers is influenced by fashion. Linen and satin uppers can be easily dyed and are popular for weddings. Jute and strawcloth are popular uppers for summer shoes, as is canvas for sneakers and sports shoes. One disadvantage of fabric uppers is that they are hard to keep clean and looking attractive. They may also wear out more quickly than leather.

■ SHOE MATERIALS: THE SOLE

Some common materials used for shoe soles are leather, neolite (hard plastic), crepe, and wood. Each material has distinct qualities. These soles are compared in Table 2. For example, leather soles are comfortable and flexible, but they are expensive and wear out faster.

■ SHOE MATERIALS: THE HEEL

Common materials for shoe heels are crepe, leather, plastic, rubber, and wood. The durability and comfort of these materials are ranked in Table 3. The most durable and comfortable heel is made of crepe. Although leather is comfortable, it wears down quickly. Today, most shoes have plastic or rubber heels.

■ SHOE CONSTRUCTION

Although you will seldom be asked how a shoe is made, information about shoe construction will help you understand why some shoes fit better, wear better, and have fewer defects than others.

As is true with most apparel and accessories products, a new shoe begins in the designer's office. After the sole is designed and priced, marketing research is conducted to determine if the shoe will sell. If the shoe appears to be profitable, the assembly-line process begins.

A cardboard pattern of the shoe is made. Dies, or steel cutting forms, are made from the patterns. There must be a die for every style and size of the shoe. In the cutting room, shoe parts are cut from large pieces of leather, man-made materials, or fabric. The pieces are tied and bundled by size and type.

■ **Table 3**
Comparison of Different Shoe Heels

Durability	Comfort
Crepe (most durable)	Crepe (most comfortable)
Rubber	Leather
Plastic	Rubber
Wood	Wood
Leather (least durable)	Plastic (least comfortable)

Experience 9-2 ■ Find the Customer Benefits

1. What are the customer benefits of leather uppers?

____

____

2. What are three customer benefits of leather soles?

____

____

3. What are three customer benefits of vinyl uppers?

____

____

____

In the stitching room, both handstitchers and computer stitchers are used. The completed upper then goes to the lasting room. The upper is pulled over a last to give the shoe its shape. The size of the last determines the size of the shoe. For example, a size 8B last is used to make a size 8B shoe. (A **last** is a wooden or plastic form that resembles a human foot.) The goal in lasting is to pull the leather evenly all the way around the last, leaving just enough material to attach the insole. A heat setter is used to make the upper conform to the shape of the last. The longer a shoe stays on the last, the better it fits and feels on the foot. After lasting, the upper actually looks like a shoe.

Next, the outersole is attached to the insole by cementing or sewing. In cement construction, uppers and soles are glued together. One well-known method of sewing the upper to the sole is known as the Goodyear welt. A small grooved ledge, which is almost always leather, runs around the front part of the shoe at sole height. Stitches through the side of the upper portion of the shoe anchor the welt to the insole, while separate stitches downward through the welt hold the sole in place. Welt-constructed shoes are sturdier than any other kind. They provide more cushioning between the inner and the outer soles, more protection from heat and cold, and are easily resoled. Welt construction is the most expensive way of attaching the sole.

Most shoe factories use heeling machines to attach the heel to the outersole. In better-quality shoes, heels are nailed as well as cemented in place. The nail pattern is designed to hold the heel firmly so that it will not give way under stress.

In the finishing room, the shoes are cleaned and finished. Sock lining and heel pads are added to the shoe. Finally, heels are dyed and uppers are polished. If necessary, laces are added.

■ SHOE STYLES, TEXTURES, FINISHES, AND COLORS

People buy shoes with a certain purpose in mind. Shoe salespeople hear customers ask for dress shoes, casual shoes, school shoes, athletic shoes, and work shoes. Many shoe stores and department stores stock sport shoes such as running shoes, tennis shoes, aerobic shoes, and fitness walking shoes. During the late summer, shoe stores receive their boot order: fashionable boots in smooth leather, vinyls, and suedes and the more functional rubbers and galoshes. Christmastime is when most customers purchase house slippers. There are a variety of shoe styles, textures, finishes, and colors to fit everyone's needs.

Styles

Shoe styles can be loosely grouped as oxfords, slip-ons, sandals, boots, and athletic shoes. **Oxfords** are low-cut, laced shoes with anywhere from one to six eyelets. **Slip-ons** are designed to fit snugly to the foot without laces or fasteners. A **sandal** is a sole piece held onto the foot by straps or thongs. The **boot** is a high-cut shoe with the upper rising to or above the ankle. **Athletic shoes** are designed to improve performance and prevent foot injury during sport activities. A professional shoe salesperson calls each shoe by its specific style name. For example, the two basic styles of oxfords are balmoral (or bal) and blucher. The **bal** is a laced shoe with a closed throat, stitched vamp and tongue. In a **blucher**, the tongue of the shoe is a part of the vamp. A brogue is a style of blucher oxford with a durable looking textured surface. A variation of the brogue is a wing tip. Saddle shoes are a type of bal oxfords. (See Figure 9-4.)

Figure 9-4 The two basic styles of oxford are blucher, represented here by the wing-tip brogue, and balmoral (bal), an example of which is the saddle.

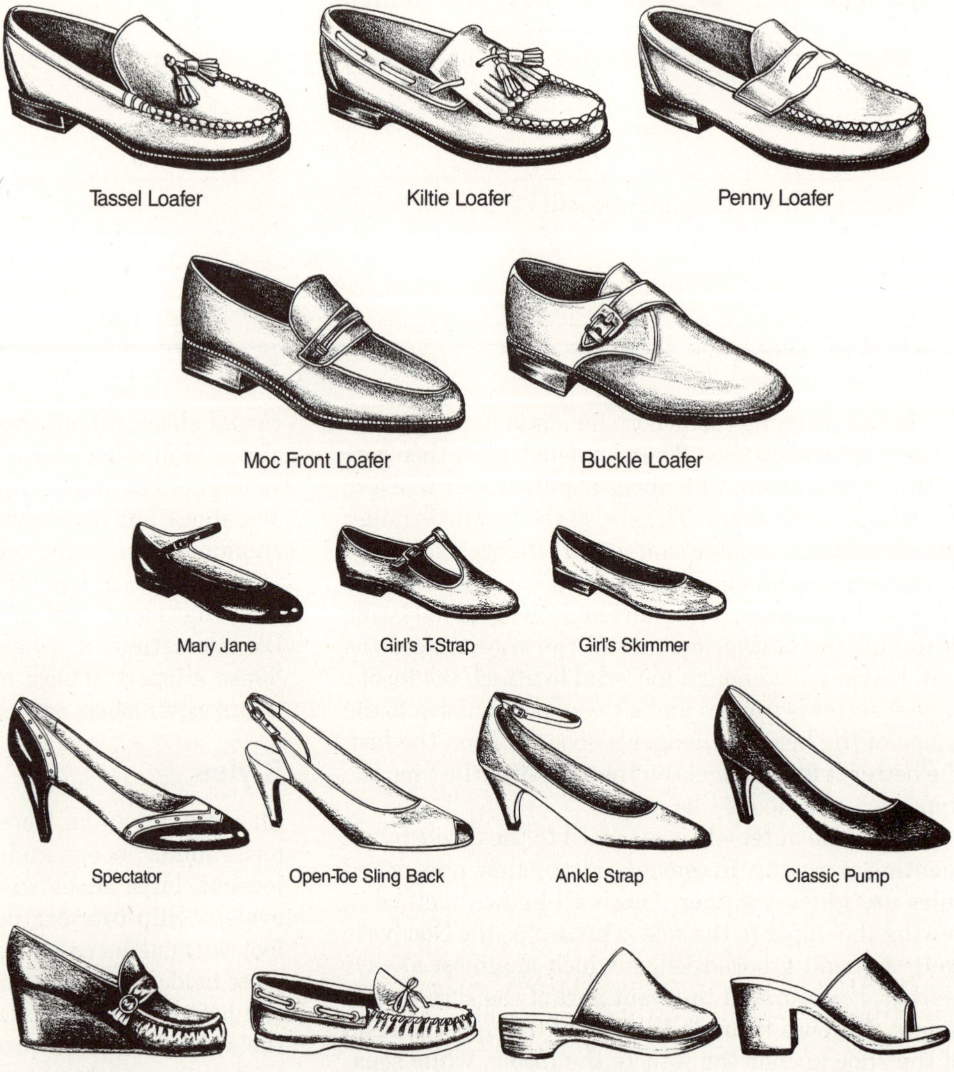

Figure 9-5 Some distinctive styles of slip-ons are shown here. © Judy Francis

Loafers and pumps are examples of slip-on shoes. Loafers are often decorated with tassels, straps, buckles, or kiltie (decorative fringe).

Pumps are available in any heel height and can be decorated or plain. A low-heel pump is less than 1½ inches high, a medium-heel pump is between 1½ and 2½ inches high, a high-heel pump is more than 2½ inches high. Some pumps have straps that fit across the instep or back of the heel. Other examples of slip-on shoes are the mule, monk, clog, and moccasin. (See Figure 9-5.)

Sandals are popular warm-weather shoes for men, women, and children. Some sandal straps wrap the leg. Thongs are one popular style of sandal. (See Figure 9-6.)

Boots are functional, fashionable, or both. Boots made of leather or synthetic materials protect the

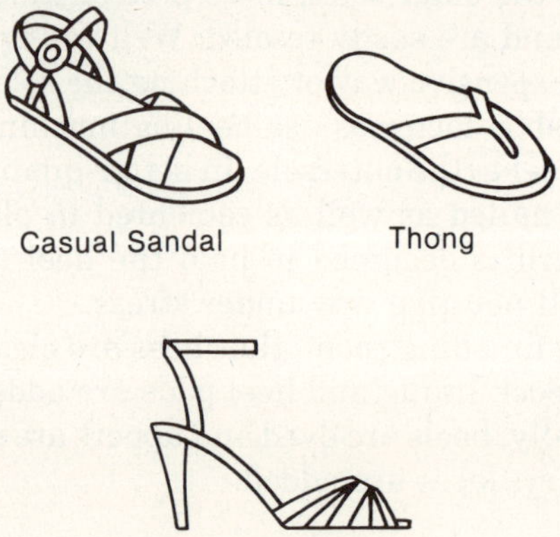

Figure 9-6 Sandals are a popular warm-weather shoe, available in dressy and casual styles.

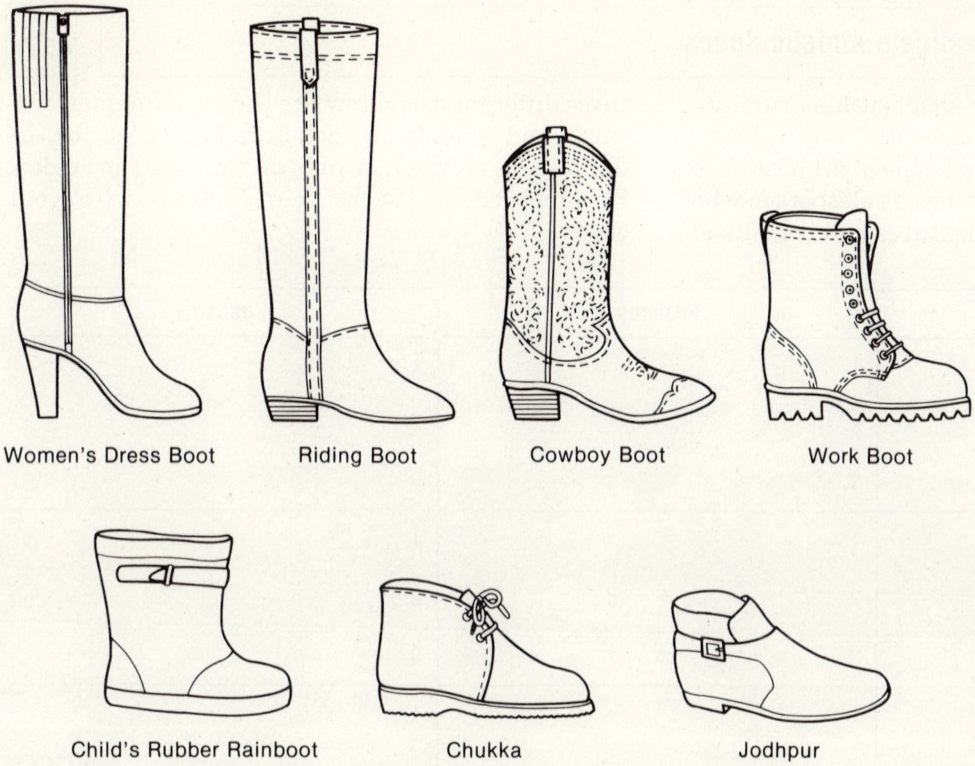

Figure 9-7 Boots are available in styles for men, women, and children; for dress and casual wear; and for protection against the weather.

foot and leg from snow and water. Work boots with thick soles and reinforced toes protect the foot from injury. Cowboy boots, with pointed toes and stacked heels (a finish that gives the illusion of thin strips of wood stacked together), are often decorated with intricate stitchery. Some are made from unusual leathers, such as snakeskin or ostrich. (See Figure 9-7.)

Athletic shoes are worn by serious athletes while they are performing their sport. They are also worn by people of all ages for casual wear. For example, athletes who run or do aerobics want shoes designed specifically for their sport so that they will be protected from injury. The shoes absorb shock and help them achieve their maximum performance. Some people wear athletic shoes because they are very comfortable and give the appearance of an active, healthy lifestyle. The original athletic shoes were sneakers, shoes with canvas uppers and rubber soles. (See Figure 9-8.)

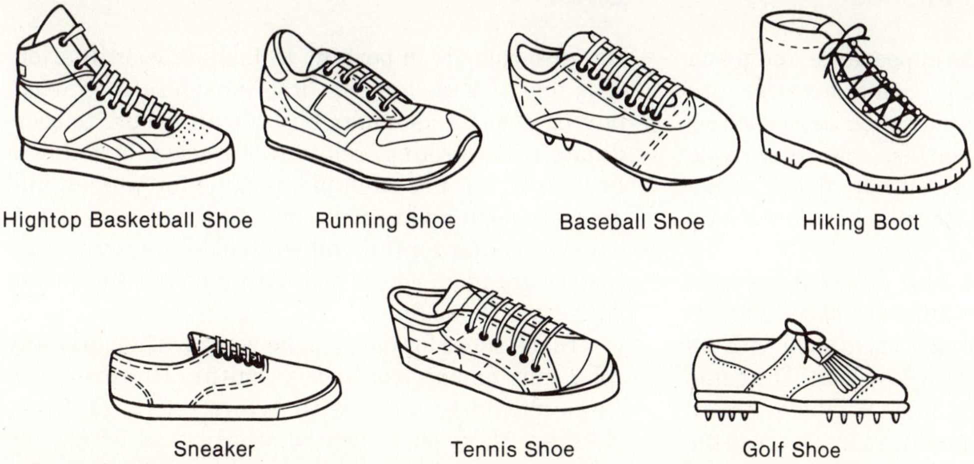

Figure 9-8 Athletic shoes are designed for particular sports, but they also serve as a fashion item. Sneakers can be worn for a variety of sports, but they do not have the specific features required by serious athletes.

Experience 9-3 ■ Investigate Athletic Shoes

Choose one type of athletic shoe, such as running shoes, aerobic shoes, or tennis shoes. Visit a sporting goods store that sells that type of athletic shoe in a variety of brands. Interview a salesperson who can explain the important features and benefits of three different brands. Write the brand names, features, and benefits for each brand and describe the differences between brands on the chart provided. Be prepared to discuss your findings with your class.

Brand	Features	Benefits
1.	1.	1.
	2.	2.
	3.	3.
2.	1.	1.
	2.	2.
	3.	3.
3.	1.	1.
	2.	2.
	3.	3.

Major differences between brands:

Surface Textures and Finishes

Surface texture refers to the appearance and touch of a shoe's upper materials. Some textures are smooth, such as kidskin. Other, such as ostrich or snakeskin, have a grained surface. Suede is made from buffing the flesh side of leather. It has a fine napped texture. Brushed leather has a coarser nap than suede.

The same style shoe but with different textures and finishes can present two different looks. For example, a patent leather skimmer style shoe with its glossy, shiny texture is a very dressy look. The same style shoe with an aniline dye finish will have a more casual look. (Aniline, a dye produced from a coal-tar base, is transparent and allows the natural grain of the leather to show.) Still other looks will result from using suede or fabric even though the basic style remains the same.

Color

Because color is important in fashion coordination, it is one of a customer's first considerations when buying shoes. Customers buy certain colors to coordinate with clothing, relate to a season, or follow a fashion trend. For example, pastels for women and girls are popular in spring and summer; darker colors are popular for fall and winter. Many times, customers are more concerned with a shoe's color than with its style or finish.

Black shoes for men can be worn with nearly any color suit or sportcoat-slacks outfit. However, tan shoes should be worn with brown, olive, or white clothing. Navy shoes can be coordinated with navy and black; gray shoes with gray, black and shades of green. Women's shoe color depends on the season and the current fashion trends. Most women try to coordinate their shoes with the color of their clothing.

Experience 9-4 ■ Choosing Shoes

Listed below are four customers who want to purchase new shoes. Read the information provided about each customer. Then, using fashion magazines, catalogs, or newspapers, locate examples of different shoe styles, textures, and colors that would be appropriate. Cut out or sketch at least five shoes for each customer. Attach your choices to a separate sheet of paper. (Try to offer each customer a wide selection of styles, colors, and textures.) Write the manufacturer, style name, texture, and color of the shoes you choose next to the selection you offer.

Customer 1 This customer is just graduating from college and is putting together an interview wardrobe. He just purchased a navy blue pinstripe suit and camel hair blazer with navy dress slacks and needs new shoes to coordinate with both outfits.

Customer 2 This customer is in high school and wants a new pair of shoes to coordinate with the clothes she wears to school, work, and out with friends. She prides herself on wearing the "latest" fashion look and doesn't like to look like everyone else. Her wardrobe includes bright colors, like hot pink, bright green, and fire-engine red.

Customer 3 This customer is a 6-year-old boy shopping with his dad. He needs school shoes that are durable and won't show wear or dirt. He is very aware of what the other boys are wearing.

Customer 4 This customer is 5 feet 11 inches tall and needs shoes she can wear to work as an officer at the local bank. She wants something neutral that will coordinate with most of her wardrobe. She wears mostly brown, taupe, and neutral colors.

■ SHOE CARE

Customers want to keep their shoes looking as attractive as the first time they wore them. By sharing your shoe care knowledge, you will help customers care for their new shoes. This additional service will give customers an added reason to buy shoes from you, and that adds up to more dollars in your pocket.

Sometime during the sales presentation, explain to customers that they can prolong the life of their shoes by "giving them a rest" between wearings. Rotating wearings allows the shoes to dry from perspiration slowly and naturally. Advise your customers to use shoe trees. Sold in pairs and shaped like a foot, shoe trees absorb the moisture in shoes and help keep the shoes in shape when they are not being worn.

The most important rule in caring for leather shoes is to keep them well-polished. Leather shoes, like skin, have pores that need to be cleaned and moisturized. Inform customers that they should apply several coats of transparent polish before wearing new shoes. This will keep the shoes looking good and will keep the leather soft and pliable. There are three main types of polish: liquids, creams, and paste waxes. Liquid polish covers scuffs well but can cause leather to crack with continual use. Shoe cream cleans and softens the leather. Paste waxes protect as well as polish.

Brush suede leather to make it look new again. Remove surface soil from suede leather by wiping it with a terry cloth towel or a wire suede brush. (The brush raises the nap and cleans the leather.) Special suede cleaners also work well. A gum eraser or the soft side of an emery board will help to take out spots. Suede should never get wet. If suede shoes do get wet, they should be dried naturally at room temperature and then brushed when dry. There are also waterproof sprays that prevent water from soaking in and ruining the nap.

Patent leather should be wiped clean with a damp cloth. A cream polish or a thin film of petroleum jelly will help make the patent leather finish shine. Prolonged sunlight will make the patent leather fade.

A vinyl shoe with a patent finish is ruined by petroleum jelly. Use all-purpose cleaner to renew the shine of man-made shoe materials. After cleaning, buff the shoe with a soft cloth.

On snowy days, leather may be exposed to salt. As soon as possible, rinse the leather shoe or boot, including the sole and heel, and allow them to dry away from direct heat. If the shoes are very wet, stuff the inside with cloth or newspaper and hand-dry the outside surface with a towel.

Washable fabric shoes that are glued to the sole should not be placed in a washing machine or submerged in hot water. Instead, stuff them with paper towels to prevent the sole from being soaked and then carefully scrub with thick, dry suds (that is, suds only). Rinse them using a damp cloth and dry them at room temperature, *never* near a direct heat source. Silk, satin, and other delicate fabric shoes should be cleaned by a professional dry cleaner.

Experience 9-5 ■ How Should I Care for These?

Circle either yes or no to answer each question.

1. Always dry wet shoes in front of a radiator or heater vent. **yes no**
2. Polishing leather shoes is only important when shoes get scuffed. **yes no**
3. Suede shoes should be brushed with a wire suede brush to keep them looking new. **yes no**
4. All fabric shoes should be dry-cleaned. **yes no**
5. Shoes made with man-made materials should be cleaned with all-purpose cleaner. **yes no**
6. Washable fabric shoes with glued soles can be washed in the washing machine. **yes no**
7. Using petroleum jelly is a good way to make all patent shoes shine. **yes no**
8. Paste waxes protect and polish leather shoes. **yes no**
9. Once leather boots get stained with salt, it is impossible to get out the stain. **yes no**

■ FITTING SHOES

Shoes are sized to the length and width of the foot. For infants and children, sizes range from 0 to 13. For adults, length begins at size 1 and continues to about size 14 or 15. Each full size is one-third inch longer than the previous full size. Each half size is one-sixth inch longer than the previous full size. Widths for shoes range from the very narrow AAAA (quadruple A) to the very wide EEE (triple E). Each width amounts to one-quarter inch. Not all shoe stores stock all these sizes. The most common size range for men is 8½ to 11 C and D. The most common size range for women is 6½ to 9 A and B.

As a shoe salesperson, you have the responsibility of properly fitting customers. To help you determine a customer's correct shoe size, many stores use the Brannock device. The **Brannock device** can take three foot measurements: heel-to-toe, heel-to-ball (arch length), and width. All three of these measurements are needed to determine the proper size shoe. There is a Brannock measuring device for children, men, and women.

Measuring

Once the customer is seated, carefully take off both shoes. When measuring infants' and children's feet, be especially gentle. Always measure both feet because one foot is usually larger than the other. The larger foot determines the shoe size needed.

To measure the customer's foot, place the customer's right foot on the Brannock device with heel firmly against the right heel cup. Then ask the customer to stand. When the customer stands, the foot spreads out and lets you see the foot's true width and length.

To determine the shoe length needed, find out the customer's heel-to-toe measurement and arch length. Determine the **heel-to-toe measurement** by locating the number at the end of the customer's longest toe. For example, in Figure 9-9, the customer's longest toe comes to the number 9. Therefore, the heel-to-toe length of the customer's foot is 9. (See Figure 9-9, letter A.) Then determine the **arch length** by placing the arch measure against the bone of the ball of the customer's foot and reading the number of the arch measure. For example, in Figure 9-9, letter B, the arch length is 8.

To determine the shoe width needed, move the width indicator against the side of the customer's foot. For example, the width of the foot in Figure 9-9, let-

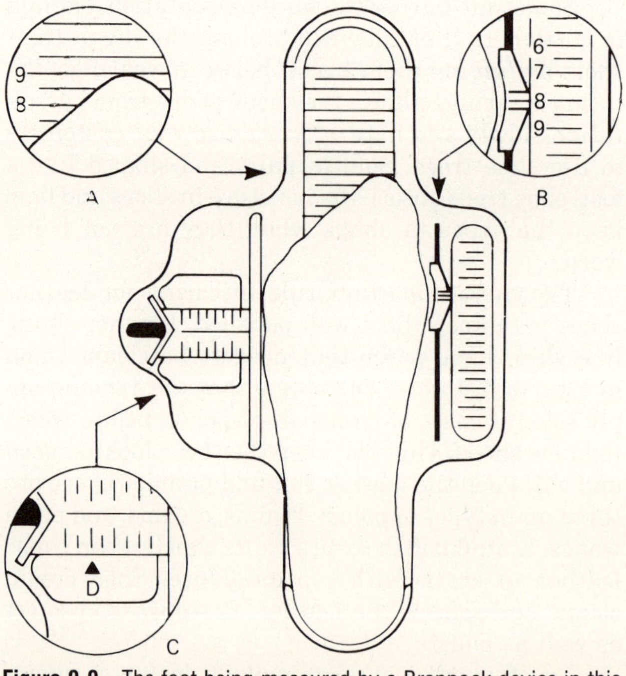

Figure 9-9 The foot being measured by a Brannock device in this illustration needs a size 9D shoe. Read this page to find out why. Courtesy The Brannock Device Co., Syracuse, NY.

ter C, is D. The customer in Figure 9-9 wears a size 9D shoe. Use this process to measure the customer's left foot.

When the heel-to-toe measurement is greater than the arch length, use the heel-to-toe measurement to determine the shoe length needed. For example, suppose a customer's heel-to-toe measurement is 10 and arch length is 9, and the foot width is C. The customer would wear a size 10C shoe. When the arch length is greater than the heel-to-toe measurement, but not more than two sizes greater, the arch length determines the shoe length needed. For example, suppose the heel-to-toe measurement is 9, the arch length is 10, and the foot width is C. The customer wears a size 10C shoe.

When the arch length is two sizes greater than the heel-to-toe measurement, subtract the heel-to-toe measurement from the arch length. Divide your answer by 2 and add this number to the heel-to-toe measurement. For example, suppose the heel-to-toe measurement is 8 and the arch length is 11, and the foot width is D. Following the formula, subtract 8 from 11. Divide the answer, 3, by 2. Add this answer, 1½, to the heel-to-toe length. The size needed is 9½D.

Measuring a foot gives you a good indication of shoe size. However, you may find that a customer does not wear the shoe size the measurements say. Shoe sizes vary by manufacturer, the shape of the last, shoe style, and heel height. A customer may wear a 9½D in one style and a 10C in another. After you have measured a customer's foot, avoid saying, "You take a 9C" or a specific size. Leave yourself an opening just in case the measured size is not correct.

You may have a customer who refuses to let you measure the feet. In situations like this, the old shoe size is a rough guide of what size is needed. Rather than disagree with the customer, bring out the size requested. Then check the shoes for proper fit.

Checking the Fit

To check the fit, watch the shoe when the customer walks. Watch to see if the shoe slips on the heel. On a flat shoe, a small amount of slippage is natural, but see if the shoe opens too much when the foot is bent. A customer should be able to move his or her heel slightly. If the heel is too snug, it will cause blisters. The topline should fit snugly and not rub against the ankle bone.

With your hand, check the shoe length. There should be about a half inch space beyond the big toe for adults. Children should be fitted with shoes ¾ to 1 inch beyond the big toe to allow room for growth.

With your thumb and middle finger, check the ball fit. The big toe should fit the ball joint pocket of the shoe. Press against the sides of the shoe. The ball of the foot and the shoe should be wide at the same place. If there are creases that form lengthwise around the ball of the foot, the shoe is too short.

Customers are the best judges of how a shoe really fits. Ask customers how shoes feel when they are sitting down and how they feel when they are walking. If the customer says the shoes are too tight, get another pair of the same shoes in a larger size. If the larger size shoes are still too tight, the shoe style may be wrong. Get another style shoe in the customer's true size. The same procedure applies to shoes that are too loose.

■ SPECIAL REMINDERS FOR SELLING FOOTWEAR

Selling shoes is very different from selling apparel. This is because only display stock is usually on the sales floor. (To prevent shoplifting, only the right shoe is placed on display, not the pair.) Customers are seated during the sales process and the salesperson often personally measures the customer to determine the correct size. In most cases, the footwear stock is then brought to the customers rather than the customers choosing the stock on their own. Let's follow some typical shoe sales to get an idea of how to sell shoes.

Invite the Customer To Be Seated

Customers often examine the many window or interior displays in the store to determine the selection of merchandise a shoe store or department carries. Salespeople should greet customers promptly, by name if possible, and invite them to have a seat.

What Kind of Shoe?

Customers usually have some idea of the type of shoe they want. Some customers take the shoe they want to try on from the interior shoe display and give it to the salesperson; others wait for the salesperson to ask questions to determine their needs. Shoe salespeople need to ask questions about what kind of shoe the customer wants, its color, its style, and the occasion for which it is to be worn. The shoe style number is located inside the shoe near the heel. The style number in the shoe will correspond with the style number on the shoe box. Possible questions are: "What will you be wearing the shoe with?" "What color did you have in mind?" "What particular style appeals to you most?" "How high a heel do you want?" As you ask these questions, take off the customer's two shoes and measure both feet. Try to have a fairly specific idea about the shoe a customer wants before going to the stockroom. Once you have located the shoes in the stockroom, try them on a customer's feet and check the fit as you explain the benefits of the shoes.

Choosing Shoes in the Stockroom

Bring out two or three variations of the type of shoe the customer wants. For example, if a woman asks for a dressy, black patent leather shoe, you may bring out a mid-high-heeled pump, a sling-back with a closed toe, and a T-strap in black patent. This gives the customer a choice. It also saves time in running back and forth from the stockroom to the selling floor.

A good way to increase shoe sales is to bring out a selection of additional shoe styles that the customer may not have mentioned. While you are in the stockroom, choose another shoe style for the customer to see. For example, bring out a popular fashion item, a traditional style in a new color, or shoes that your store is trying to close out (eliminate from stock). These other styles should be shown later in the sale.

Before coming out of the stockroom, open all the shoe boxes to be sure both shoes are in the box. If the shoes are dirty or dusty, wipe them off so they will make a more favorable impression.

Trying on Shoes

Present the shoe selection to your customer. With a little showmanship, one-by-one, open the shoe boxes of the requested style. Put the lid under the box and support the right shoe on the side of the box. This lets the customer see the various shoes you have to offer. It also builds customer interest.

While putting shoes on the customer's foot, you should be seated on the shoe fitting stool. The customer's foot should be resting on the diagonal part of the stool.

First put on the shoe the customer requested. If the shoe is to be tied, lace it prior to putting it on the foot. Hold the shoe under the arch or shank. To ease the customer's foot into the shoe, slip the front part of the foot in the shoe. Place your shoehorn at the customer's heel and slide the foot into the shoe. If the shoes are sling-backs, you do not need to use your shoehorn.

Tie the laces or adjust the shoe strap so that the shoe fits firmly. Check the fit of the one shoe. If it seems to fit properly, try on the other shoe. Again check the fit.

Ask the customer to stand and take a few steps. Observe the shoes as the customer walks. A salesperson cannot always see an incorrect fit. Be sure to ask the customer how they feel. Get the customer to walk over to a mirror to see how the shoes look.

CHAPTER 10

Accessories

Swatch has turned an inexpensive Swiss watch into a fashion phenomenon. Since the fall of 1983, Swatch has bypassed jewelers, the traditional outlets for Swiss watches, in favor of department stores. It pushed watches as fashion accessories—and American teenagers and their parents started wearing two, three, even four watches at once!

What's unusual is that Swatch never meant to be a fashion company. Swatch's owner, SMH (Switzerland's largest watchmaker and supplier of watch movements), only wanted to keep its factories operating at full production. The advent of inexpensive electronic watch movements had transferred dominance in the watch market to Japan. But Swatch found a new market with its fun, "funny" watch.

The Swatch is made on a fully automated production line, and its parts are welded, not screwed, together. That construction makes the watch almost impossible to repair, but it also means that far fewer things can go wrong with it. And, most important, it means that the watches can be sold for $35 each and still provide a healthy profit to retailers and SMH.

Swatch is a good example of an accessory item that made a big fashion statement. Like all accessories, the Swatch watch lets its customers express their individuality and creativity.[1]

Most customers need help in selecting the right fashion accessory. They have questions about how to coordinate it with their clothing—and how to care for it after it's worn. To help you answer customer questions, this chapter will discuss three types of accessories in detail: bags, belts, and fashion jewelry.

HANDBAGS

Handbags, also called purses, pocketbooks, and bags, have long been an important accessory in the well-dressed person's wardrobe. Many business people use leather or vinyl briefcases or portfolios to carry their papers to and from the office. In addition, tote bags have become a "must have" item for school, work, shopping, or the beach. Today, both men and women use bags to carry their personal belongings, such as keys, wallets, glasses, appointment calendars, and business papers. Although handbags vary in size, color, style, and material, the manufacturer is often instantly identifiable because of the unique logo or name printed on the bag. For example, Anne Klein's lion, Liz Claiborne's triangles, and Gucci's unique initials add status to the leather goods and handbags they produce.

Parts of a Handbag

A handbag consists of many parts, including outer material, filler, frame, fastener, lining, and handle. The outer material may be leather, such as glove-tanned cowhide. Many bags are made from fabric, such as woven cotton or canvas. Vinyl handbags are very popular and less expensive than leather.

Filler is the material or substance that helps the bag hold its shape. The frame also helps to give the bag a distinct shape. Handbag fasteners may be snaps, turn locks, latches, buckles, drawstrings, or zippers. The lining and interlinings give the inside of the bag an attractive, neat appearance by hiding the filler. The handle, which is used to carry the bag, comes in a variety of sizes—small for holding in the hand, medium length to fit over the arm, and shoulder length.

Handbag Styles

Some popular handbag styles are shown in Figure 10-1. Some of these styles are as follows:

Envelope Resembles an envelope with a top flap. This style may or may not have handles. Some shoulder bags are envelope styles. Clutch purses and business portfolios (designed to carry papers) may also be envelope-styled.

Box Shaped like a square, oval, or rectangular box. This bag is rigid and does not expand. It may have small, medium, or shoulder-length handles. An example of a box is a briefcase.

Pouch Has a top closing, a frame, or a top handle. This bag is usually soft and is pleated or gathered to a top opening. Examples of this bag are the duffle and tote.

Care of Handbags

Customers want to know how to keep their handbags looking attractive. Explain that handbags that are not overloaded will last longer than bags that are too full. Overloading a bag weakens seams, stretches the handbag out of shape, and pulls the material away from the frame or handle.

Vinyl handbags should be wiped clean with a damp cloth. Suede handbags should be brushed with a suede brush. Leather handbags should be polished with a good leather cream. Fabric handbags should be cleaned by brushing, spot cleaning, or dry cleaning. The handbag lining should be kept clean by brushing or vacuuming.

Stored fabric handbags should be filled with paper to keep their shape, and wrapped in plastic. Vinyl or leather handbags should be wrapped in tissue paper.

Color, Line, and Design of Handbags

To achieve a fashion look, handbags should either blend with the wearer's clothing or provide a pleasing contrast. Many people choose their handbags to coordinate with their shoes so that the textures, colors, and designs of both harmonize. The right handbag can make a person look "put together"; the wrong handbag can ruin the best looking ensemble.

When helping customers select the right handbag, remember that the handbag should be scaled to the size of the wearer. When a handbag and wearer are not in scale, (1) a small person looks dwarfed by an oversized bag, (2) a tall person looks even taller with a small bag, and (3) a large person makes a tiny bag look as if it were a child's.

A small person should carry a small- to medium-sized handbag. A large person should carry a medium to large handbag. You should also be aware

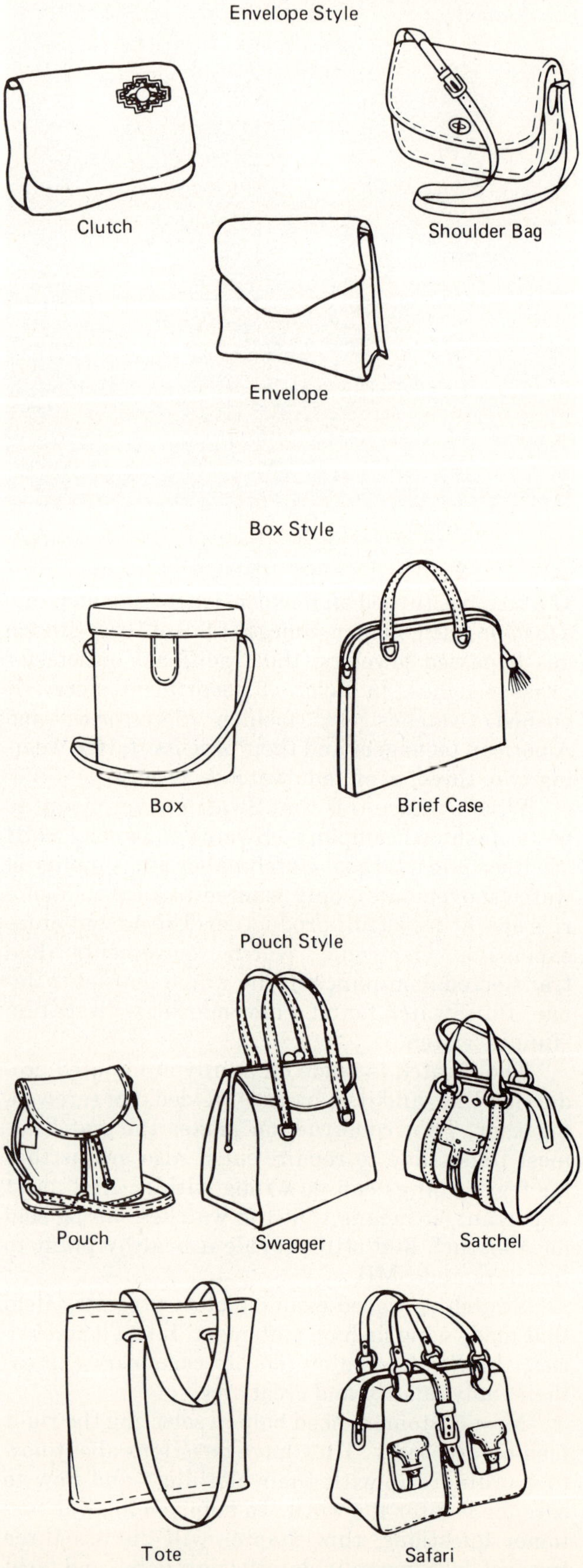

Figure 10-1 You should be able to recognize and name these handbags, which are grouped into the three most popular styles.

that shoulder bags can call attention to the wearer's hips depending where they sit on the body. For example, a larger tote bag that is parallel with a person's hips can make that person look broader.

Although a wide variety of handbags are available, many customers buy only one bag to coordinate with all their clothing. Customers who buy only one or two bags for their wardrobe should choose bags in neutral tones, such as taupe, navy, or black.

Small Leather Goods

Handbag salespeople, department managers, and buyers must also be familiar with small leather goods, including wallets, credit card cases, passport cases, appointment books, cigarette cases, keycases, and eyeglass cases. These small items which are available in vinyl, fabric, and leather, are often coordinated to give the customer a total look.

Experience 10-1 ■ This Style Is Called a...

1. Using a fashion magazine or newspaper, locate pictures of 10 different bags for men, women, and/or children. Cut out the pictures and attach them to a separate sheet of paper. Label the style of each.
2. Identify the style of handbags your classmates are carrying. Write the style and the classmate's name on a separate sheet of paper. Examine each classmate's body build and decide whether each student has chosen the most attractive size and style handbag. If not, suggest a size and style that you think would be more appropriate. Write your comments on your paper.

■ BELTS

Belts may be functional or decorative. They may be used to hold up pants or add a splash of color or interest at the waist. Fashion trends determine the types of belts worn.

Belt Materials

The most popular belt materials are leather, vinyl, and fabric. Belts are made by cutting the material to the desired length and width, stitching the edges, adding buckles, and punching holes. Some belts also have a backing.

Belt Styles

Belt styles change with fashion. The most popular style is a $\frac{3}{4}$ to $1\frac{1}{8}$ inch leather belt with buckle. Some belts are 5 inches wide and give the appearance of a cummerbund. Stretch belts are popular to wear with casual, sporty looks. A **self-belt** is made out of the same fabric as the clothing with which it is worn. Some leather belts are reversible. For example, they may be black on one side and brown on the other. (See Figure 10-2.)

Belt Sizes

Belts are sized according to waist measurements. Women's belts are sized 22 to 38 inches. Men's belts are sized from 28 to 50 inches. Belts are also sized small, medium, large, and some belts are one size fits all.

To measure a customer for a belt to wear with slacks, place a measuring tape through the belt loops of the customer's slacks or skirt. If the customer is buying the belt for a dress, wrap the measuring tape around the customer's waist.

Care of Belts

When belts are not used, they should be rolled or hung. Leather belts may be polished with a cream or leather polish. Patent leather belts should be wiped with a damp cloth. For fabric belts, follow the care instructions for the entire garment if the belt does not have a leather, vinyl, or cardboard backing. Fabric belts with these backings should be dry-cleaned so that the belt keeps its shape and appearance.

Color, Line, and Design of Belts

Belts can make a waistline appear smaller, higher, or lower. They can make a tall person appear shorter. They can add interest and give a "different" look to an ordinary outfit.

Belts that match clothing do not attract as much attention as belts that contrast with clothing. Narrow self-belts are especially becoming for customers with broad hips and thick waists. Belts in a contrasting color or design call attention to the waist and, if the customer is short, tend to cut the customer's height.

Customers of average height and slender, defined waists can wear almost any style and color belt. Many short people wear thin belts in the same color as their outfit (1 inch width or less). Tall people often use wide belts in contrasting colors to cut their height by dividing their body proportions horizontally.

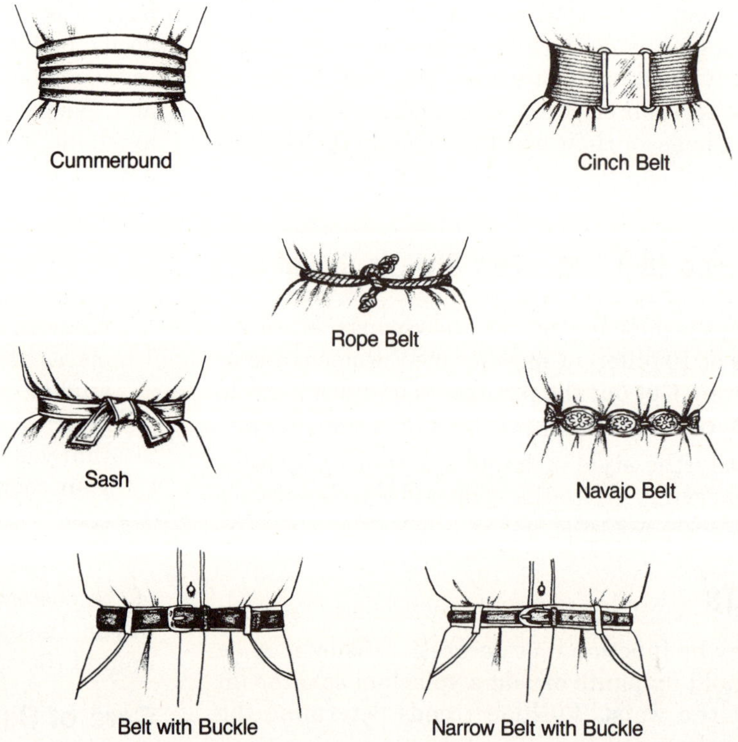

Figure 10-2 Be able to identify these belt styles. © Judy Francis.

Long-waisted people may place the belt above the waist to make them appear more typically proportioned. They can also wear wide belts, which make their waistlines appear higher. Another way to shorten the torso is by matching the belt to the color of the pants or skirt being worn. Large buckles or belt styles that curve upward are also flattering for long-waisted people. Likewise, short-waisted people should match their belt color with the color of the garment worn above the waist. They should avoid belts wider than 2 inches and large belt buckles. Belts that are in the shape of a V or slope downward are also good choices for short-waisted customers.

A popular alternative to wearing a belt is suspenders. Suspenders, also called braces, attach to the slacks with button ends or tabs and are adjustable to fit the customer's body. They are usually worn with pleated trousers or slacks and are available in different fabrics, such as silk or nylon, and a wide variety of patterns, including polka dots, stripes, paisley, or solids. Suspenders move the focal point away from the waist and to the person's chest.

Another alternative to wearing a belt is wrapping a colorful scarf around the waist. A scarf around the waist can add interest to an otherwise plain garment. Of course, people who do not want to emphasize their waists should avoid this waist-treatment.

Experience 10-2 ■ Around the Waist

Belts attract attention to a person's waist. Using a catalog, fashion magazine, or newspaper, locate three examples of belts you would suggest for each of the following customers. Attach your recommendations to a separate sheet of paper. Be prepared to explain your choices.

1. Customer A has a slim waist and is of average height and weight.
2. Customer B has a large waist and broad hips. She is of average height.
3. Customer C is tall and is very long-waisted.
4. Customer D is very short and of average weight.
5. Customer E is of average height but very short-waisted.

FASHION JEWELRY

Fashion jewelry is worn to decorate or add interest to an outfit. It differs from fine jewelry in that it is not made from precious or semi-precious metals and gems, such as gold or rubies. It is made from base metals, such as aluminum or tin. Often, these base metals are covered with a thin coating of gold or silver to give the jewelry a richer look. Fashion jewelry is also made from plastic and wood.

Jewelry Styles

Fashion trends greatly influence the look of fashion jewelry each season. Sometimes the styles are delicate; sometimes they look heavy. Customers may refer to the following style terms when buying jewelry:

Barette Clamplike device that holds hair in place

Bracelet Arm bands of various types

- *Bangles* Round hoops of varied width that slip over the hand
- *Charm bracelet* Bracelet with tiny curios of metal or stone that dangle from a metal chain
- *Identification bracelet* Bracelet with the wearer's name engraved on a piece of metal

Brooch Ornamental pin

Charm Small ornament hung from bracelet or chains

Collar pin Straight, bar, or looped pin worn between collar points

Cuff links Ornaments on stems that are used to fasten French cuffs

Earrings Ornaments worn in the ears with screw backs or clips or posts and butterflies or backings for pierced ears.

- *Ball earrings* Earrings round in shape, like a ball
- *Button earrings* Round, flat earrings that fit against the earlobe
- *Drop earrings* Earrings with a pendant or hoop that dangles
- *Hoop earrings* Earrings circular in shape that resemble a hoop

Fob chains Chains that connect pocket watches to a watch pocket

Key chains or rings Chains or rings that hold a group of keys together

Necklace Jewelry worn around the neck

- *Amulet* Pendant worn on a neck chain
- *Bib* Multistrand necklace that hugs the base of the neck
- *Choker* Single or multistrand necklace that fits the neck snugly
- *Collar* Flat necklace that fits the neck just atop the collarbone
- *Lavaliere* Pendant or ornament that hangs from a chain or cord
- *Locket* Small case hanging from a chain, cord, or ribbon that holds a picture or other mementos
- *Matinee-length necklace* Pearl or bead necklace 30 to 35 inches long
- *Opera-length necklace* Pearl or bead necklace 48 to 120 inches long
- *Rope* Long chain or bead necklace, matinee or opera length

Ring Ornament worn on the finger

- *Cluster ring* Ring with stones set in groups
- *Cocktail ring* Ring in fancy shapes and fitted with stones
- *Dinner ring* Long, narrow ring usually worn on fingers of the right hand
- *Signet ring* Ring with a crest or initials

Stud Small button or ball on a stem; used instead of buttons on a formal shirt

Tie bar Bar with hinged back worn to hold a tie in place

Tie tack Small ornament set on a short, straight pin that pierces the tie and holds it in place

Watch Timepiece worn on a bracelet or chain

Color, Line, and Design of Jewelry

When choosing jewelry colors, it is easier to design an outfit around two colors rather than three or more. If an ensemble is a solid color, jewelry in a contrasting color may give the outfit added appeal. Jewelry for clothing that is multicolored should repeat one of the colors in the garment.

Jewelry should be chosen in relation to a customer's figure, personality, and age. A petite person will look most attractive wearing small jewelry, whereas a large person can successfully wear bolder, chunkier styles.

Necklaces call attention to the neck, throat, and face. Bracelets and rings call attention to the hands and fingers. Earrings call attention to the face and eyes. Customers who do not want to emphasize a particular feature or part of their body should refrain from wearing certain jewelry items.

When helping customers choose jewelry, keep the shape of the face and body in mind. Long necklaces that make a V are more slenderizing and a better choice than large, round beads for a person with a short, thick neck. A pin centered on a customer's sweater or dress will have a slenderizing effect on the face and neck because it draws the eyes downward. Two pins, one on each side, will add breadth to a thin person. A pin worn high on the shoulder will make a short person appear taller or a heavy person appear more balanced.

When helping a customer choose earrings, suggest round earrings for the person with a long face to add width to the face. Likewise, the round-faced person should look more attractive when wearing drop earrings, which add length. In some stores, earrings for pierced ears are not returnable; in others, they may not even be tried on. Know the policy in your store.

People with heavy arms should wear wide bracelets, not delicate ones, or several narrow ones. People with thin arms should not wear bulky bracelets or watches; these will make the arms look thinner. Large hands can wear large rings. Small hands need delicate rings.

The purpose of jewelry is to decorate the body, to make it look more attractive and to provide a way to express each customer's unique personality. Choose jewelry that emphasizes each customer's best features and minimizes the most undesirable ones. Customers will appreciate your suggestions of what jewelry styles and colors help them look their best.

Experience 10-3 ■ Different Looks...Different Jewelry

1. Look through a fashion magazine or catalog and locate a woman's basic clothing outfit that could be worn to work, school, and a social event. Cut out or copy the outfit you chose and attach it to a separate sheet of paper.
2. In magazines, catalogs, or newspapers, locate fashion jewelry that gives this outfit three different looks—work, school, and social event. Cut out or copy the pictures and attach them to your paper. Label the jewelry "work," "school," or "social event," depending on the occasion.

■ OTHER FASHION ACCESSORIES

Other fashion accessories sold in clothing and accessory stores include gloves, hats, wigs, hairpieces, hair bows, scarves, shawls, handkerchiefs, and sunglasses. Each of these accessories may fill a functional need. For example, scarves may be worn around the neck for warmth, hair bows may hold back the hair. However, when customers are buying these accessories, they give most consideration to how that item looks on them. If they feel the item is unattractive, customers will refuse to buy it.

When selling these accessories, find out the purpose the customer has in mind for the item. Color is very important when helping customers select wigs, hair accessories, scarves, or shawls. Color and shape are both important considerations when helping customers choose sunglasses. When working with these accessory items, suggest a color and style that will highlight the customer's best features.

Chapter Note

1. *The New York Times,* August 16, 1987, p. 4F.

UNIT 3

CASE

Units Stores

Units is a chain of women's clothing specialty stores that sells a *unique* product in *unique* stores. In its first three years of operation, the company, which is a division of J. C. Penney, grew to 140 stores in 35 states.

Product and Price

Much of Units' success can be attributed to an innovative product. Rather than producing a different line each season, Units offers a core collection of interchangeable knit separates. The designs, which are derived from a simple square, are modular—each piece interfaces with the other to create a multitude of fashion looks. The same pieces that build a "dress-for-success" office look can create casual weekend attire or sophisticated after-five wear. For example, a wrap design called 28SSR can be tied as a crop jacket, worn reversed, or worn as an unbuttoned cardigan jacket. (See Figure 1.)

In addition to the core collection, the company offers new silhouettes on a seasonal basis that provide customers with fashion options. For example, Units introduced new dress and jacket styles, such as the chemise, the princess, the bolero, and the 32-inch long jacket in the Fall 1988 season. All new designs are consistent with its philosophy of allowing the customer to adjust every fashion look to suit her individual style and the occasion.

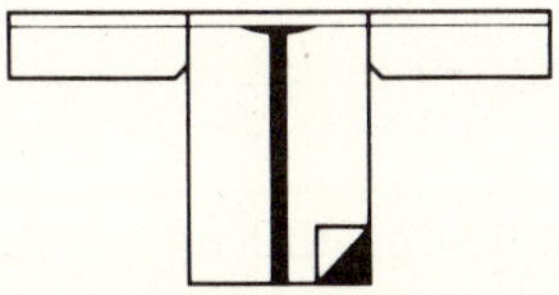

Figure 1 This wrap, 28SSR, can be worn as an unbuttoned cardigan or can be tied in the front or back to create several different looks. It can be coordinated with other unit modules in the same or contrasting colors. Courtesy Units.

The assortment of silhouettes is appropriate for many figure types. Layering and the addition of tubular belts or bandeaus allow the wearer to shift the proportion of an outfit and create the shape that is most attractive for her.

All the garments are made in the U.S. of 50 percent cotton/50 percent Fortrel® polyester interlock knit jersey. This lightweight fabric is comfortable to wear and easy to drape and fold. The fabric is especially well suited to thermal layering, offering additional warmth in cooler climates. It is ideal for the business traveler or vacation-goer; it is wrinkle-resistant and can be hand- or machine-washed.

All Units' separates are free sized, meaning that one size conforms to and flatters most figure types. Petite sizing is also available. This sizing strategy ensures that customers do not experience out-of-stock problems in their particular size.

All basic styles are available in five colors—black, white, red, navy, and ecru. Additional colors and patterns that are right for the season are introduced twice each year. Examples of fashion colors include amethyst, rose quartz, sapphire, winter jade, and goldenrod.

In addition to a peplum, cowl, sash, and bandeau—accessories which are used to change the shape or proportion of an outfit—the company also offers scarves and shawls. No other accessories are sold.

Certainly, part of the attraction of Units clothing is its price. Items in the core collection range from $6 for a bandeau to $48 for a jumpsuit. Most tops cost $25 to $35 each. In addition, customers receive extra value for their money—the versatile design of a Units separate lends itself to becoming part of many different fashion looks.

As evidenced by the rapid growth in sales, women like these simple designs that allow the wearer to create her own personal style. They enjoy having fashion options that are versatile, comfortable, and affordable.

Retail Stores

Most Units stores are located in malls on corridor intersections. Although they are only about 800 square feet, the store design, like the clothing, is unique. The square motif typified in Units' clothing is repeated in the cabinetry, benches, gridwork ceiling, and the store windows. A white-on-white scheme adds to the overall sense of order and balance. Dressing rooms have no mirrors. This encourages shoppers to view the clothing at full-length store mirrors and receive help from one of the store's stylists.

The clothing is not hung on racks or displayed on fixtures. Rather, all merchandise is packaged neatly in zip-lock plastic bags and tucked into modular cubicles. Sketches of the clothing are placed in the top square to help customers make their selection.

Each store has about 12 stylists, who assist customers with their selection of merchandise. When working, the stylists are required to wear Units clothing. This helps customers visualize how many clothing options are available with the separates. When not working with customers, the stylists are busy coordinating new looks for each other. This attracts the attention of shoppers who are passing by the store and encourages them to come in and examine the clothing more closely.

Advertising

Like most smaller specialty shops, Units rarely advertises. Although ads are run in local newspapers about twice a year, the company does encourage its stylists to send promotional brochures to customers on their mailing list. The stylists also hand out these promotional pieces and their business cards to customers who enter the store. Corporate executives also make appearances on local television shows demonstrating the versatility of Units designs.

Target Market

The primary target market for Units is women aged 30–45. Because the baby boomers are getting older, this is the largest female market segment. (The second largest market segment is women over age 50.) Research regarding the target market reveals that:

- The majority of these women work outside the home.
- An increased amount of their discretionary income is being used to set up first time homes, for home electronic items, and for children's clothing. (This is due to the number of women who are marrying later and delaying childbirth.)
- One in three women in the United States wears size 16 or above.
- These women are very concerned about aging and looking older.

Competition

There is competition in free-sized modular clothing. Multiples by Sandra Garratt and The Harley Farber Singles System are sold in many department stores, such as Macy's and Bullock's. The stores selling these lines use a similar promotional strategy as Units; the clothing is not displayed on racks but inside clear plastic bags. Salespeople are specially trained to sell and coordinate the different pieces for the customer. They also wear the clothing while working. Brochures that illustrate the different fashion looks are distributed to in-store customers. Prices are comparable.

Other stores, such as Dayton-Hudson's Mervyn's and Target, also offer their own brand of modular clothing in bright fashion colors.

1. From its research, Units knows that its target market likes to wear knit modular clothing for casual wear. However, many of the working customers in their market do not feel that knits are appropriate for the office. Therefore, they buy more tailored designs for career wear. What could you suggest that Units do to alter this market segment's attitude toward wearing Units designs for business?

____

____

____

____

____

2. Units is not having difficulty attracting younger customers (teenagers and those in their early 20's) into their shops. Younger customers like the opportunity of using their creativity to create fashion looks that express their own style. But

this is a shrinking market and the company would prefer to get a stronger hold on its target market, women between the ages 30–45. What suggestions could you make that would attract more customers in the target market to shop at Units?

____

____

____

____

____

3. Identify an additional market, within the primary target market, whose needs and wants could be satisfied by Units. Outline any changes in the marketing mix that would be required to attract this market sub-segment.

____

____

____

____

____

4. What are the advantages and disadvantages of having salespeople serve as stylists? Do you think this service meets the needs of its target market?

____

____

____

____

____

5. Identify two additions to the Units' product mix that meet the following criteria: (1) Is consistent with the company's philosophy of providing customers with versatile, affordable modular knit clothing; and (2) would increase sales. Give the rationale for your answer.

____

____

____

____

____

UNIT 4

Marketing Skills

Marketing fashion merchandise is different from marketing other products. Because a perfectly functional clothing item will be rejected when it is "out of fashion," speed is very important. Being first with the merchandise that customers want can provide retailers with large profits.

This unit discusses the marketing skills involved in the retailing of clothes and accessories, including buying merchandise from suppliers and having it delivered to the stores. It also reviews how fashion merchandise is promoted, displayed, and sold. The special management skills required by a store or department manager, such as recruiting, hiring, and evaluating employees and financial responsibilities are also discussed.

After reading this unit, you will be ready for a job in apparel and accessories marketing. You can look forward to an exciting career with many rewards. Good luck!

CHAPTER 11

Buying and Pricing

In 1962, when Dress Barn started, it was only an idea. Elliot Jaffe, then a Macy's buyer, believed that there was a market for quality-labeled women's fashions at 20 to 50 percent below department store prices. Today, the Dress Barn is an off-price retailing chain with over 300 stores, mostly on the East Coast and in the Midwest.

It was tough getting started. Jaffe was often forced to buy at full wholesale prices, which put a squeeze on his own profit margins. Few top-line manufacturers were willing to see their first quality clothes on Dress Barn's off-price racks, especially when they were already selling to department stores. Some insisted that their labels be cut out of garments sold by the Dress Barn. But Jaffe would see that just enough of the label was left for customers to know they were getting name-brand clothing at a discount.

The sales increased and the number of stores grew. By the midseventies, Dress Barn had 18 stores. It had become an important customer of manufacturers like Liz Claiborne, Ciao, Jones New York, and Calvin Klein Sport. Jaffe paid his bills on time—something that isn't always true of a fashion retailer—and he didn't ask for the usual concessions demanded by department stores, such as cooperative advertising allowances, return privileges, or profit guarantees. By giving Jaffe 20 to 40 percent discounts, manufacturers found they could fill otherwise unused manufacturing capacity or clear out unwanted stock late in the season.

A major part of Dress Barn's success is due to the selection and pricing of merchandise. Each of the eight buyers has experience with a major retailer. Each is responsible for selecting and pricing about $25 million of merchandise a year. These highly paid professionals can earn over $90,000 a year when their sales and profit exceed company plans.[1]

This chapter, which includes many of the tasks Dress Barn buyers perform, is important to anyone who wants to pursue a retail buying career. To be a successful buyer, you must know how to plan your purchases, negotiate with vendors, and monitor and react to what is selling.

PLANNING YOUR PURCHASES

"Good morning. I just dropped in to discuss any questions you may have before you leave on your spring buying trip," says Amy Davis, your merchandise manager at Malone's Department Store. You are the new buyer of the Junior Sportswear Area. Your department sells casual pants, shirts, sweaters, and jackets to young adults.

"It's a bit frightening to be going to New York on your first big buy," Ms. Davis says. "You're responsible for a large sum of money and your success as a buyer depends on the buying decisions you make. I'll be going on the trip with you, but I also have to supervise four other buyers. That's why I want to be certain that you are as well organized and prepared as possible."

Preparing a Six-Month Merchandise Plan

As your manager hands you the six-month merchandise plan (see Figure 11-1), you remember the first time you saw the form. It was over two years ago. You were right out of college and had just joined the company as a merchandise trainee. You attended many meetings trying to understand what

Malone's SIX MONTH MERCHANDISING PLAN

Department Name *Junior Sportswear* Department No. *175*

	PLAN (This Year)	ACTUAL (Last Year)
Workroom cost		
Cash discount		
Season stock turnover		
Shortage reserve		
Operating expenses		
Operating profit		

SPRING 199–		FEB.	MAR.	APR.	MAY	JUNE	JULY	SEASON TOTAL
SALES	Last Year	54,000	63,000	56,000	53,000	54,000	49,000	329,000
	Plan	58,000	68,000	60,000	57,000	58,000	52,000	353,000
	Percent of Increase							
(A)	Revised							
	Actual							
RETAIL STOCK (BOM)	Last Year	200,350	279,100	243,600	237,450	213,300	220,500	1,394,300
	Plan	215,200	301,250	261,000	255,350	229,100	234,000	1,495,900
	Revised							
(B)	Actual							
MARKDOWNS	Last Year	14,000	11,000	6,000	5,500	9,750	6,900	53,150
	Plan (dollars)	12,600	9,900	5,400	4,950	8,800	6,200	47,850
	Plan (percent)							
	Revised							
	Actual							
RETAIL PURCHASES	Last Year	132,750	27,500	49,850	28,850	61,200	89,300	389,450
	Plan	144,050	27,750	54,350	30,750	62,900	100,900	420,700
	Revised							
	Actual							
PERCENT OF INITIAL MARKON	Last Year							
	Plan							
	Revised							
	Actual							

Comments

All numbers rounded to nearest #50

Merchandise Manager *A. Davis* Buyer *Your Name*

Controller *B. Bosley*

Figure 11-1 This is an example of a six-month merchandise plan that would be prepared by a Junior Sportswear buyer.

each column meant and how it was to be completed. You remember hearing, "Our store, like other clothing stores, prepares a six-month merchandise plan twice a year, once for the spring-summer season, February through July, and once for the fall-winter season, August through January. This plan helps buyers control the amount of money they invest in inventory. The six-month plan shows the amount of sales planned, the amount of stock needed in order to achieve those sales, and how much merchandise we need to purchase to achieve the required stock level and planned sales. This plan helps buyers control the amount of money they invest in inventory. When followed, it assists buyers in earning a profit for the store. Notice that it includes information about sales, retail stock, markdowns, and retail purchases. All the numbers on the form represent dollars."

Sales All the activities of a store are dependent on sales. Therefore, when a store does its planning, it forecasts its sales first. To determine what sales will be for the coming year, a buyer looks at last year's sales figures, as well as other previous years' figures. Usually, there is an upward or downward trend. For example, if sales have been increasing by 5 percent every year for the last five years, they will probably continue to increase by 5 percent next year. Of course, this depends on whether or not conditions during the coming year are similar to conditions in the past.

After many hours of examining past sales records, you and Ms. Davis forecast the planned sales for the coming year. The planned sales are listed on the second line of the six-month merchandise plan. (See Figure 11-1, letter A on page 149.)

Stock There must be adequate stock to achieve the planned sales. This direct relationship between the amount of stock in a store and the planned sales is called the **stock-sales ratio.** It tells buyers that if they plan to sell a certain amount of merchandise, they must have a certain amount of stock in the store. The stock-sales ratio is determined over a period of time. The ratio is calculated by dividing the actual stock, in dollars, by the actual sales. For example, a department that, during the month of April, has a stock or inventory of merchandise amounting to $20,000 and sales amounting to $10,000 has a stock-sales ratio of 2.

$$\frac{\$20{,}000}{\$10{,}000} = 2$$

The historical monthly stock-sales ratios for the Junior Sportswear area are shown in Table 1.

■ Table 1
Junior Sportswear: Monthly Stock-Sales Ratios

Month	Ratio	Month	Ratio
January	3.64	July	4.50
February	3.71	August	4.42
March	4.43	September	4.21
April	4.35	October	4.18
May	4.48	November	4.28
June	3.95	December	4.92

You used your stock-sales ratios to determine the dollar volume of merchandise you need in stock at the beginning of every month. For example, the six-month merchandise plan shows that the planned sales for March are $68,000. The stock-sales ratio for that month is 4.43. By multiplying 4.43 by $68,000, you determine that you need planned retail stock amounting to $301,240 for the beginning of the month (BOM). This is rounded to the nearest fifty, $301,250. (See Figure 11-1, letter B.) By multiplying the stock-sales ratio for each month by the planned sales for that month, buyers can determine the retail stock needed at the beginning of each month. The stock-sales ratios help buyers keep their stock in balance with their sales.

Markdowns The next section of the six-month merchandise plan lists last year's and this year's planned markdowns for the Junior Sportswear Area for the spring season. Markdowns reduce the dollar value of inventory in the store or department, just as sales do. No buyer can avoid taking markdowns, as they are a necessary part of merchandising. Customers expect sales at certain times of the year. Markdowns move out damaged, out-of-season, and slow-selling merchandise and make room for new purchases. By planning markdowns, buyers can use them as a tool to increase sales.

Last year, the markdowns in the Junior Sportswear Area were high. Poor buying decisions were made and markdowns were necessary to move out the merchandise. This year, a 10 percent reduction in markdowns is planned.

Retail Purchases The retail purchases section of this six-month merchandise plan shows the dollar amount of merchandise that a buyer may purchase for one month without going over the planned inventory level. The formula for computing planned purchases is:

Planned purchases = planned sales
+ planned EOM inventory
− planned BOM inventory

For example,

Planned purchases for February
= $58,000 + $301,250 − $215,200
= $144,050

Notice that the end of month (EOM) inventory for February is equal to the beginning of month (BOM) inventory for March.

Open-to-Buy Although the open-to-buy figures are not listed on the six-month merchandise plan, they are derived from retail purchases. In the merchandise plan, buyers are allotted certain amounts of money for retail purchases. When buyers place orders for any new merchandise, they decrease the total amount of money available for retail purchases. Because most merchandise is not received the same month it is ordered, buyers have to keep track of what they have spent and what is still available to spend. **Open-to-buy** refers to the available amount of money a buyer can spend at any particular time.

Buyers can calculate their open-to-buy amount by using a formula:

Open-to-buy = planned purchases
− merchandise on order

For example, in February, the planned purchases for Junior Sportswear Shop equal $144,050. There are outstanding orders for February totaling $67,500. This merchandise has already been ordered but not yet received. The open-to-buy for February equals

$144,050 − $67,500 = $76,550

When buyers spend more money than there is available, the store or department is in an "overbought" situation. Overbought departments do not have the proper balance between stock and sales. Buyers in overbought departments or stores are usually in trouble.

Retail Versus Cost All figures on the Junior Sportswear six-month merchandise plan are expressed in retail, not cost, amounts. **Retail** is what the store charges for the merchandise. **Cost** is what the store pays for the merchandise it has for sale.

In reality, the buyer does not have $144,050 to spend on purchases or an open-to-buy of $76,550 in February. To determine the actual amount of money the buyer can spend, or the open-to-buy at cost, multiply the open-to-buy at retail by the store's markup complement. The **markup complement** is equal to 100 percent minus the markup percentage. Suppose, for example, that the buyer marks up all merchandise by 52 percent. The open-to-buy at cost for February would equal $32,400.

Open-to-buy at cost = open-to-buy at retail
× (100 percent − markup percentage)
= $76,550 × (100 percent − 52 percent)
= $76,550 × 48 percent
= $36,744

Completing the Merchandise Plan The six-month merchandise plan is one of the buyer's most important tools. Because it shows the amount the buyer should spend on new stock to achieve the planned sales, the merchandise plan keeps the buyer from overspending and not achieving planned profit for the store.

There are spaces in the merchandise plan for revisions. When a store or department does not meet its planned sales, or exceeds its planned sales, the buyer can adjust the other figures on the plan. At the end of each month, the buyer records the actual sales, retail stock, markdowns, and planned purchases on the form. This helps to determine how closely the planned figures are to the actual figures.

Experience 11-1 ■ How Much Can I Buy?

Complete the fall merchandise plan for the fashion jewelry department on page 152. The planned sales have already been forecast. You estimate that you can reduce last year's monthly markdowns by 15 percent. Use the stock-sales ratios listed to the right. The retail stock BOM for February equals $17,300. When completing the merchandise plan, round off each number to the nearest $50.

Month	Ratio
August	3.72
September	3.69
October	3.41
November	3.85
December	3.30
January	3.40

Malone's

SIX MONTH MERCHANDISING PLAN

Department Name *Fashion Jewelry* Department No. *325*

	PLAN (This Year)	ACTUAL (Last Year)
Workroom cost		
Cash discount		
Season stock turnover		
Shortage reserve		
Operating expenses		
Operating profit		

FALL 199–		AUG.	SEP.	OCT.	NOV.	DEC.	JAN.	SEASON TOTAL
SALES	Last Year	3500	4000	5000	5500	6500	4500	29,000
	Plan	4000	4600	5600	6100	6900	4750	31,950
	Percent of Increase							
	Revised							
	Actual							
RETAIL STOCK (BOM)	Last Year	13,000	14,750	17,050	21,200	21,450	15,300	102,750
	Plan							
	Revised							
	Actual							
MARKDOWNS	Last Year	350	400	450	400	350	700	2,650
	Plan (dollars)							
	Plan (percent)							
	Revised							
	Actual							
RETAIL PURCHASES	Last Year	5250	6300	9150	5750	350	7500	
	Plan							
	Revised							
	Actual							
PERCENT OF INITIAL MARKON	Last Year							
	Plan							
	Revised							
	Actual							

Comments

All numbers rounded to the nearest $50

Merchandise Manager *C. Greene* Buyer ______________

Controller *B. Shultz*

Experience 11-2 ■ How Much Can I Spend?

Compute the open-to-buy at retail and open-to-buy at cost for the following merchandise items. Assume the store's markup is 50 percent.

Merchandise Item	Planned Purchases	Merchandise on Order	Open-to-Buy at Retail	Open-to-Buy at Cost
Boys' elasticized belts	$17,500	$ 7,380		
Boys' polo shirts	27,200	12,500		
Boys' socks	9,460	4,800		
Boys' T-shirts	3,600	1,220		

Analyzing Unit-Control Records

Once a buyer has determined planned sales and needed inventory levels, the buyer must decide what merchandise to purchase. Most buyers pay careful attention to what customers have purchased in the past.

Buyers use **unit-control records** to keep track of sales and control stock assortments. The information in unit-control records, also commonly called style/stock status reports, is expressed in units of merchandise. These records tell buyers the total history of a particular merchandise item or merchandise style. They show the number of units purchased, sold, ordered, returned, and in stock. Unit-control records help predict buying trends. They help buyers identify what items are "hot" and what items are, in merchandising language, "dogs." Buyers who know how to analyze these records can learn many facts about their customers.

You remember your manager telling you, "There is no standard set of unit-control records. We receive a computer report that lists all the merchandise we have in stock or on order and a brief history of the merchandise. Some stores still use manual reports; a worker has to record the sales, purchase, and return information by hand. No matter how the information is prepared, the purpose of the unit-control records is the same. They help you plan your purchases and control your inventory."

Reading Unit-Control Records "Let's look at the first two entries on the report. From the style description, we can see that both are sweater styles—a two-pocket classic cardigan and a long-sleeve pullover with collar. (Refer to Figure 11-2, letter A, on page 154.) Both sweaters are from the same vendor, #302, called Looking Good. This is a small manufacturer of sweaters and knit tops that we just started to use. (See letter B.)

"Reading across the report, we see that we bought and received 64 cardigans and 64 pullovers. (See letter C.) Of the 64 cardigans, we bought 24 red, 20 black, and 20 green. (See letter D.) We have sold only 17 of the cardigans (see letter E), or 27 percent of the total. The red cardigans have the best sell-through, 8 out of 24, or 33 percent. (See letter F.) Two of the green sweaters that sold were returned by customers (see letter G) and one of the black sweaters was returned to the vendor (see letter H), probably due to manufacturer's defect. We still have 48 of these sweaters available for sale. (See letter I.)

"We purchased the same number of pullovers as cardigans (see letter C). The cardigans arrived on November 12, the pullovers arrived on November 15. (See letter J.)

"So far, at the end of six weeks of selling time, we have sold 38 pullovers. (See letter K.) Again, red is the best color, we have sold about 80 percent of them (19 sold ÷ 24 received = 79 percent). (See letter L.) White pullovers are also doing well; 13 out of 20, or 65 percent, are gone (see letter M). We have sold only 30 percent of the black pullovers during the six weeks.

"The cardigan is priced to sell at $90, the pullover at $100 (see letter N). Neither has been marked down; if they had, the lower price would be listed in the column labeled current retail (see letter O).

"Although the styles are not identical, the two sweater styles can be compared. They are displayed next to each other, and we believe the customers for both styles are the same.

"Even though the pullover is $10 more than the cardigan, it has a better sales record. Our customers seem to want the pullover style with a special collar treatment that is so popular this season. We have

UNIT-CONTROL RECORD

Dept 175 Better Sportswear Date 12/20/9- Page 8

Vendor #	Style #	Style Name	Cost Initial MU%	Current Retail 1st Retail	Receipts	Sales	Customer Returns	Returns to Vendor	On Hand	On Order
Last action and date										
302	13242	2 Pkt	45.00	90.00 (O)	64	(E) 17	2	1	48 (I)	
		Cardigan	50%	90.00						
	13242-1	Red			24	(F) 8			16	
(B)	13242-2	Black			20 (D)	6		(H) 1	13	
	13242-3	Green		(N)	20	3	2 (G)		19	
Receipt 11-12 (A)										
302	13334	L/S	50.00	100.00	64	(K) 38			26	
		Pullover	50%	100.00						
(J)	13334-1	Red			24	(L) 19			5	
	13334-2	Black			20	6			14	
	13334-3	White			20	(M) 13			7	
Receipt 11-15				(C)						

Figure 11-2 This is an example of a unit-control record. Buyers use this record to keep track of sales and control stock assortments.

always thought that our customers are fashion-conscious, and this selling history supports the idea. In fact, they are willing to pay $10 more for the fashion look. We need to remember how important fashion is to our customers when we make future purchases. Although we should continue to buy classic styles, like the cardigan, we should purchase a greater percentage of trendy styles than in the past.

"When we analyze the color selection, it's obvious that red is best. That isn't surprising. This report is for the week before Christmas and red is traditionally a popular color. That's why almost 38 percent of the total sweater purchase was red (48 red sweaters out of 128 total = 37.5%).

"Green was the weakest color. Only three were sold, and two of those were returned. Green was a popular fashion color this season and we bought it to add a Christmas look to the selling floor. We should continue to experiment with small quantities of other fashion colors next season. Green is a difficult color for some women to wear; perhaps that is why it didn't sell well.

"If we look at the number of returns to vendor, we can conclude that the sweaters were well-made, only one was sent back to the manufacturer (see letter H). Also the customer return rate was low, only two, or 1.5 percent of the total purchases (see letter G). This vendor manufactures a quality product and we will continue to do business with them."

Ms. Davis then gave you the following guidelines for analyzing unit-control information:

1. Identify the strongest price lines. (A **price line** is a specific range of prices at which an assortment of merchandise is regularly offered for sale.) The price lines show the amount of money customers are willing to spend. When buying goods, choose merchandise that fits those lines.
2. Identify the best-selling styles and colors. Some merchandise styles and colors stay popular for long periods of time. Others are out of fashion in a season. Use the unit control records to identify which items are basic stock items, or classics. These items, which have a consistent customer

demand, should always be part of the merchandise assortment. Clothing styles and colors change fast, so unit-control records do not show the exact fashion goods customers will purchase. However, the records show if customers are fashion leaders or followers. If last year the "in" look was cropped pants, and buyers couldn't keep cropped pants in stock, then the customers are the fashion leaders.

3. Identify the best-selling sizes. If sizes are included on the report, analyze them carefully. The best-selling sizes change very slowly over time. They should be purchased in the largest quantity. Sizes that rarely sell should be dropped or purchased in very limited quantities.
4. Identify which merchandise had to be marked down. For example, the green cardigan sweaters will need to be marked down to clear the stock. If past records show that a particular color or style of merchandise must always be marked down, buy a limited merchandise assortment of that color or style.

Experience 11-3 ■ Reading the Records

Analyze the unit-control record for men's hosiery below. Answer the following questions.

1. What color is most popular with customers?

____

2. What style is most popular with customers?

____

3. Have there been any markdowns taken on men's hosiery?

____

4. If you could buy only one style and color of men's hosiery, what would you buy? Explain your answer.

____

UNIT-CONTROL RECORD

Dept 403 Better Sportswear Date 3/11/9- Page 3

Style Description

Last action and date

Vendor #	Style #	Style Name	Cost Initial MU%	Current Retail 1st Retail	Receipts	Sales	Customer Returns	Returns to Vendor	On Hand	On Order
365	0234	wool argyle	3.00	$6.25	576	378	69	69	198	300
	0234-1	black	52%		108	82	10	10	26	70
	0234-2	gray			108	43	14	14	65	70
	0234-3	brown			90	24	7	7	66	40
	0234-4	maroon			120	84	16	16	36	40
	0234-5	navy			150	145	22	22	5	80
360	04137	acrylic argyle	2.75	5.50	780	399	6	-	387	420
	04137-1	gray	50%		240	164	2	-	78	150
	04137-2	navy			360	193	-	-	167	190
	04137-3	maroon			180	42	4	-	142	80

Examining Customer Requests

When customers ask for merchandise that the store does not stock, sales are lost. To minimize lost sales, smart buyers find out what merchandise customers couldn't find in the store. Because buyers don't work regularly with customers, they rely on salespeople to tell them about customer requests for merchandise. Many buyers ask their salespeople to use customer want slips to record information about unavailable styles, colors, and price ranges that customers wanted. (See Figure 11-3.) Buyers should examine requests for merchandise when deciding what to purchase before a buying trip.

Identifying Current Fashion Trends

Unit-control records show what has sold in the past. However, they cannot predict the big fashion news for the future. Buyers must turn to fashion designers, reports, and magazines to find out the latest fashion trends.

To keep up with the current trends, buyers read trade journals and magazines. Trade journals, such as *Women's Wear Daily* and the *Daily News Records,* predict fashion trends and spotlight best sellers. *Vogue, Harper's Bazaar, W, Elle, Glamour, Mademoiselle, Seventeen, Gentlemen's Quarterly,* and *M* emphasize fashion news. These periodicals help buyers know the types and styles of clothing and accessories that will be sold by merchandise suppliers. They help buyers form an idea of new items to purchase.

Evaluating Economic Conditions

Buyers should realize that economic conditions in their areas influence what sells and what doesn't. For example, the income level and unemployment rate of an area affect clothing and accessory purchases. New businesses moving into a town will increase the demand for clothing and accessories. An anticipated layoff or closing of a major business makes people hesitant about buying unnecessary merchandise.

Newspapers, television news, and radio reports are good sources of information about local economic conditions. They help the buyer determine the type and quantity of merchandise to buy and its price ranges.

Buyers who use demographics, psychographics, and behavioristic methods to segment their target market (as described in Chapter 2) are in a better position to determine how economic conditions will affect current and potential customers. For example, suppose the town's biggest employer decides to move its operation to another state. Many local jobs will be lost. People will either transfer out of the area or look for new employment. This change in economic conditions will not have much impact, however, if a store's target market consists of individuals who are retired.

Determining Competition

Buyers should consider the competitive conditions in their market areas when planning their purchases. They should know their competition—the other stores where their customers make clothing and accessory purchases. They should know which stores carry the same merchandise but at a lower price.

To keep informed of these competitive conditions, buyers need comparison shopping information. Either the buyers or their assistants must visit other stores that serve the same customers. After these visits, buyers sometimes complete a comparison shopping report, which lists facts about the merchandise, price, and promotional atmosphere of the competition.

Malone's

CUSTOMER WANT SLIP

Store 12 Dept. No. 6 Salesperson Nels Laarsen Date 2/13/—

Item Requested by Customer	Style No.	Size	Color	Price	Action Taken (To Be Completed by Buyer)
Rugby Shirt	—	12	blue/green	—	
Tank top	8338	medium	rust	$18.00	
Suede vest	2111	large	beige	$96.00	

Figure 11-3 A typical example of a customer want slip prepared by salespeople for buyers.

Preparing Stock Plans

After buyers have determined the money available to spend and have examined past customer purchases, customer requests, current fashion trends, local economic conditions, and local competition, they are ready to prepare stock plans. **Stock plans** describe the merchandise items the buyers should purchase. Ms. Davis discusses two types of stock plans.

"Clothing and accessory buyers must purchase two kinds of merchandise, basic stock and fashion goods," Ms. Davis explains. "**Basic stock** items have a consistent customer demand in a complete range of sizes. These items are stocked in the best-selling price lines, sometimes in an entire color assortment. Some basic stock items are seasonal and some are popular all year long. For example, in the summer, there is a consistent demand for tank tops. Every winter, turtlenecks in a variety of colors and sizes are basic stock items. Standard-rise blue denim jeans have high sales all year long.

"**Fashion goods** have a shorter selling cycle than basic stock. They are in high demand for short periods of time and then lose their customer appeal. Most of the pants, shirts, jackets, and sweaters in the Junior Sportswear Department are fashion goods."

Basic Stock Lists Buyers prepare basic stock lists that show the strong sellers in the store or department. They use these lists to determine the type and quantity of basic stock they need to purchase. Basic stock lists include a description of the merchandise, merchandise vendor, style, size, color, and price. They also include the minimum and maximum quantity of merchandise that should always be on hand in the store. (See Figure 11-4.)

When determining what to purchase, buyers should examine the sales record of their basic stock. Items that are not selling well should be dropped. Items that have become consistent sellers and that are not included on the basic stock list should be added. Basic stock lists show buyers exactly what type of merchandise customers are purchasing. They help simplify the buying decision. By examining the sales of basic stock items, buyers can buy the merchandise customers have purchased in the past and will purchase in the future.

Model Stock Plans Buyers prepare model stock plans to determine the kinds of fashion merchandise they should purchase. Ms. Davis explains more about model stock plans: "A model stock plan is not as specific as a basic stock list. Instead of identifying a particular merchandise item by brand or style, the model stock list may describe merchandise by type (such as blouses), size (such as two size 8, four size 10, four size 12, and so on), and price line (such as $20 to $49 and $50 to $80)." (See Figure 11-5.)

Buyers develop their model stock plans from information in the unit-control records, which show price, color, and size preferences of past customers. They also estimate where their customers fit in the fashion cycle, either as fashion leaders or followers.

Fashion goods change so quickly that each season clothing and accessory buyers are faced with different merchandise selections. By preparing model stock plans, buyers take some guesswork out of buying merchandise.

BUYING MERCHANDISE

Most buyers make several trips to different market areas each year. In large organizations, some buyers travel to foreign fashion centers, such as Paris, Milan, London, and Tokyo, to negotiate the purchase of imports for their specific target customer. Most large retail chains maintain a foreign buying office in the large fashion capitals of the world. For example, May Department Stores operates a year-round buying office in the increasingly important Hong Kong market. This office is involved in selecting merchandise for all divisions of May Department Stores, including May Company in Los Angeles, G. Foxx in Connecticut, and Famous Barr in St. Louis.

The majority of buyers do not travel abroad. Rather they visit the major clothing and accessory fashion centers in the United States: New York, Los Angeles, and Dallas. Buyers also visit regional markets such as Atlanta, Boston, Chicago, Denver, Pittsburgh, and Seattle. Unlike the major markets, little design and production takes place near the regional markets.

Each of the above markets has market weeks, or scheduled periods throughout the year during which producers and their sales representatives introduce new lines for the upcoming season to retail buyers. There are several market weeks for each area, such as men's or women's sportswear, each year. Buyers usually plan their market trips to coincide with the scheduled market weeks. This allows them to see the largest assortment of new goods and place orders to receive the new merchandise in a very efficient way.

When arriving at "market," many buyers visit their **resident buying offices** before going to any merchandise supplier, or vendor, showrooms. These buying offices, located in major fashion centers, keep buyers informed on current marketing information and fashion trends. The offices provide the companies that subscribe to their services with facts about merchandise suppliers, hot items, and deliveries.

pg 6

Malone's

Basic Stock List

Store 2 Dept 507 Effective Date 4/15/--

Vendor: Action Wear

Item Description	Style	Size	Color	Retail	Minimum Quantity	Maximum Quantity
Athletic Shirts	U6304	Small		8.50		
			Teal		6	26
			Red		8	28
			Gray		8	28
			Royal blue		8	28
			Black		8	28
			White		10	30
		Medium		8.50		
			Teal		8	28
			Red		10	30
			Gray		10	30
			Royal blue		8	26
			Black		10	30
			White		14	34
		Large		8.50		
			Teal		8	28
			Red		10	30
			Gray		10	30
			Black		10	30
			White		14	34
		X-large		8.50		
			Teal		8	26
			Red		8	28
			Gray		8	28
			Royal blue		8	28
			Black		8	28
			White		10	30
Crew-Neck Shirts	U6694	Small		9.50		
			Gray		10	30
			Black		10	30
			White		16	36
			Yellow		6	26
		Medium		9.50		
			Gray		14	34
			Black		14	34
			White		20	40
			Yellow		10	30

Figure 11-4 Here's a basic stock list that shows the strong sellers.

Model Stock Plan

Item	Style	Price	Color	Sizes S	M	L	XL
Sweatshirts	Puff-Prints	29.99	White	4	6	6	4
			Red	6	8	8	6
			Black	6	8	8	6
			Yellow	4	6	6	4
	Solids	26.00	White	4	6	6	5
			Purple	6	8	8	6
			Red	6	8	8	6
			Gray	6	8	8	6
	Quilted	26.00	White	4	6	6	4
			Jade	6	8	8	6
			Red	6	8	8	6
			Gray	4	6	6	4

Figure 11-5 Model stock plans are not as detailed as basic stock lists, as this example shows. Compare it with the basic stock list on page 158.

There are so many merchandise vendors to buy from that buyers must carefully choose the ones they want to visit. At this busy time, your choice of vendors should depend on the merchandise assortment you want to purchase. Vendors with whom you have done business in the past, who produce goods that appeal to your customers, and who deliver on time should be at the top of your list.

Most successful buyers suggest that you see all your scheduled merchandise lines before you write out an order. This will enable you to evaluate all merchandise that interests you and determine the best buy at the best price.

Classifying Goods

Let's suppose you're in the vendor's showroom to see a junior's sportswear line. The merchandise may be attractively displayed on hangers or grouped on stock racks. Some vendors use live models and others display the goods on mannequins. Some vendors give buyers an order form that shows a picture and description of the merchandise available.

When examining merchandise in vendor showrooms, try to classify all the merchandise seen. If the vendor gives you a pictured order form, you may assign numbers to the merchandise that you would consider buying next to the pictures on the form. For example, you may assign a 1 to merchandise that you think would be an excellent buy. You assign a 2 to merchandise that you think would be a good buy. Goods that you would never consider buying would be given no rating at all.

If the vendor does not provide you with a pictured order form, you will need to write down the style number and some styling characteristic of the goods you like. To take notes about the merchandise, you will need an excellent knowledge of various styles and fabrics.

Evaluating the Merchandise Salability

When deciding what merchandise to buy, try to determine if customers will pay the price your store must charge. Suppose, for example, you are considering the purchase of a line of blouses that cost $32 each. If your store marks up all its merchandise by 50 percent, the retail price of these blouses will be $64. If you know your customers will not pay more than $50 for a blouse, buying the blouses would be a mistake.

Another way to evaluate the salability of merchandise is by comparing the new merchandise with similar items that have retailed for the same price at your store. For example, if you are considering the purchase of blouses that are supposed to sell for $40 each, compare these blouses with other $40 blouses that you have sold. Ask yourself if your customers purchased the blouses in that price range or if the price was too high or low. Determine the quality of the blouses: examine how they are made and check the fabric.

If you think the blouses are overpriced, ask yourself one more question. Is there something unique about the blouse, some style characteristic, that makes you feel customers would pay the price asked? If so, and if the blouse fits into your store's image and price lines, then you may want to include it in your selection.

Another factor buyers must keep in mind is the competition in the local market. If another store is going to sell the item at a lower price or can promote the goods better, it may have an advantage over yours. Buyers often ask vendor salespeople about the purchases of other stores in their areas to find out about competition.

Experience 11-4 ■ Making Your Selection

Assume you are a buyer for the women's coat department. Your target market consists of upscale customers. They prefer natural fabrics that wear well and are very fashionable.

1. Choose three local stores that sell lightweight casual jackets, appropriate for your target market to wear in the spring and fall. (You are going to assume that these stores are your merchandise vendors.)
2. Visit each store. Examine their assortment of women's lightweight jackets.
3. While at each store, identify and write a detailed description of the jackets you believe your customers will particularly like. Use a separate sheet of paper.
4. Without looking at the price tags, decide what retail price you think is fair to charge for the jackets you described in no. 3 above. Then look at the price tags and record the real prices of the jackets on your paper.
5. Assume your open-to-buy will allow you to purchase only one of the jackets you described. Choose the one jacket that you believe will best appeal to your target customer. Write the reason you chose that particular jacket on your paper.

Determining the Ordering Terms

When you find merchandise that fits into your buying plan, you should ask the vendor about the ordering terms. Ask questions such as:

When will the goods be delivered? If goods cannot be on your selling floor when you need them, they are useless.

What shipping terms can I get? Because freight costs are increasing, they represent a sizable portion of the total cost of the merchandise. If a vendor quotes you **free on board (FOB) shipping point** or **FOB selling point,** your store is responsible for paying all transportation charges from the shipping or selling point to your store. Title to the goods passes to your store at the time of shipping, so your store assumes all risks of loss or damage to the merchandise. **FOB destination** means the vendor pays the entire transportation charges to the buyer's location. Title to the goods passes to the store when the merchandise reaches its destination.

What discounts can I get? **Discounts** are deductions off the regular price of the merchandise. Buyers may negotiate **quantity discounts** when they buy in large quantities or **promotional discounts** or allowances to help pay for advertising the merchandise. Stores receive **cash discounts** for paying the bill within a specified time before the bill is actually due. Some women's apparel vendors offer 8/10EOM payment terms. That means if the bill is paid within 10 days of the invoice date, an 8 percent deduction off the total price will be allowed. The total amount is due at the end of the month. If, for example, an invoice for $7,000 is dated March 15 and has payment terms of 8/10EOM, and if it is paid by March 25, the total amount due is $6,440 ($0.08 \times \$7,000 = \$560$; $\$7,000 - \$560 = \$6,440$). Men's and children's wear vendors usually offer 2/10EOM payment terms.

Can I reorder? If the goods turn out to be fast-selling items, you will want to be able to reorder. Clothing vendors have a limited amount of fabric. When all the fabric is used, they cannot fill reorders.

Are there any promotional aids? Some vendors offer promotional help. Promotional aids, such as point of purchase signs, fashion videos, celebrity appearances, or training clinics help sell the merchandise.

Balancing the Stock

When buyers have seen the vendors' merchandise assortments and are deciding which merchandise they will buy, they may find that the items they like are all very similar. Before writing any orders, buyers should examine their model stock plan so as to buy the clothing their customers like. By balancing the stock, buyers provide a complete assortment of merchandise, in various styles, colors, and sizes, to satisfy all their target customers.

When buying new merchandise, buyers must pay special attention to size assortment because rarely can a customer substitute one size for another. On the average, most customers wear the medium-sized ranges. As a buyer of men's sport shirts, sized S, M, L, XL, you would probably buy more medium and large than small or extra large. Following this same principle, some manufacturers package goods in sized assortments, such as two small, four mediums, four large, and two extra large.

Unit-control records provide buyers with an accurate idea of the size ranges of customers. After examining the records, buyers may find that their customers are larger than average and must, therefore, purchase larger sizes.

Determining the Order Quantity

There are so many fashion goods to choose from that it is impossible for buyers to predict accurately which items will be the best sellers. Therefore, they purchase new merchandise items, styles, and colors in small quantities. By carefully watching which items are selling, they determine which items to reorder.

In the beginning of a season, the merchandise assortment is usually **broad and shallow.** That means there are a variety of styles, colors, and price lines (broad) with a small number of items for each size (shallow). As some styles and colors stand out as the strong sellers, buyers reorder the same item or another item in a similar style and color. At the peak of the selling season, the merchandise selection should include only the most popular items and the depth of the stock (in sizes and colors) should be increased. This assortment is referred to as **narrow and deep.** Once the peak of the selling season has passed, buyers try to sell out the remaining stock to make room for the next season.

Sometimes buyers find an item that they are sure will be a big seller. If there are no reorders on the item, which is the case when buying imports, buyers may be daring and order a large enough supply for the expected customer demand. This is a risky situation for buyers.

Although they are a "star" if their decision was a profitable one, buyers who fail to anticipate customer acceptance of an item may be left with thousands of dollars of markdowns and a big reduction in gross margin. (**Gross margin** is the difference between the net sales for a period and the net cost of merchandise sold during that period.)

The total quantity of merchandise the buyers can purchase depends on the open-to-buy amount and demand within a season. Many buyers do not like to spend all their open-to-buy at one time. They want some in reserve so that they can reorder hot items or buy special promotion goods.

Writing the Order

Stores usually require buyers to use the store purchase order form when buying goods. (See Figure 11-6 on page 162.) Buyers complete the order form by:

1. Writing in the name and address of the vendor (letter A).
2. Writing the date the order is written, the date the goods are to be shipped, and the date the order is canceled because of late shipment (letter B).
3. Writing in the terms and dating agreed upon by the vendor and buyer (letter C).
4. Writing in the quantity and description of the goods ordered (letter D).
5. Writing in the size breakdown of the goods (letter E).
6. Writing in the unit cost and total and the unit retail and total (letter F).
7. Signing the order (letter G).

Following Up the Order

When buyers return from market, it is their responsibility to make sure the orders are delivered at the right time, in the correct quantity, and in good condition. Buyers may have to telephone or write vendors to find out why delays in deliveries occurred. Usually, buyers indicate a delivery date on their purchase orders. If the merchandise does not arrive at the store by that date, the store does not have to accept the merchandise.

Buyers can also contact vendors when shipments arrive with more or less merchandise than was ordered. They work with vendors who back order certain items. When vendors **back order** merchandise, it means that they cannot fill the entire order at the time and are sending only a partial shipment. The vendor will send the rest of the order as soon as the ordered merchandise becomes available. This can cause a big problem for buyers. For example, if a vendor ships only the jackets, without coordinating skirt or pant, sales will suffer. Many buyers hold one part of a coordinating outfit until the rest has arrived. Of course, if all the merchandise has not arrived by the delivery date specified on the purchase order, the order may be canceled.

Sometimes the merchandise samples in vendor showrooms are not of the same quality as the goods shipped to the store. When shipments of faulty fabric or poorly sewn garments are received, the buyers must contact the vendors and arrange for merchandise returns and refunds to the store. These refunds are called **chargebacks.**

Merchandise that is returned to vendors is subtracted from the store's inventory records and unit-control records. If this subtraction is not made, then the records won't agree with later inventory counts. Improper recording of merchandise makes the managers believe that there was a merchandise shortage caused by shoplifting or employee theft.

Purchase orders are a contract and usually can-

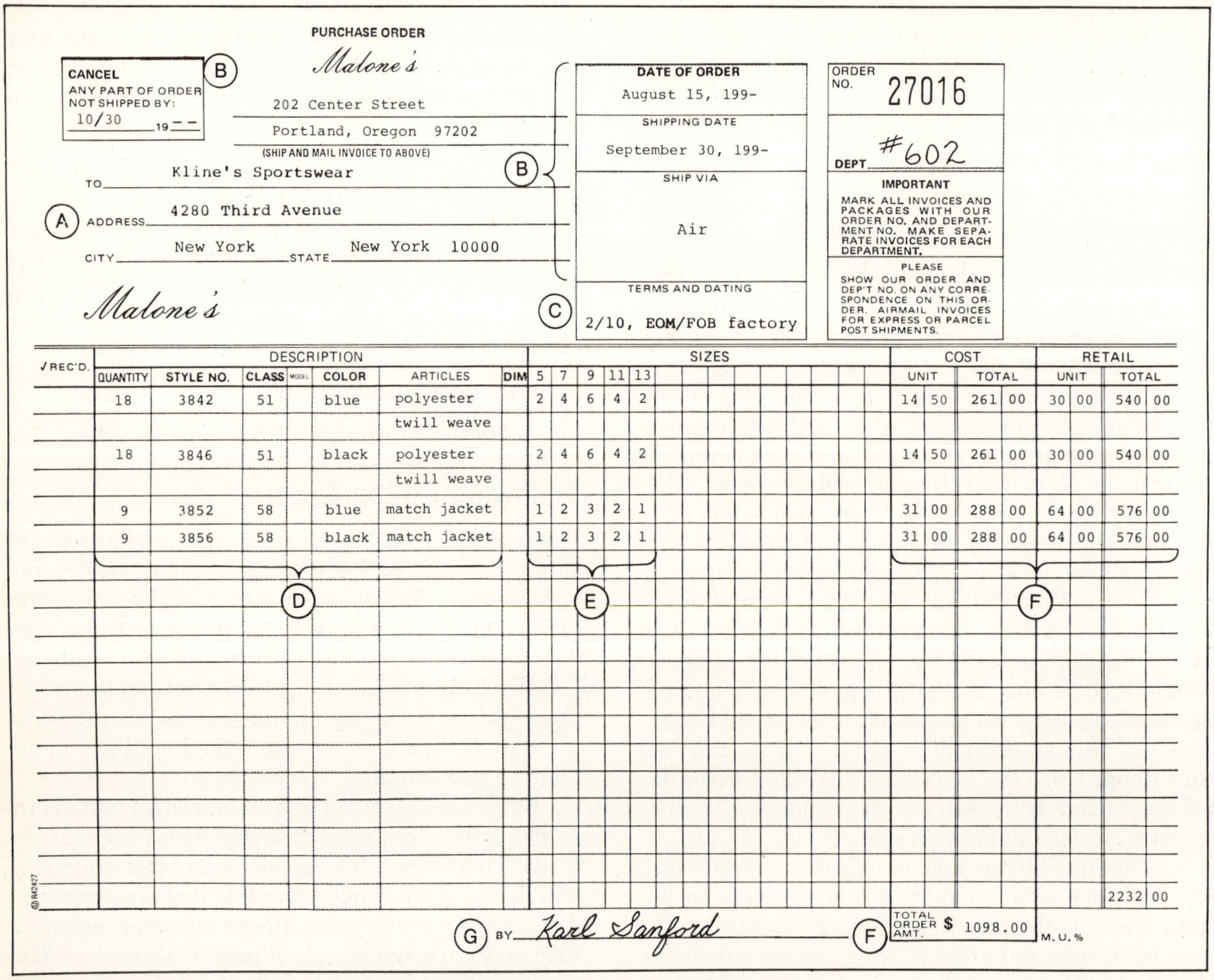

PURCHASE ORDER

Malone's

202 Center Street

Portland, Oregon 97202

(SHIP AND MAIL INVOICE TO ABOVE)

CANCEL ANY PART OF ORDER NOT SHIPPED BY: 10/30 19-- (B)

TO Kline's Sportswear

(A) ADDRESS 4280 Third Avenue

CITY New York STATE New York 10000

(B)

Malone's

DATE OF ORDER: August 15, 199-

SHIPPING DATE: September 30, 199-

SHIP VIA: Air

(C) TERMS AND DATING: 2/10, EOM/FOB factory

ORDER NO. 27016

DEPT. #602

IMPORTANT: MARK ALL INVOICES AND PACKAGES WITH OUR ORDER NO. AND DEPARTMENT NO. MAKE SEPARATE INVOICES FOR EACH DEPARTMENT.

PLEASE SHOW OUR ORDER AND DEP'T NO. ON ANY CORRESPONDENCE ON THIS ORDER. AIRMAIL INVOICES FOR EXPRESS OR PARCEL POST SHIPMENTS.

√REC'D.	QUANTITY	STYLE NO.	CLASS	MODEL	COLOR	ARTICLES	DIM	5	7	9	11	13	COST UNIT	COST TOTAL	RETAIL UNIT	RETAIL TOTAL
	18	3842	51		blue	polyester twill weave		2	4	6	4	2	14 50	261 00	30 00	540 00
	18	3846	51		black	polyester twill weave		2	4	6	4	2	14 50	261 00	30 00	540 00
	9	3852	58		blue	match jacket		1	2	3	2	1	31 00	288 00	64 00	576 00
	9	3856	58		black	match jacket		1	2	3	2	1	31 00	288 00	64 00	576 00
																2232 00

(D) (E) (F)

(G) BY Karl Sanford

(F) TOTAL ORDER AMT. $ 1098.00 M. U. %

Figure 11-6 Here's an example of a purchase order buyers use. Each of the lettered items is explained in the text.

not be canceled without the consent of both the vendor and the store. Therefore, if the store buyer wants to cancel an order, the buyer must ask the vendor for approval. If the vendor cannot meet all the requirements of the order, such as a certain delivery time, then either the buyer or vendor may cancel the order.

■ PRICING MERCHANDISE

There are no standard rules for setting the retail price of clothing and accessory merchandise. Buyers, who compute the selling prices, must be certain each item can be sold at a price that covers their costs and earns a profit. Stores add a markup, also called markon, to the wholesale cost of merchandise. The dollar **markup** equals the difference between the retail price and the cost of the merchandise sold:

$$\text{Dollar markup} = \text{retail price} - \text{cost}$$

Two variations of this formula are:

$$\text{Retail price} = \text{cost} + \text{dollar markup}$$

$$\text{Cost} = \text{retail price} - \text{dollar markup}$$

All buyers should be able to compute the dollar markup, cost, and retail price of any merchandise item. Buyers can compute these amounts by using the formulas shown above. For example, if a merchandise item costs the store \$20 and the store sells it for \$42, the dollar markup equals \$22.

Buyers should also be able to compute the retail markup percentage and retail price when only the cost and the retail markup percentage are known. Buyers use the following formulas for these computations:

$$\text{Retail mark up percentage} = \frac{\text{dollar markup}}{\text{retail price}}$$

$$\text{Retail price} = \frac{\text{cost}}{\text{100 percent} - \text{retail markup percentage}}$$

For example, the merchandise that cost the store \$20 and sold for \$42 has a retail markup percentage of 52.4 percent.

$$\text{Retail markup percentage} = \frac{\$22}{\$42}$$

If the store purchases an item for \$22 and wants to maintain a 50 percent markup, the retail price is \$44.

$$\text{Retail price} = \frac{\$22}{100\% - 50\%} = \frac{\$22}{50\%} = \$44$$

The actual merchandise prices set by buyers are determined by such factors as the type of product, customer demand, store image, store expenses, competitors' prices, and profit required. When buyers find a "hot" item that their customers can't get enough of, they may be able to increase the markup percentage a few points and earn additional profit. Of course, the competition may also sell the "hot" item, and most customers will gravitate to the store with the best price. Some of the many pricing policies followed by clothing and accessory buyers are:

- *Predetermined Markup* Some stores determine the amount of markup needed to pay for operating costs and provide a profit, such as 48 percent or 51 percent. These stores initially mark up all their merchandise by this predetermined markup percentage. Many clothing stores add a **keystone** markup, which is a 50 percent markup, to all their merchandise items. The keystone markup means that the retail price is double the cost price. If a scarf that costs the store \$16 has a keystone markup, it will retail for \$32.
- *Leader Pricing* Some stores follow a leader pricing policy. Prices of some merchandise items are set lower than usual to attract customers to the store. The reasoning behind this pricing policy is that once customers are in the store, they will also purchase other, more profitable merchandise.
- *Odd-Ending Pricing* Stores that set prices that end in odd numbers, such as \$15.99 or \$27.97, are following an odd-ending pricing policy. This policy usually suggests lower prices and bargains.
- *Even-Ending Pricing* Stores that set prices that end in even numbers, such as \$25 or \$38.50, are following an even-ending pricing policy. Even-ending pricing suggests high-quality merchandise.
- *Price Lining* Many clothing stores follow a price lining approach when pricing merchandise. They set up distinct price points or ranges of prices. They then price all merchandise to fit into the predetermined price points or price ranges that appeal to their target customers. For example, a store may sell gloves at three specific price points: \$19.99, \$24.99, and \$29.99. Or it may sell handbags in three price lines: \$24 to \$32, \$40 to \$60, and \$75 to \$125. All gloves or handbags would fit into one of the store's price points or lines.

Experience 11-5 ■ What's the Price?

On a sheet of paper, compute the following:

1. Dollar markup on the following merchandise:
   a. Ties: cost = \$9.50; retail price = \$20.00
   b. Calfskin belt: cost = \$15.60; retail price = \$32.50
2. Percentage of retail markup on the following merchandise items:
   a. Earrings: cost = \$22.00; retail price = \$56.00
   b. Socks: cost = \$2.25; retail price = \$4.50
3. Retail prices of the following merchandise items: (Assume your store rounds its prices to the nearest dollar.)
   a. Slippers: cost = \$11.75 per pair; desired markup percentage = 45 percent
   b. Nightgowns: cost = \$13.00 each; desired markup percentage = 55 percent

Retail Price Changes

The original retail price is not always the price at which merchandise is actually sold. Because customer demand for merchandise items fluctuates, the retail price of clothing and accessories goes up and down. Salespeople are responsible for marking the retail price changes on price tickets. They record the price changes on the appropriate store retail price change forms.

Markdowns Suppose it is early June. The buyer wants to move out the swimsuits and beachwear to make room for the coats. Your store carries swimsuits at three price lines: $36 to $49, $50 to $79, and $80 to $120. The buyer decides to take 25 percent off the selling price. (A markdown of 25 percent is a typical initial markdown for clothing and accessories.) To make the price changes, the salesperson slashes the $36 price and writes $26.99. (Even though a 25 percent markdown of an item priced $36 would be $27, it is this store's policy to round the markdown price to the nearest 99 cents ending in the customer's favor.) The $50 suits are reduced to $36.99 and the $80 suits are reduced to $59.99. (See Figure 11-7.)

It is important that all price changes are recorded on a retail price change form, similar to the one in Figure 11-8. Follow the letters on this form to see how to keep track of total inventory dollars in a department. When price changes are not properly recorded, the store does not have the necessary information to determine accurately the amount of profit earned.

To complete the retail price change form, you do as follows:

1. Write the symbol of the type of price change (markdown, MD). (See Figure 11-8, letter A.)
2. Enter the date the price change is made. (See letter B.)
3. Enter the department number as shown on the price ticket. (See letter C.)
4. Enter the style of the item from the price ticket. (See letter D.)
5. Write the description of the merchandise. (See letter E.)
6. Enter the class of the item from the price ticket. (See letter F.)
7. Write the retail price shown on the ticket before the price change. (See letter G.)
8. Write the new retail price after the markdown. (See letter H.)
9. Write the difference in the present and revised price. (See letter I.)
10. Count the number of items with price changes. (See letter J.)
11. Multiply the quantity by the difference and record the total in the extension column. (See letter K.)
12. Total the extension column. (See letter L.)
13. Write the reason for the price change. (See letter M.)
14. Sign the retail price change form. (See letter N.)
15. Have the buyer sign the form to indicate approval of the changes. (See letter O.)

Markups Suppose one of your basic stock items is a camel-colored vinyl shoulder bag that retails for $24. When the buyer went to market in April, the handbag vendor issued price increases on all the vinyl models. The new cost to the store is $13.75. Your store keystones, or doubles, the cost of every item and follows even price lines, such as $25.00 and not $24.95. You compute the new selling price of the shoulder bag to be $28 ($13.75 × 2 = $27.50, rounded up to $28.00.) The price tickets of the purses in stock must be marked up $4.

Clothing and accessory stores do not record price increases on price tickets by slashing out the old price and showing the new price. This causes too much customer resistance, so stores issue new price tickets. This price change is also recorded on the retail price change form.

Markdown Cancellations A markdown cancellation restores the original price after a special sale markdown. For example, on Saturday your store participated in the town's annual Sidewalk Sale. The buyer marked down tennis shirts from $28 to $19. Your store, like many other clothing shops that run limited-time sales, attached a sale ticket to the shirts and did not remove the original price ticket. Now that the sale is over, you must remove the sale price

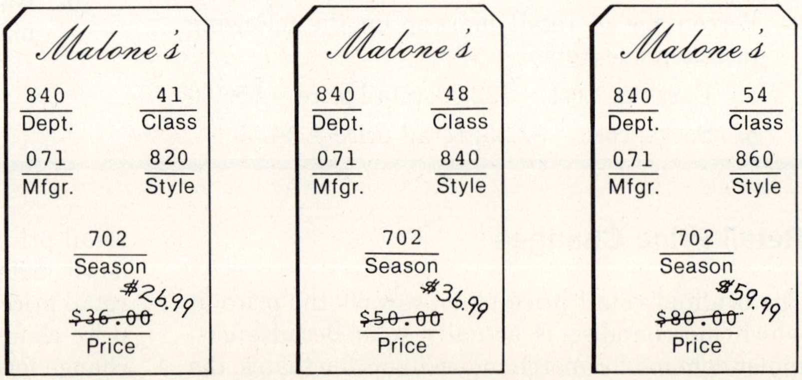

Figure 11-7 These three price tickets reflect typical midsummer markdowns.

Malone's

RETAIL PRICE CHANGE

TYPE OF PRICE CHANGE	SYMBOL
1. MARKDOWN	M D
2. MARKUP	M U
3. MARKDOWN CANCELLATION	M D C
4. MARKUP CANCELLATION	M U C

INDICATE SYMBOL: MD (A)

M 7215

DEPARTMENT NO. (C) 840 DATE July 5, 199- (B)

STYLE	DESCRIPTION	MDSE CLASS	RETAIL PRICE PRESENT	RETAIL PRICE REVISED	RETAIL PRICE DIFFERENCE	QUAN	EXTENSION
820	Maillot swimsuit	41	36 00	26 99	9 01	8	72 08
840	Maillot swimsuit	48	50 00	36 99	13 01	7	91 07
860	Maillot swimsuit	54	80 00	59 99	20 01	12	240 12
(D)	(E)	(F)	(G)	(H)	(I)	(J)	(K)
						TOTAL PRICE CHANGE →	403 27 (L)

REASON FOR PRICE CHANGE: (M) Summer Stock Clearance

PREPARED BY (N) K. Brown

APPROVAL (O) *Olive Hansen* BUYER

Figure 11-8 This is an example of a retail price change form. Each of the lettered items is explained in the text.

tag and record the markdown cancellation on the retail price change form. The present price of $19 is revised to $23.

Instead of taking markdown cancellations, many stores now use point-of-sale (POS) markdowns. Rather than mark the merchandise at a lower price for a temporary time period and then re-mark it when the sale is over, these stores place a sign next to the merchandise. The wording on the sign is usually similar to "Take an extra 25% off." When a customer buys an item that is marked down that day, the salesperson enters the point-of-sale markdown into the register. At those stores with laser scan guns, the markdown is automatically recorded on the sales receipt when the scanner reads the price ticket. POS markdowns eliminate the need for markdown cancellation paperwork and greatly reduce the number of hours required to set up a sale for merchandise that is temporarily marked down.

Markup Cancellations The buyer purchased 12 dozen orlon pullover sweaters. They were retailed at $32 each and were a sellout. The buyer reordered another 20 dozen of these sweaters. After two weeks of selling on the floor at $32, the buyer issued a markup cancellation that permanently lowered the price to $24.99 to increase sales and liquidate the goods.

You record the new price by slashing through the original price and writing $24.99 on the price ticket. The markup cancellation, which shows the revised price of $24.99, is also recorded on the retail price change form.

Experience 11-6 ■ Making Changes

Your store has had a difficult time moving sequin party tops. The buyers have already taken two markdowns, but there are still five of these tops on the sale rack. You are instructed to take 50 percent off the original price. Using today's date, complete the retail price change form on page 166. (Round the markdown price to the nearest dollar ending in the customer's favor.) A salesperson will record the new price on the price ticket later.

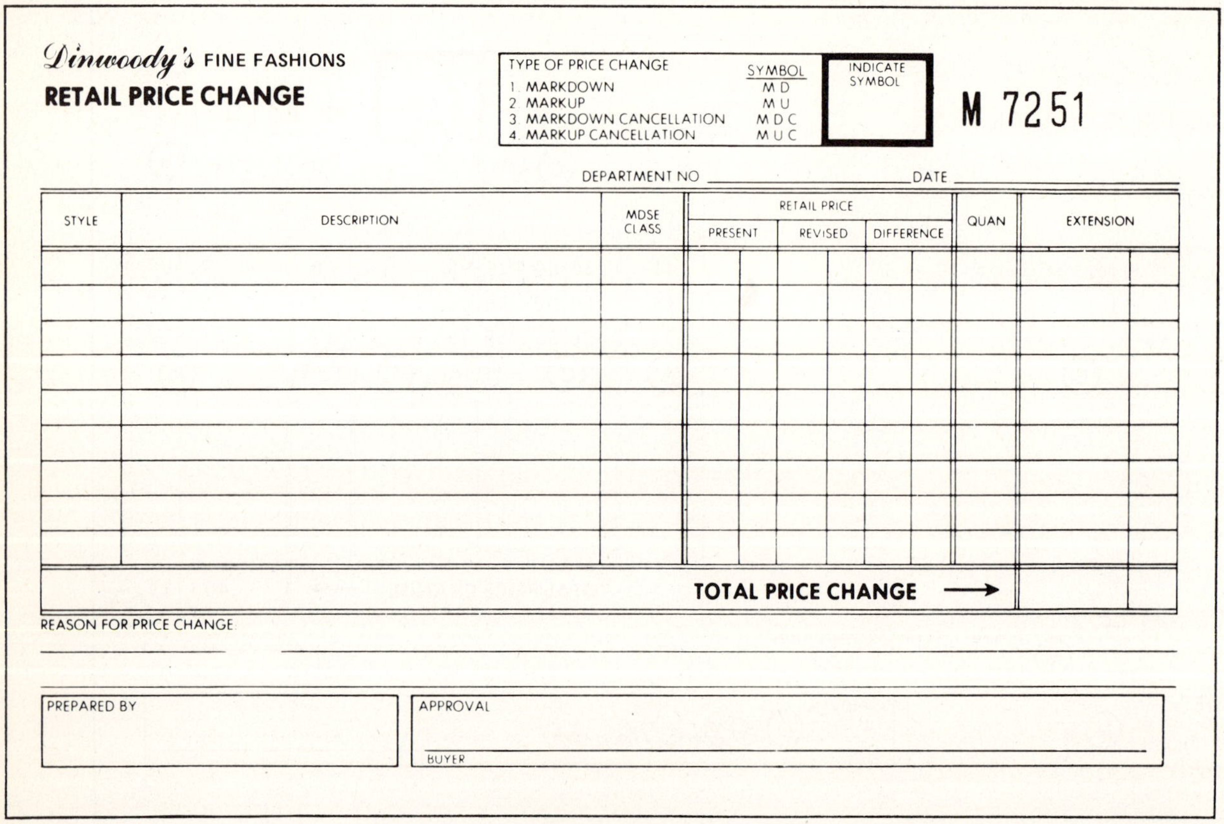
Dinwoody's FINE FASHIONS
RETAIL PRICE CHANGE

TYPE OF PRICE CHANGE	SYMBOL
1. MARKDOWN	M D
2. MARKUP	M U
3. MARKDOWN CANCELLATION	M D C
4. MARKUP CANCELLATION	M U C

INDICATE SYMBOL

M 7251

DEPARTMENT NO ______ DATE ______

STYLE	DESCRIPTION	MDSE CLASS	RETAIL PRICE			QUAN	EXTENSION
			PRESENT	REVISED	DIFFERENCE		
TOTAL PRICE CHANGE →							

REASON FOR PRICE CHANGE

PREPARED BY

APPROVAL

BUYER

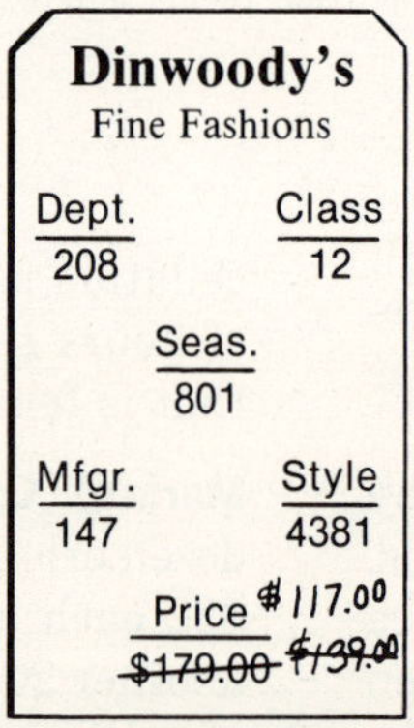
Dinwoody's
Fine Fashions
Dept. 208
Class 12
Seas. 801
Mfgr. 147
Style 4381
Price ~~$179.00~~ ~~$139.00~~ $117.00

■ WATCHING SALES

A buyer's work isn't finished once merchandise is on the sales floor. The buyer must also carefully watch sales. Merchandise that shows strong sales needs to be reordered as soon as the need for more stock is discovered, also slow-selling merchandise needs special attention. Buyers must carefully monitor inventory levels so that there is an adequate supply of merchandise to meet customer needs.

Reordering Basic Stock

Many times basic stock lists include the minimum amount and maximum amount of merchandise that should be on hand at all times, called **model stock.** These amounts are determined from past sales. As soon as the store's actual supply of merchandise reaches or falls below the minimum quantity, an order for more merchandise should be made.

The amount of merchandise to be reordered is computed by subtracting the quantity on hand and the quantity on order from the model stock. Suppose that there are only three packages of men's T-shirts, size 40, on the sales floor. There is no merchandise in reserve stock in the stockroom. Three dozen packages are on order but have not yet been received. (The buyer must reorder in even dozens.) The model stock for this item is six dozen. The buyer should reorder three dozen (thirty-three rounded to the nearest dozen) more T-shirts in size 40. Notice the following:

Model stock (72) − quantity on hand (3) − quantity on order (36) = quantity to be reordered (36) (33 rounded to the nearest dozen)

Reordering Strong Sellers

Sometimes an item hasn't been included on the basic stock list but has shown strong, consistent sales. When the stock supply of this item gets low, the buyer must determine how much to reorder. To do this, the buyer uses these formulas:

1. Minimum stock = reserve + (delivery period × weekly sales)
2. Maximum stock = minimum stock + (reorder period × weekly sales)
3. Quantity to order = maximum stock − (stock on hand + stock on order)

For example, suppose you sell an average of eighteen medium-sized rugby shirts during a week. Your reorder period is every four weeks. Your delivery period for the shirts is two weeks. You think you need six extra shirts in stock just to take care of an unexpected increase in sales. At this time, you have a total of sixteen shirts in stock and 48 on order. In this situation, you would reorder 50 shirts.

Minimum stock = 6 + (2 × 18) = 42

Maximum stock = 42 + (4 × 18) = 114

Quantity to order = 114 − (16 + 48) = 50

Let's suppose these shirts are prepacked in a certain color assortment, such as six to a package in red, blue, and green. You have to order nine packages to meet your merchandise requirements. This amount is computed by dividing the quantity needed, 50, by the number of shirts to a package, six.

By following the reorder formula, which is based on weekly sales figures, you will be able to keep your stock in balance with your sales.

Experience 11-7 ■ Ordering, Ordering, and Reordering

1. Using the following information, determine the quantity of basic stock you should reorder.

Merchandise	Quantity on Hand	Quantity on Order	Model Stock	Reorder Quantity
a. Knee socks—navy	13 pair	0	48 pair	
b. Gold button earrings	1 pair	5 pair	18 pair	
c. Men's white shirt, 15½/33	0	12	24	
d. One-size-fits-all women's white turtlenecks	3	12	48	
e. Black leather clutch handbag	5	6	18	
f. 12-month white-snap undershirts (2 per package)	9	18	36	

2. Using the following information, determine the quantity of merchandise you should reorder.

Merchandise	Weekly Sales	Reserve Quantity	Reorder Period (weeks)	Delivery Period (weeks)	Stock on Hand	Stock on Order	Reorder Quantity
a. Men's socks	10 doz.	20 doz.	2	2	3 doz.	0	
b. House slippers	15	18	4	3	14	24	
c. Men's black belts	7	12	3	6	16	24	
d. Women's knit tank tops	12	24	1½	2	18	30	

Promoting Slow Sellers

Just as a buyer cannot know which items will be the best sellers, a buyer cannot predict unpopular items. To overcome poor turnover and low profits, buyers must stimulate consumer demand for poor-selling items. To stimulate sales, a buyer does those things that cost the store the least amount of money first, such as moving merchandise on the sales floor. Other ways to promote slow sellers are creating interior displays, training the salespeople, lowering the retail price, or transferring the goods to another store in the chain.

Moving Slow Sellers Around An item may not be selling well because of its location on the sales floor. Customers cannot see merchandise that is hidden from view, and, therefore, will not buy it. The first way many buyers try to stimulate sales is by moving the goods to different locations in the department.

Display Some items are poor sellers because customers are not aware that the store or department carries the goods. A buyer may use signs or interior displays to call customers' attention to the merchandise. Depending on the type of item, the display may feature merchandise on top of racks, counters, or shelves. Chapter 14 discusses display in more detail.

Sales Training Some items may not be selling because the sales staff has not received training in the features and benefits of the goods. It is the buyer's responsibility to be certain that salespeople know the features of their merchandise and how to translate these features into benefits for the customer.

Lower Prices Some items are poor sellers because they are priced too high. Customers who consider the price too high will often purchase goods if the price is lowered. Many buyers use a keystone (or 50 percent) markup at retail. If the goods fail to sell at the original retail selling price, the buyers mark twenty or twenty-five percent off the price. If the goods still do not sell the buyers mark the goods down again, usually to one-third and later to one-half of the original retail price. After multiple markdowns, many buyers have special sales. The goods are placed on a clearance table and sold at a very low price.

Merchandise Transfers Buyers for stores that have more than one location may consolidate merchandise assortments that have been "broken" to stimulate sales. By bringing all the remaining sizes and colors together at one location, the buyer provides customers with a better selection from which to choose. Usually, the merchandise has already been marked down.

To initiate the consolidation, the buyer or assistant buyer prepares the merchandise transfer document (see Figure 11-9, on page 169.) by recording:

1. Date (see letter A)
2. Where the merchandise is transferred from (store and department) (see letter B)
3. Where the merchandise is being transferred to (store and department) (see letter C)
4. Merchandise class number (see letter D)
5. Merchandise style number (see letter E)
6. Description (see letter F)
7. Merchandise class in the receiving store and department (see letter G)
8. Retail price of the transferred item (see letter H)
9. Reason for transfer (see letter I)
10. Signature of the person initiating the transfer document (see letter J)
11. Signature of the person approving the transfer (see letter K)

When the sending store receives the form, the manager is responsible for assembling the merchandise that is to be transferred, counting the pieces that are being sent, packaging the merchandise for shipping, and completing the quantity sent and extension columns on the document. (See letters L, M, and N.) The merchandise is then taken to the dock or receiving area, where it is picked up for delivery to another store.

Transfer documents may also be initiated by a manager who receives a customer request for merchandise to be sent to another store. This happens most often when one store has already sold a particular size that a customer wants to buy.

■ EVALUATING THE BUYER'S PERFORMANCE

Most buyers are evaluated annually for salary increases. Companies often base the buyer's evaluation on four criteria: sales, gross margin, markdowns, and turnover. For the buyer to receive the highest increase, they must surpass their company's goals in each of these specific areas. Because we have already mentioned sales, gross margin, and markdowns, we will turn our attention to turnover.

Stock Turnover

Stock turnover is the number of times that an average stock of merchandise has been turned into sales during a given period. It reflects the rate at which merchandise is placed on the selling floor, sold, and additional new goods are purchased.

A good turnover rate is important for many reasons. It increases the open-to-buy, allowing the buyer

Malone's
MERCHANDISE TRANSFER

DATE September 13, 19__ (A) T 86270

TRANSFER FROM	TRANSFER TO
(B) LOCATION: ☑ 1 Portland, Oregon ☐ 2 Eugene, Oregon ☐ 3 Spokane, Washington — DEPT 420	(C) LOCATION: ☐ 1 Portland, Oregon ☐ 2 Eugene, Oregon ☑ 3 Spokane, Washington — DEPT 420

MDSE CLASS Send Dept	STYLE	DESCRIPTION	MDSE CLASS Rec Dept	UNIT RETAIL	QUAN	EXTENSION
70	8256	Fur-trimmed leather jacket	7	250 00	12	3000 00
14	3475	Imitation leather bomber jacket	14	95 00	14	1330 00
(D)	(E)	(F)	(G)	(H)	(L)	(M)
		TOTAL TRANSFER →				4330 00 (N)

REASON FOR TRANSFER: Consolidation (I)

PREPARED BY: Allen Smithstone (J)   APPROVAL: Denise Martin (K) BUYER   MDSE. RECEIVED BY:

FORM A 402 (REV. 8/74)   MOORE BUSINESS FORMS INC. M   MERCHANDISE CONTROL COPY

Figure 11-9 Merchandise transfers, which sometimes stimulate sales of goods in other branches of a store, are recorded on a form such as this.

to purchase new merchandise that will add "freshness" to the current merchandise assortment. A constant flow of fresh merchandise will cause customers to return to the store more often, and they will be more inclined to buy the new items at full price—which is much more profitable—than the older items. New merchandise also creates an exciting visual presentation and a renewed enthusiasm among the sales people. For the company, a good turnover rate provides an increased profit. Also, the money that is generated from sales can be used to pay expenses.

A good turnover rate is not, however, the highest rate possible. If the rate is too high, there will be constant out-of-stock situations. Customers will not find the styles, colors, and sizes they want and may decide to shop elsewhere. A high turnover rate will also result in high handling and billing costs because the buyer will be placing many orders to fill in the stock needed.

Calculating Turnover

The buyer or manager computes the turnover rate for the store or department by following this formula:

$$\frac{\text{Net sales}}{\text{Average stock}} = \text{stock turnover}$$

Suppose you want to compute the turnover rate for the spring season, February through July. Before you can compute the turnover rate, you must first compute the average stock. The **average stock** is determined by adding all the BOM (beginning of the month) stock plus the EOM (end of month) inventory for the end of the period divided by the number of inventory figures used. For example, if the inventory figures are:

Feb. 1	3,000
March 1	4,500
April 1	4,200
May 1	4,800
June 1	5,500
July 1	3,800
July 31	2,700

Then, the average stock for the period is:

$$(3{,}000 + 4{,}500 + 4{,}200 + 4{,}800 + 5{,}500 + 3{,}800 + 2{,}700) = 4{,}071$$

The net sales for this 6-month period are \$11,700. Therefore, the stock turnover rate for this period is:

$$\text{Stock turnover} = \frac{11{,}700}{4{,}071} = 2.87$$

Average turnovers for clothing and accessories are published annually by the National Retail Merchants Association in its magazine, *Stores*. If you learn to use sales information and your study of fashion trends to predict future sales, and if conditions in your market are favorable, you may meet or beat these typical rates.

Experience 11-8 ■ Calculate the Rate

1. Compute the stock turnover rate for the six-month period August to January. The net sales for the six months is $13,500. The monthly inventory levels are:

Aug. 1	2,400
Sept. 1	2,800
Oct. 1	3,400
Nov. 1	4,650
Dec. 1	5,500
Jan. 1	2,700
Jan. 31	2,100

Stock turnover rate:______

2. The turnover rate for last year's August to January period was 3.2. What conclusion can you draw when you compare last year's turnover rate with this year's rate?

____

____

Chapter Note

1. Alyssa A. Lappen, "Macy's Buyer Makes Good," *Forbes*, February 22, 1988, pp. 54, 56.

CHAPTER 12

Physical Distribution

Levi Strauss and Company began using bar-coded labels on its shipping cartons in July 1988. The purpose of the labels, which are similar to the bar coding on items at the grocery store, is to shorten the time it takes retailers to process incoming goods and move them to the selling floor.

The system works as follows: Levi transmits a packing slip to the retailer by computer. The slip identifies the carton by its code and describes the contents, including style, color, and size. When the shipment is received by the retailer, the code on the carton is scanned. The retailer's computer matches the code on the carton with the already-received packing slip and identifies the merchandise, which can then be speeded to its destination.

The goal of marking products with bar codes and relaying information electronically is to make manufacturing, distribution, and store operations faster and more efficient. Previously, all these tasks had to be done manually, which was time consuming and did not always result in accurate information.[1]

Bar coding and computer linkups between manufacturers and retailers are two of the exciting innovations that are affecting the way clothing and accessory companies do business. The goal of any of these computerized systems is to keep track of inventory, speeding its movement through the system. These innovations help buyers to reduce their inventory levels without hurting their sales, which, of course, increases turnover.

Whether or not you work in a company that uses computers extensively, it is very important that you understand how the inventory control system operates. This chapter will discuss the basics of every inventory control system: taking inventory and receiving, checking, and pricing merchandise.

■ PERPETUAL INVENTORY VERSUS PERIODIC INVENTORY

Dollar control and unit-control records can be based on perpetual or periodic inventory systems. If dollar control records are based on the **perpetual inventory system,** the records will perpetually, or continually, show the retail value of merchandise on hand. If the dollar control records are based on the **periodic inventory system,** the records will be updated only when an inventory is taken.

Unit-control records based on the perpetual inventory system will always show the merchandise items in stock. Under a periodic inventory system, the only way to determine the quantity of merchandise units on hand is by making an actual stock count. Under this system, there is no attempt to track the merchandise; it assumes that the difference between the merchandise received and the merchandise on hand was sold.

Retail Inventory Method

Most apparel and accessory stores use an accounting system known as the retail inventory method. Stores using this inventory method record all transactions at the retail price, including sales, merchandise purchases, markdowns, transfers, and returns-to-vendors. The inventory control systems discussed in this chapter will use the retail inventory method.

■ THE INVENTORY CONTROL SYSTEM

How do apparel stores establish and maintain an inventory control system? There are many variations, especially with computers to keep track of inventory levels. Let's examine how a fictional store, called Michael's, handles its inventory control system. Twice a year, in January and again in July, this medium-sized clothing store's employees count and record all merchandise in stock. Suppose you are a salesperson at Michael's and you are part of the inventory team.

Taking Inventory

Your manager, Helen Wilson, explains the inventory procedures to you and the other salespeople. "We are going to use the team method to take inventory," Helen says. "A team is made up of two persons—a caller and a lister. Each team will be assigned to a particular area and section of the store and will be responsible for counting all the merchandise in that section. Accuracy is most important in taking inventory. If you overlook any merchandise or call out or write down the wrong information, your errors will affect the store's financial records."

You have been assigned the job of lister in department 380. This department sells clothing for boys, sizes 8 through 18. "As lister," Helen explains, "you are responsible for recording information on an inventory sheet." (See Figure 12-1.)

Part of the inventory sheet will be completed for you by your manager. As lister, you are to complete the rest, including writing the names of the caller and lister and a general description of the merchandise, such as pant, shirt, belt.

When taking inventory, the caller must read and call out information on the price ticket. (See Figure 12-2.) Because Michael's takes inventory by class, the caller and lister must be sure the merchandise being counted belongs to the same department and class listed on the inventory sheet. The department number represents a grouping of similar merchandise, such as department 180 for socks, department 182 for underwear, and department 185 for robes. The class number represents a specific kind of merchandise in the department, such as class 40 for dress socks and class 50 for athletic socks.

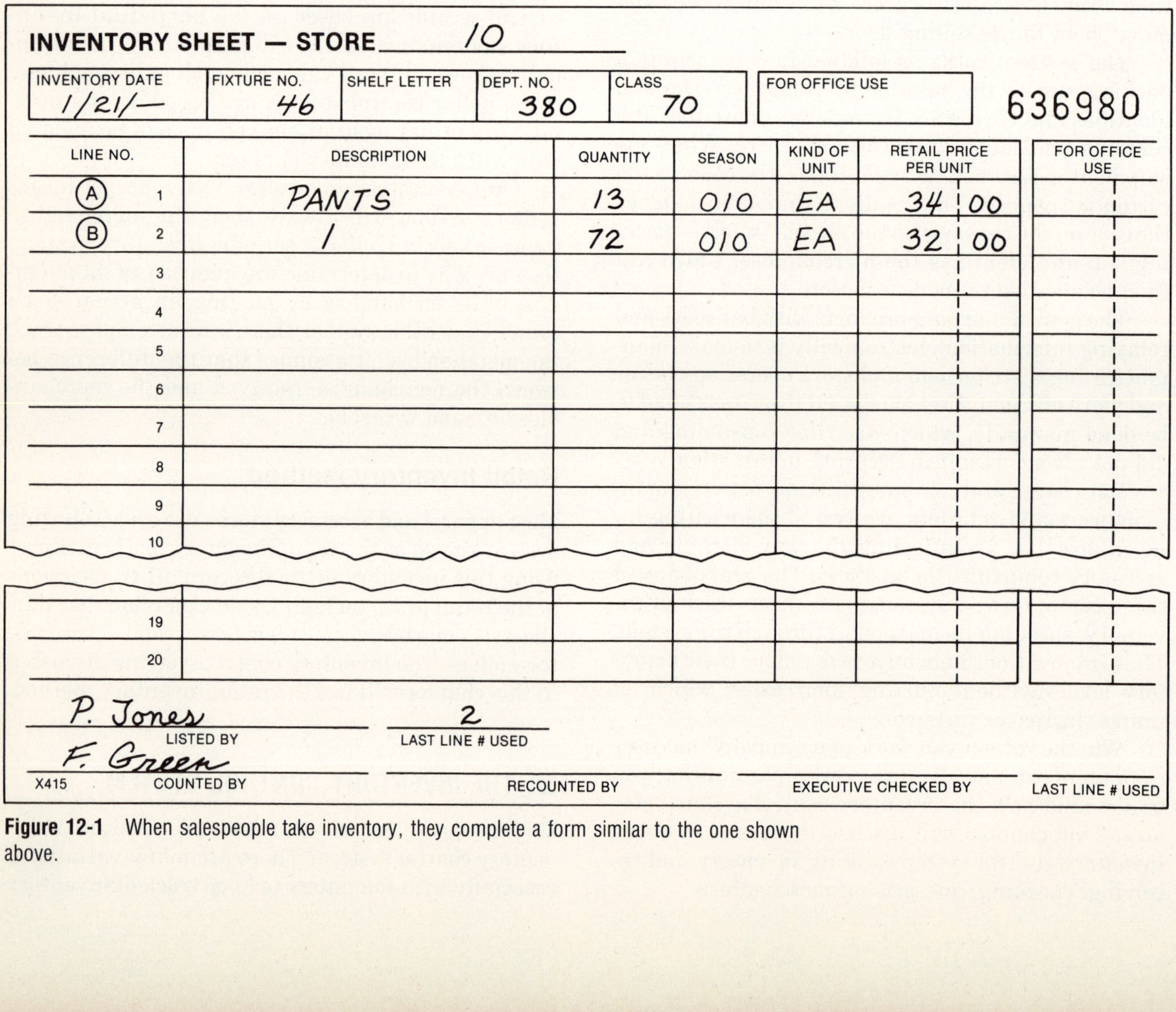

INVENTORY SHEET — STORE 10

INVENTORY DATE	FIXTURE NO.	SHELF LETTER	DEPT. NO.	CLASS	FOR OFFICE USE
1/21/—	46		380	70	

636980

LINE NO.		DESCRIPTION	QUANTITY	SEASON	KIND OF UNIT	RETAIL PRICE PER UNIT	FOR OFFICE USE
(A)	1	PANTS	13	010	EA	34 00	
(B)	2	/	72	010	EA	32 00	
	3						
	4						
	5						
	6						
	7						
	8						
	9						
	10						
	19						
	20						

P. Jones — LISTED BY

2 — LAST LINE # USED

F. Green — COUNTED BY

X415 RECOUNTED BY EXECUTIVE CHECKED BY LAST LINE # USED

Figure 12-1 When salespeople take inventory, they complete a form similar to the one shown above.

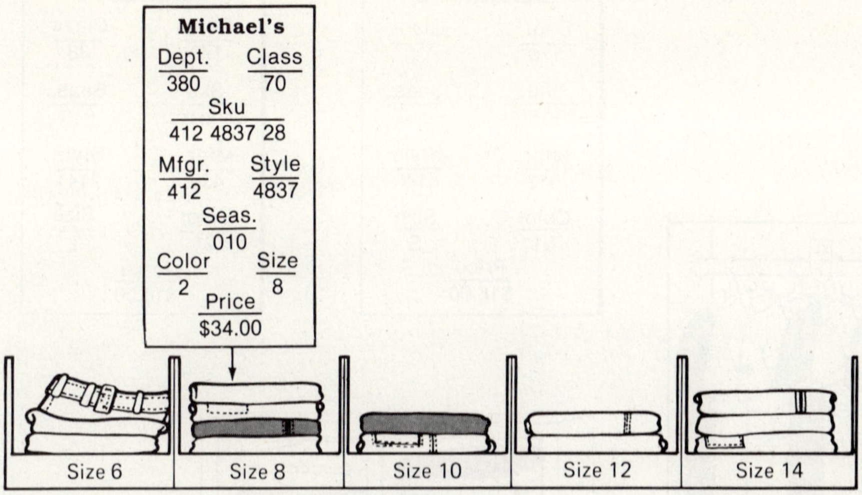

Figure 12-2 The top portion of the tiered counter you and Lenny are to inventory is shown at the right. If all the pants shown had the same department, class, season, and price, what's the total inventory?

Some retailers assign stockkeeping units (SKUs) to merchandise and keep sales and stock records according to these SKU numbers. An **SKU** identifies a unique merchandise item or group of items by its specific characteristics (such as manufacturer, style, size, and color) which distinguish it from all other merchandise. The SKU number is printed on the price ticket and various other reports.

The quantity is the total number of units with the same department, class, season, and price. Kind of unit refers to the way the merchandise is sold; write "each" if the merchandise is sold in individual units, like a blouse; write "pair" for a pair of gloves, and "package" for a package of three T-shirts. The season, which is also included on the price ticket, identifies when the merchandise was received by the store. Stores set up their own season code. At Michael's the first digit represents the year the merchandise arrived at the store; the second and third digits represent the month the merchandise arrived. For example, 902 is the season code for merchandise that arrived in February 1989. The price is the lowest retail price, after any markdowns. (*Note:* If you were taking a unit-control stock count, you would also be concerned with the manufacturer, style, color, and size of the merchandise.)

The first item you inventory is boy's denim pants. The pants are sized on a tiered counter. Lenny Hackett, your caller, starts in the top left-hand section of the counter and calls off the department, class, season, and price. After checking each price ticket to be sure all the merchandise belongs to the same department and class and has the same price, Lenny counts the pants.

You notice that Lenny follows a certain pattern when counting the merchandise. After he has finished counting, he calls out: "Pants, a total of 13, season 102, retailing at $34.00." You record this information on your inventory sheet. (See Figure 12-1, letter A.)

The next item for your count is boy's corduroy pants. Lenny says, "Pants, a total of 72, season 102, retailing at $32.00." You record the information on line 2 of the inventory sheet. Because the description is the same as listed on the prior line, you draw a line in the column to show the items are alike. (See Figure 12-1, letter B.)

As you continue taking inventory, you notice that all the merchandise has been organized for easy counting. Your manager has made sure all merchandise with the same department and class numbers is together on the merchandise fixtures. She also checked to be sure all merchandise had correct price tickets reflecting the current selling price of the items. This inventory preparation helps the caller and lister teams take fast, accurate counts.

Experience 12-1 ■ Counting the Goods

You're taking fall inventory and have been assigned to the men's sportwear area. Following the proper inventory procedures, count the merchandise shown and complete the inventory sheet on page 174. The hanging shirts are $18.00 each. The folded shirts are $16.50 each. You have already checked the price tickets and found that all the hanging and folded shirts have the same department, class, and season.

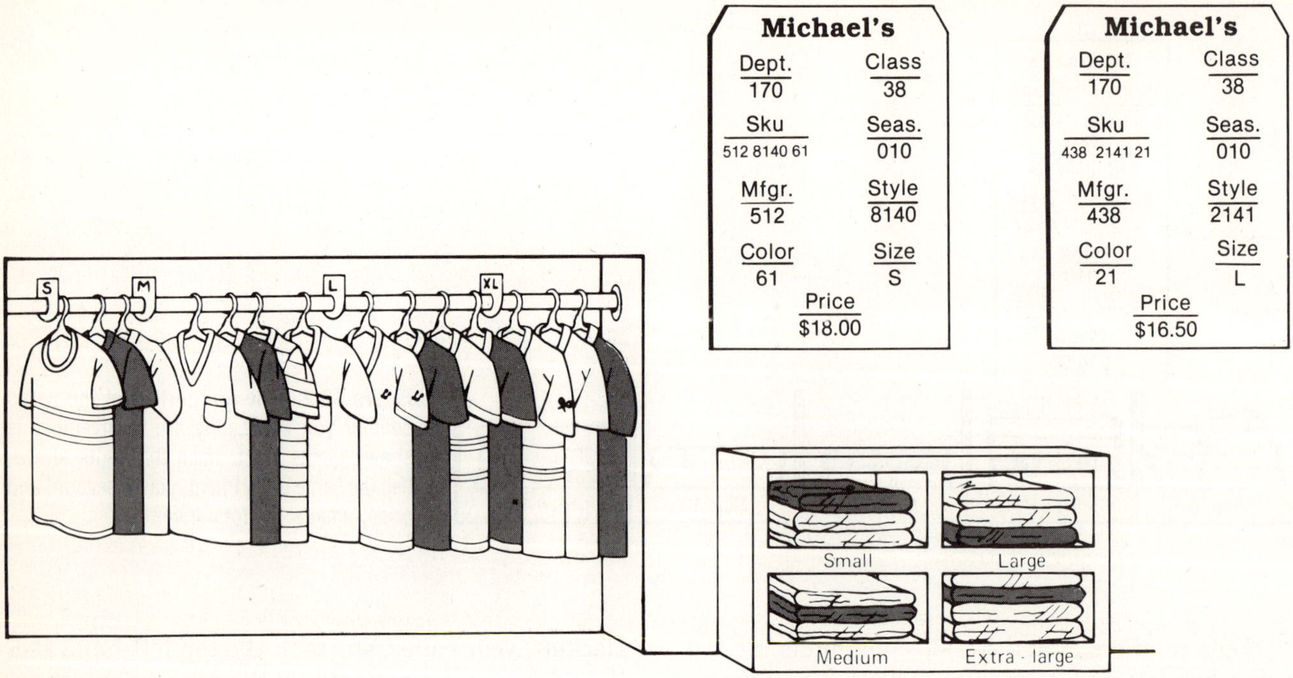

INVENTORY SHEET — STORE 10

INVENTORY DATE	FIXTURE NO.	SHELF LETTER	DEPT. NO.	CLASS	FOR OFFICE USE

636980

LINE NO.	DESCRIPTION	QUANTITY	SEASON	KIND OF UNIT	RETAIL PRICE PER UNIT	FOR OFFICE USE
1						
2						
3						
4						
5						
6						
7						
8						
9						
10						

Setting Up and Maintaining Inventory Control Records

Both dollar control and unit-control records require similar types of information. For example, both need an initial stock count to determine the dollar value and type of merchandise on hand. To keep the records accurate and up to date, both types of inventory control records also need information about all merchandise that has entered the store and left the store. They also need information about any changes in the merchandise price.

Most stores have prepared many kinds of forms to keep track of store inventory. See Table 1.

Dollar Control Records To prepare dollar control records, which are concerned only with dollar values, the total quantity of merchandise counted is multiplied by its price. Notice the example given on page 175.

■ **Table 1**
Inventory Control Forms

	Information Recorded	Form Used
Incoming merchandise—Added to Inventory	Merchandise orders	Purchase order
	Merchandise arrivals	Receiving record
	Customer returns	Merchandise return or credit slip
	Transfers from other stores	Transfer form
Outgoing merchandise—Subtracted from Inventory	Sales	Sales slip or stub of price ticket
	Transfers to other stores	Transfer form
	Returns to vendors	Vendor claim form
Other—Added or Subtracted from Inventory	Retail price changes	Retail price change form
Other—Subtracted from Orders	Canceled orders	Order cancellation form

Class	Quantity	Price	Total
70	764	$24.99	$19,092.36
74	612	$48.00	$29,376.00
82	831	$36.50	$30,331.50
			$78,799.86

To keep the dollar control records current and up to date, the value of all incoming merchandise is added to the dollar value of stock. All outgoing merchandise is subtracted from the stock. Markdowns and markup cancellations are subtracted from stock; markups and markdown cancellations are added to stock.

Suppose, for example, that the total dollar value of the initial inventory in the boys' department is $77,895. During one month of business, $23,364 of merchandise was added to the store inventory. In the same month, $29,655 of merchandise was subtracted from the store inventory. The current dollar value is $71,604.

Initial stock count		$77,895
Merchandise added to stock:		
Merchandise received	$22,380	
Customer returns	304	
Transfers from other stores	680	+ 23,364
Merchandise subtracted from stock:		
Sales	24,835	
Returns to vendors	740	
Markdowns	3,650	
Transfers to other stores	430	– 29,655
Dollar value of inventory		$71,604

Inventory control records must reflect accurately the merchandise in stock. Any errors in counting, recording, and math can affect merchandise and profit decisions.

Experience 12-2 ■ How Much Is in Stock?

Using the following information, determine the dollar value of merchandise at the end of August in the shoe department at Michael's.

Stock count on August 1: $48,342

Merchandise transferred to other stores: $632

Merchandise transferred to your store: $1,431

Sales: $22,658

Merchandise received from vendors: $15,438

Merchandise returned to vendors: $341

Markdowns: $2,385

Merchandise returned by customers: $2,060

Dollar value of merchandise: $______

Unit-Control Records As mentioned in Chapter 11, "Buying and Pricing," there are many different types of unit-control records. Some records are prepared by computer; others are prepared by hand. The information recorded on all unit-control records, which tells about orders, sales, returns, purchases, and merchandise transfers, is basically the same.

Suppose you are responsible for the unit-control records at Michael's, which are prepared by hand. You must record the information as follows (see Figure 12-3):

Use the buyer's purchase orders to record information about merchandise orders. Record the merchandise department, class, manufacturer, vendor, delivery date (completion date), and the date the order was written on the form. (See letter A.) Also record the style, color, and sizes of the merchandise ordered. (See letter B.) You record this information by marking a small dot in each square to show a certain size, color, and style is on order. For example, Figure 12-3, letter C, shows that five medium black turtlenecks are on order.

Use the receiving record to record information about what merchandise is actually received. Information listed on the receiving record may differ from the purchase order because vendors do not always send all the merchandise that was ordered at one time. Sometimes they have to back order items that are not yet available. You record information about items that are received by making a straight line in the appropriate squares. (See letter D.)

Use the price tickets from sold items to record information about what merchandise is actually sold. Suppose that whenever a sale is made in your department, one section of the price ticket is torn off the item and placed in a box next to the cash register. Use these tickets to record sold merchandise. Make a diagonal line to represent each sold item. (See letter E.)

Use merchandise transfers, merchandise returns to vendors, and purchase order cancellations forms to record transactions that reduce inventory. Circle the items that are leaving your inventory and indicate why these items are being removed from stock. Remember *t* means transfer and *v* stands for merchandise returned to vendors; a circled dot with a *v* stands for a canceled purchase order. (See letter F.)

Use customer return slips to record customer returns Customer returns increase inventory. When a customer returns an item that can be sent to the selling floor, erase the sold mark. By doing this, the unit-control records show the item is again available for sale.

Use retail price change forms to record price changes of merchandise Cross out the old price and write in the new retail price on the unit-control record. (See letter G.)

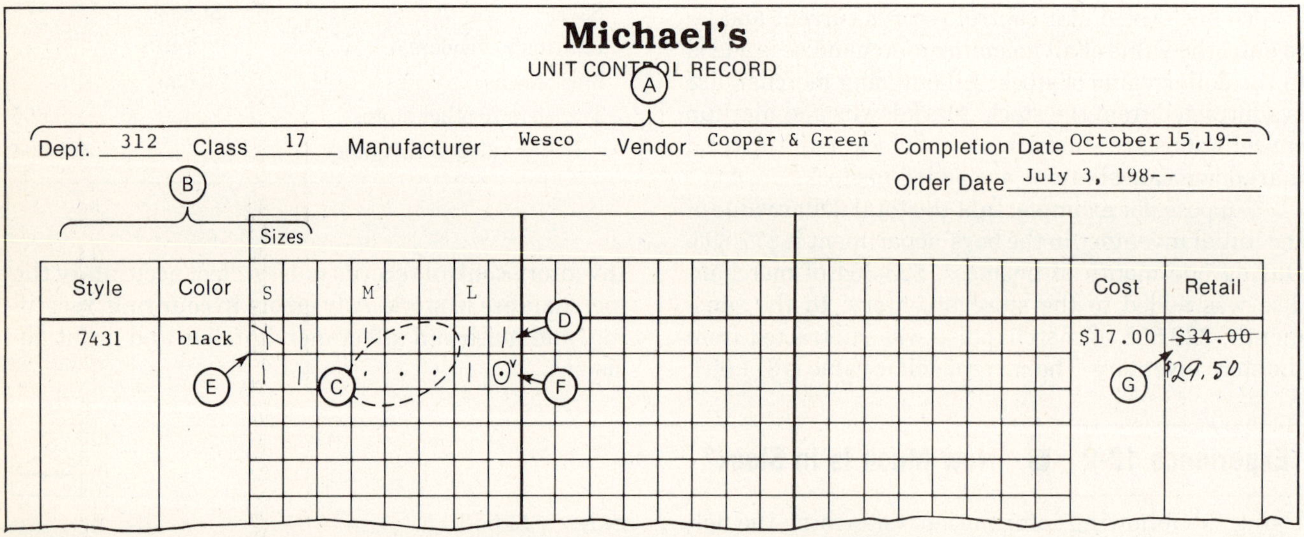

Figure 12-3 What types of information were used to fill out this unit-control record? (See the discussion above for the answer.)

Experience 12-3 ■ Completing the Records

Using the following information, complete the unit-control record shown below.

1. On February 4, you purchased an assortment of short-sleeve broadcloth shirts in white, blue, and

pink. The shirts, which are sized S, M, L, and XL, are to be delivered on April 15. You purchased two small, four medium, four large, and two extra large shirts in each color. The manufacturer and vendor are Aristocrat Shirt Company. The style number is 8241; the department number is 811, the class number is 72. The cost of the shirts is $9.50. The retail price of the shirts is $22.00.

2. The shirts arrive on April 8. Two size medium blue shirts are missing.

3. During the week of April 12 to 18, you sell all the medium and extra large pink shirts, one small and two large pink shirts, three large blue shirts, and two medium and one large white shirts.

4. One customer returns a medium pink shirt for a large pink shirt. You return the medium shirt to the selling floor.

5. Another branch store calls and asks for a transfer of one extra large white shirt.

UNIT CONTROL RECORD

Class ______________ Manufacturer ______________ Vendor ______________ Completion Date ______________

Order Date ______________

Style	Color	Sizes	Cost	Retail

Analyzing Inventory Control Records

Inventory control records provide management and buyers with important information about sales, shoplifting and employee theft, and merchandise reorders. As a clothing and accessory employee, you can use the information from inventory control records to prepare financial reports and determine what merchandise should be reordered or closed out. These records can also help you pinpoint some problems with your store's security procedures.

Determining Units Sold From Stock Counts Inven tory control records, used in conjunction with stock counts, allow a store owner, manager, or buyer to determine how many units were sold in a given period. To compute what has sold in the past:

- Physically count and record all stock on hand at the beginning of a period.
- Add to this amount all the merchandise that has arrived during the period. (This includes all merchandise from vendors, transfers from other stores or departments, and customer returns.)
- Subtract all merchandise that has left the store, such as returns to vendors and transfers to other stores.
- Physically count and record all stock on hand at the end of the period.

(*Note:* Because this is a unit count, it is not necessary to subtract markdowns or any changes in price.)

For example, suppose you want to know the sales of men's slacks during February. On the first of the month, the store would take an actual count of all men's slacks in the store. It would add to that amount the merchandise received and subtract the merchandise that left the store and the quantity still on hand at the end of the month. The following procedure would be used to calculate the units solds.

Merchandise on hand, February 1		1,240
Amount of merchandise arriving during February		
From vendors	749	
From transfers	18	
From customer returns	24	+ 791
Amount of merchandise leaving during February		
By returns to vendors	82	
By transfers	111	− 193
Merchandise on hand February 29		− 1,138
Units sold during February		700

Experience 12-4 ■ How Many Did We Sell?

Using the following information, compute the unit sales during May for the men's basic underwear department at Michael's.

Stock count, May 1: 1,274

Stock count, June 1: 1,349

Merchandise received from vendors: 308

Merchandise returned to vendors: 54

Merchandise received from transfers: 17

Merchandise transferred out of store: 24

Merchandise returned by customers: 11

Units marked down: 380

Unit sales in May:______

Using Inventory Records to Reorder Merchandise

One of the most important functions of inventory control records is to help buyers determine what merchandise should be reordered. Unit-control records, which specifically show merchandise sales, markdowns, and customer returns for certain items, can help buyers identify what the customer demand is for merchandise. Fast-selling merchandise or merchandise that has a strong sales record should be in sufficient supply to meet customer requests. Reorders should be placed before all merchandise is gone, or future sales will be lost. Slow-selling merchandise or merchandise that must be marked down before customers will buy it often indicates low customer demand. Buyers probably will not want to purchase more merchandise that, in the past, has been significantly marked down. Buyers may also hesitate to purchase a certain item if customer returns have been high.

Using Inventory Records to Determine Shortage

Most clothing stores take a physical inventory of their total stock twice each year, once in January and once in July. After the dollar value of the counted merchandise is computed, it is compared to the inventory records. If the value of the merchandise counted is less than the records, or "book," show, there is a shortage. For example, suppose the physical inventory in the knit hat and glove department is $17,439. The book inventory is $17,842. The difference, $403, is shortage.

Shortage can result from dishonesty, paperwork errors, or waste. For example, dishonest employees may let friends take two items when charging for one or take money out of the cash register. Customers who shoplift also contribute to shortage.

Taking Basic Stock Counts To be sure that the store is never totally out of basic stock items, some stores have salespeople take basic stock counts once a week, every 2 weeks, or once a month. The salespeople count the items listed on the basic list and record the on-hand stock.

The "on-hand" basic stock figure is usually given to the assistant buyer or buyer, who determines the amount of merchandise to reorder. Sometimes an experienced salesperson computes the reorder. Computing basic stock reorders is discussed in Chapter 11, "Buying and Pricing."

■ RECEIVING, CHECKING, AND PRICING MERCHANDISE

How merchandise is received, checked, and priced at a clothing and accessory store is another part of inventory control. If merchandise is not received and checked properly, the inventory control records will show inaccurate information. Incorrectly marked price tickets can seriously affect the store's sales and profit.

Receiving Merchandise

To make sure that all incoming merchandise is entered into the inventory control system, many stores require all new merchandise to follow certain receiving procedures. These stores realize that tight control of the merchandise must begin as soon as the goods arrive at their receiving docks.

Jon Trentman supervises the receiving, checking, and pricing of merchandise at Michael's. He is just beginning to explain the receiving procedure to Sandra, a new employee: "We receive many shipments of new merchandise each week. These shipments are delivered to our store as soon as they arrive in town. Some of the merchandise is delivered by motor carriers, such as P.I.E. Nationwide or Yellow Freight systems. Sometimes, shipments are sent by air, with companies such as Emery Air Freight, and then delivered to our store by truck. We also receive deliveries of packages that weigh between 1 and 50 pounds from companies such as United Parcel Service. All these trucks bring the shipments to our receiving dock, which is located at the back of the store."

Receiving and Delivery Schedules Clothing and accessory stores receive many merchandise deliveries from vendors. At stores that have more than one unit, there is usually much merchandise being transferred from store to store. At companies with a regional distribution center, merchandise is sent to the stores within the region each week. Some clothing stores also provide delivery service of merchandise to their customers.

To coordinate the flow of merchandise in and out of stores and their distribution centers, clothing and accessory stores establish a receiving and delivery schedule. This schedule, which shows when merchandise will be picked up and delivered, allows for the efficient movement of merchandise. Managers, salespeople, and buyers refer to the schedule to know when new merchandise shipments and transfers are expected. Managers use these schedules to staff the store with enough stock workers to unload large shipments.

Every morning, a Michael's delivery truck visits each of the five local Michael's stores. This truck picks up merchandise that is to be transferred to other stores and drops off the transfers and newly-received merchandise shipments.

Receiving the Freight Bill After a driver pulls a truck up to the receiving dock, the driver hands the store receiver a freight bill. Jon tells Sandra, "As a receiver, you should check to be sure the merchandise being delivered really belongs to our store. Our store name should be written where it says consignee."

Unloading the Shipment Many receivers help the driver unload a merchandise shipment, especially when the shipment is large. Safety procedures should be followed at all times.

Checking the Freight Bill The freight bill shows the transportation charges due. Usually, clothing and accessory stores must pay the freight cost for shipping merchandise. Jon continues his explanation of receiving procedures for Sandra. "After the boxes are unloaded, you must count the number of boxes that are actually received. Then compare the counted quantity to the quantity listed on the freight bill. If the numbers agree, sign the bill and date it. If the numbers disagree, write the actual number received and sign your name and date. If all the boxes are not received, a copy of the freight bill is sent to the transportation company as evidence of the shortage. You are to keep one copy of the freight bill; the driver keeps the others." (See Figure 12-4.)

"Sometimes," Jon says, "I open a box or two and examine the merchandise before signing the freight bill. If any of the boxes have been opened or if the

Fast Transit Inc.
3248 Soda Drive
Los Angeles, CA 90053
Phone 213-492-1083

FREIGHT BILL 138114
PRO NUMBER S 6295S
DATE 2/17/--

CONSIGNEE	DESTINATION	
Michael's	1211 Main Street	Miami, FL 33100

SHIPPER	POINT OF ORIGIN	SHIPPER'S NO.
Martin's of California	78 Los Angeles Blvd., Los Angeles, CA 90015	720

NUMBER OF PACKAGES, ARTICLES, AND MARKS	WEIGHT	RATE	FREIGHT CHARGES
5 Boxes, sweaters NMFC 49880	65	$74.35	$74.35 Collect
RECEIVED 4 BOXES, 1 BOX CRUSHED Jon Trentman 2-27- / Andrew Craig, driver			
TOTAL PIECES 5 — *ICC REGULATIONS REQUIRE PAYMENT OF THIS BILL WITHIN 7 DAYS*	TOTAL WEIGHT 65	$74.35	← PAY THIS AMOUNT

Figure 12-4 When checking freight bills, such as the one at the left, it is important to compare the actual quantity of the shipment with what appears on the freight bill.

tape or seal of any box is broken, I always open the box and check the condition of the merchandise. Opened boxes are a clue that pilferage or theft may have taken place.

"If any of the boxes are crushed or damaged, be sure to write that information on the freight bill. (See Figure 12-4.) Make it a habit to inspect the merchandise inside the damaged boxes while the driver is still there. Have the driver initial your notation about damage on the freight bill. The accounting office will file a claim with the transportation company for the damaged merchandise.

Completing the Receiving Record Michael's, like most clothing and accessory stores, keeps a record of all inventory shipments on a form called a receiving record, "key/rec," or apron. "This preprinted number in the top left-hand section of the form is our control number," Jon says. "It is used for all future identification of that shipment. (See Figure 12-5.) When a shipment is received, the receiver locates a copy of the buyer's purchase order. These orders are filed by the vendor. Then the receiver completes the top line of the receiving record by writing:

1. The department number from the purchase order (see letter A).
2. The store number, also from the purchase order (see letter B).
3. The name of the shipper or the company that sent the merchandise (see letter C).
4. The place the merchandise was shipped from (see letter D).
5. The name of the company that transported the merchandise (see letter E).
6. The freight bill number (see letter F).
7. The number of boxes of merchandise received (see letter G).
8. The weight of the shipment according to the store (see letter H).
9. The weight of the shipment according to the transportation company or carrier (see letter I).
10. The cost of transporting the merchandise (see letter J).
11. The date the merchandise was received (see letter K). This information is easy to find on the freight bill. (See Figure 12-4.)

Prioritizing the Shipments Once a shipment is unloaded, it must be processed, that is, checked, priced, and sent to the selling floor. When multiple shipments are waiting to be processed, a priority order should be established. The shipment with top priority is the merchandise that is featured in soon-to-be-run advertisements. Next pri-

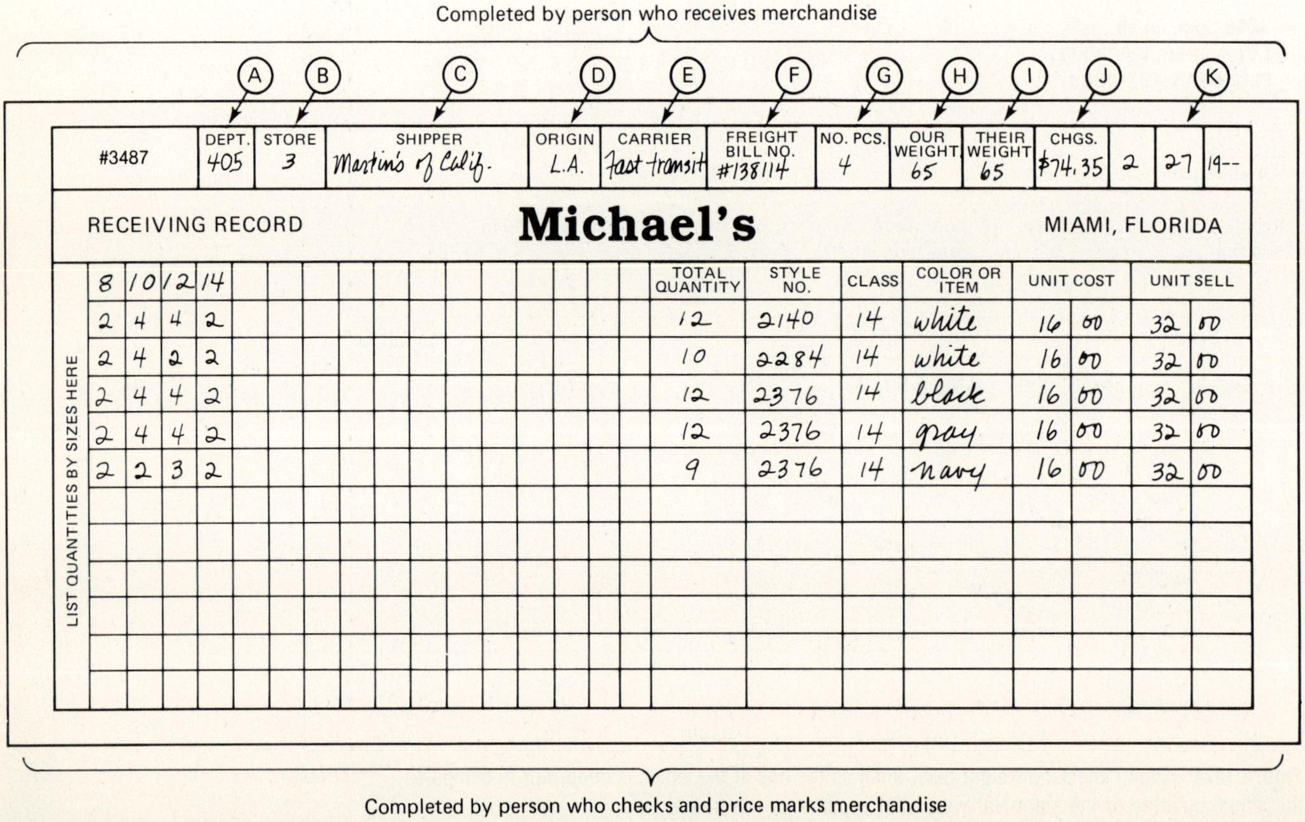

#3487	DEPT. 405	STORE 3	SHIPPER Martin's of Calif.	ORIGIN L.A.	CARRIER Fast transit	FREIGHT BILL NO. #138114	NO. PCS. 4	OUR WEIGHT 65	THEIR WEIGHT 65	CHGS. $74.35	2	27	19--

RECEIVING RECORD **Michael's** MIAMI, FLORIDA

LIST QUANTITIES BY SIZES HERE

8	10	12	14	TOTAL QUANTITY	STYLE NO.	CLASS	COLOR OR ITEM	UNIT COST	UNIT SELL
2	4	4	2	12	2140	14	white	16 00	32 00
2	4	2	2	10	2284	14	white	16 00	32 00
2	4	4	2	12	2376	14	black	16 00	32 00
2	4	4	2	12	2376	14	gray	16 00	32 00
2	2	3	2	9	2376	14	navy	16 00	32 00

Figure 12-5 From where does the information on the top line of a receiving record come?

ority goes to fashion items because of their shorter selling life than basic stock. High-fashion clothing stores often negotiate with vendors for first shipment so they can be first to offer the latest fashions to their customers. Shipments that contain missing coordinate pieces should also be given high priority. For example, suppose the jackets and blouses of a coordinate group have already arrived, without the skirts, pants, and sweaters. When the matching pieces arrive, they should be rushed through processing. The group will sell better when customers can choose from all pieces. Low processing priority is given to basic stock reorders, also called fill-in stock. There should be a sufficient quantity of this merchandise available for sale because basic stock is ordered on a regular schedule.

Checking New Merchandise

"After the new shipment has been recorded on the receiving record," Jon continues, "open the boxes with a box cutter. Be careful to cut only the edge of the cartons so that the merchandise is not damaged.

"There will be a packing slip with each merchandise shipment. (See Figure 12-6.) This form shows the actual quantity and description of merchandise shipped. Count the merchandise in each shipment and compare it to that listed on the packing slip. If the actual quantity received agrees with the quantity on the packing slip, write OK and initial the packing slip. If the quantities do not agree, write the quantity received and your name. (See Figure 12-6, letter A.)

"Once the merchandise is counted and checked to be sure it is not defective, send the freight bill, packing slip, and purchase order to Accounts Payable. If you have indicated on the packing slip that we have not received the full order, the company will pay only for the merchandise that we received. The buyer will follow up on undelivered merchandise.

"Also, if any of the merchandise in the shipment is defective, contact the buyer. The buyer must call or write the vendor and receive permission to return the defective merchandise. When permission is given, the buyer will tell Accounts Payable to deduct the cost of the defective merchandise from the bill."

PACKING SLIP

From: MARTIN'S OF CALIFORNIA
78 Los Angeles Boulevard
Los Angeles, California 90015

To: Michael's
1211 Main Street
Miami, Florida 33100

Customer's Order Number 741

QUANTITY ORDERED	DESCRIPTION	QUANTITY SHIPPED	QUANTITY BACK-ORDERED
12	White pullovers, style #2140	12	
12	White cardigans, style #2284	10	2
12	Black tunics, style #2376	12	
12	Gray tunics, style #2376	12	
12	Navy tunics, style #2376	~~10~~ 9	2

A — Only 9 shipped, Jon Trentman

Figure 12-6 If the quantities received in a shipment do not agree with those shown on the packing slip, write the quantity received and your name.

Experience 12-5 ■ Comparing the Merchandise With the Packing Slip

1. A shipment of blouses you received is shown on page 182. Compare the actual shipment with the packing slip below it. Record any problems on the packing slip and initial it.
2. Suppose you are to pay the vendor for the shipment. You check the buyer's purchase order and find that ten solid blouses (two 8s, three 10s, three 12s, and two 14s) and ten patterned blouses (two 8s, three 10s, three 12s, and two 14s) were ordered. Each costs $12.50. What would you pay the vendor? (Assume that your store doesn't pay for back ordered merchandise.)

____

PACKING SLIP

From: Shirt-Tale
2900 Creek Road
Dallas, Texas 75201

To: Michael's
1211 Main Street
Miami, Florida 33100

Customer's Order Number 328

QUANTITY ORDERED	DESCRIPTION	QUANTITY SHIPPED	QUANTITY BACK-ORDERED
2	Size 8, Solid white tie blouses	2	
3	Size 10, Solid white tie blouses	2	1
3	Size 12, Solid white tie blouses	3	
2	Size 14, Solid white tie blouses	2	
2	Size 8, Classic plaid shirts	2	
3	Size 10, Classic plaid shirts	3	
3	Size 12, Classic plaid shirts	2	1
2	Size 14, Classic plaid shirts	2	

Price Marking Merchandise

Once buyers determine the initial retail price at which clothing and accessories will be sold, they record the price on their copy of the purchase order. When the shipment arrives, the purchase order is pulled and, using that information, the price tickets are prepared. The season code printed on the ticket depends on the date the shipment arrives.

Many retailers are printing UPC codes, similar to those on grocery store items, on their price tickets. These codes are read by laser scanners and greatly speed the entry and increase the accuracy of the information recorded into the POS system.

Selecting Price Tickets Michael's uses plastic connector tickets, pin tickets, string tickets, gummed tickets, and wedge tickets to price mark its merchandise. The kind of merchandise determines the type of ticket used. For example, a gummed ticket would be a poor choice on a hanging blouse, but it is the best choice for boxed wallets. (See Figure 12-7.)

The plastic connector makes it almost impossible for dishonest people to switch price tickets. A special machine shoots the plastic connector, which is attached to the price ticket, into the garment. It is used on a wide variety of merchandise.

Because this ticket makes a small hole in the garment, it should not be used on delicate garments or leather unless it is attached to the garment's label or inside lining.

Most stores attach the plastic connector tickets to the left side of the merchandise. On blouses, the ticket is attached to the left sleeve on the inside seam

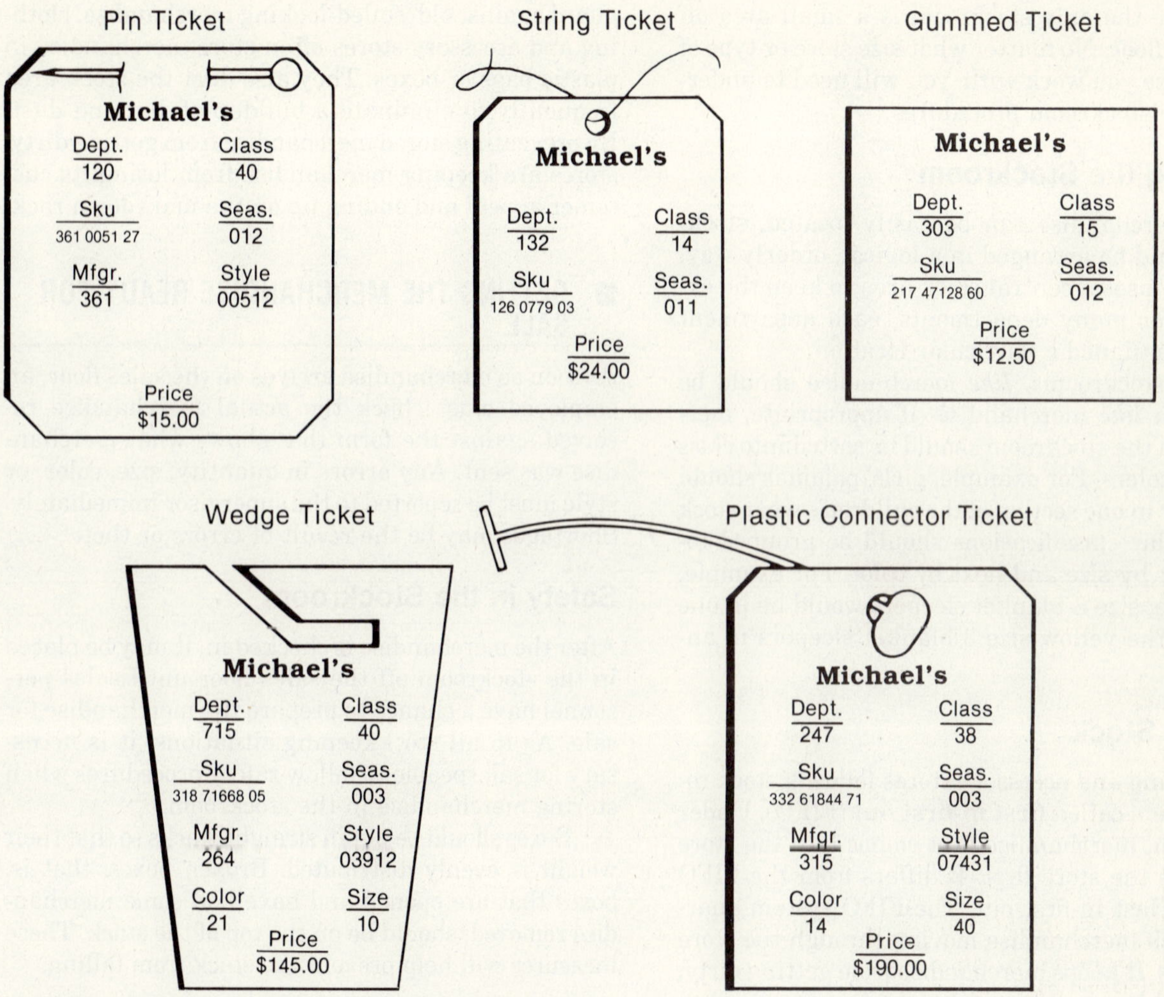

Figure 12-7 Some common price tickets are shown here.

above the cuff. On slacks, it is attached to the left side of the waistband, on the inside seam. On short-sleeve or sleeveless dresses, it is attached where the seams meet under the left arm. When merchandise is hung on racks and other merchandise fixtures, all the price tickets are hanging on the outside, in easy reach of the customers.

Some stores use string tickets or wedge tickets rather than the plastic connector ticket. They may attach the string ticket by sewing it to the waistband of skirts or slacks at the seams. String tickets are attached to handbags at the handle and to jewelry at the clasp. For lingerie, many stores use pin tickets so as to not snag the garment. Gummed tickets are attached to earrings, boxed merchandise, and perfume. The wedge tickets are used for leather and vinyl coats.

■ STORING MERCHANDISE

Buyers often purchase large quantities of basic stock merchandise. Because of the lack of space, only limited quantities of this merchandise may be placed on the sales floor. Large quantities of the extra merchandise is kept in a warehouse; smaller quantities are stored in a stockroom. When the stock on the sales floor begins to sell, merchandise from the warehouse or stockroom is used to replenish the missing items.

Sometimes merchandise arrives too early or is left over at the end of the season. For example, a large quantity of swimwear may be left over at the end of the summer. To make room on the selling floor for the winter coats, these swimsuits may be kept in a storage location until next year's swimsuit promotion.

Occasionally, buyers instruct salespeople to keep merchandise in the stockroom until the day before an ad runs. This practice ensures that a sufficient supply of the merchandise is available for the ad.

You may work at a clothing store that has a large distribution center where all merchandise shipments are received, checked in, price marked, and when necessary, stored. Or you may work for a very small store where the only stockroom is a small area off the selling floor. No matter what size store or type of merchandise you work with, you will need to understand basic stockroom procedures.

Arranging the Stockroom

So that merchandise can be easily located, stockrooms should be arranged in a logical, orderly way. If the store uses a central stock area to keep the extra stock for many departments, each department should be assigned a particular location.

In all stockrooms, *like* merchandise should be stored with *like* merchandise. If appropriate, merchandise in the stockroom should be sorted into class sizes, and colors. For example, girls' pajamas should be together in one section of the children's wear stock area. Similar classifications should be grouped together first by size and next by color. For example, all the blue, size 3 blanket sleepers would be in one group, all the yellow size 3 blanket sleepers in another.

Rotating Stock

Most clothing and accessory stores follow a stock rotation system called **first in-first out** (FIFO). Under this system, merchandise that comes into the store first leaves the store first. It differs from the LIFO system, or last in-first out. The FIFO system guarantees fresh merchandise moving through the store at all times. It keeps merchandise from getting dirty, fading, and losing its appeal to customers.

To use the FIFO system, refer to the season code on the price tickets that dates the arrival of the merchandise. For example, merchandise coded 907, or July 1989, is older than merchandise coded 007.

Following the FIFO system, all merchandise should be stocked so that the oldest items are in the front and the newest items are in the back. For example, men's undershirts with a season code 907 should be displayed in front of the same shirt with a season code 911.

Merchandise on the sales floor should also follow this stock rotation system. When additional undershirts are needed to replenish the ones that have sold, they should be placed behind any shirts with earlier season codes.

Keeping Store Merchandise Clean

Dust, dirt, and fading from sun or age can make merchandise lose its fresh look and customer appeal. To guard against old, soiled-looking merchandise, clothing and accessory stores often store merchandise in plastic bags or boxes. They also dust the stock area frequently to eliminate a buildup of dirt and dust. By preventing stored merchandise from getting dirty, stores are keeping merchandise from losing its customer appeal and ending up on the markdown rack.

■ GETTING THE MERCHANDISE READY FOR SALE

As soon as merchandise arrives on the sales floor, an employee must check the actual merchandise received against the form that shows what merchandise was sent. Any errors in quantity, size, color, or style must be reported to the supervisor immediately. Shortages may be the result of errors or theft.

Safety in the Stockroom

After the merchandise is checked in, it may be placed in the stockroom off the sales floor until sales personnel have a chance to prepare the merchandise for sale. As in all stockkeeping situations, it is necessary for salespeople to follow safety procedures when storing merchandise in the stockroom.

Boxes should be put in straight stacks so that their weight is evenly distributed. Broken boxes, that is, boxes that are opened and have had some merchandise removed, should be on the top of the stack. These measures will help prevent the stack from falling.

Changing Clothes Hangers, Buttoning, Belting, and Dusting

Salespeople should make merchandise look as attractive as possible so that customers will want to purchase it. Sometimes, this involves changing the hangers from the wire or plastic hangers sent by the vendor to the store's own hangers. Sometimes, salespeople add a foam padding insert to the hanger so that merchandise does not slip off its hanger or lose its shape. Dresses or blouses should be buttoned and belted before going on the sales floor; plastic, leather, or boxed merchandise should be dusted. Some stores steam out the wrinkles with a hand-held steam iron.

Chapter Notes

1. Holly Haber, "Levi's to Start Sticking On the Bar-Coded Labels," *Women's Wear Daily,* July 25, 1988; and Michael G. Ker, "Codes Speed Manufacturing," *Apparel Industry Magazine,* July 1987, pp. 54, 56.

CHAPTER 13

Promotion

It's not enough to purchase the right merchandise in the right quantities and offer it at the right price. To be successful, a clothing or accessory store must let customers know it has exciting merchandise and encourage them to come into the store and, ultimately, buy.

That's what Mel and Patricia Ziegler did when they started the Banana Republic Travel and Safari Clothing Company in 1978. Appealing to a market of affluent, well-educated, and well-traveled customers, they offered old military surplus clothing from around the world and new safari clothes deliberately outside the mainstream of fashion. These easy-to-wear, natural-fiber clothes were offered in sunproof colors only: khaki, gray, white, olive drab, faded indigo, and dusty plum.

When the store started, they couldn't afford much promotion. Nevertheless, they printed a catalog featuring Mel's descriptions (he's a former newspaper reporter) and Patricia's drawings (she was an artist). They distributed the catalog in the store and through classified ads. The catalog was written as the Zieglers' own travel journal and described the mysteries and surprises one might encounter while wearing safari clothing. It included witty and often irreverent descriptions of the clothes featured and read more like a novel than a retail advertisement for clothing. (See Figure 13-1.)

Although the catalog is no longer written, the Banana Republic's design and interior display continue to carry out the theme of travel. Some shops are designed as jungle outposts with thatched huts and bamboo walls. Other shops create a travel image by using old jeeps as display props. The clothing and accessories look right at home displayed on top of wooden crates with names of exotic destinations such as Zaire, Cairo, and Chad printed on them.

Banana Republic, which is now owned by The Gap, has successfully created and coordinated its promotion mix to appeal to its customers.[1] Other stores may use a different promotion mix to influence customers to stop and shop. One of your responsibilities as a clothing and accessory manager will be to design and coordinate your own promotion plan.

■ THE PROMOTION MIX

Promotion is the communicating of information about products, services, image, and ideas to influence purchase behavior. Promotion consists of advertising, visual merchandising, personal selling, sales promotion, special events, and publicity.

Advertising is any paid nonpersonal message that promotes a product, service, or idea by an identified sponsor. It's considered a nonpersonal message because it is one-way communication from the business to its customers. The ad in the Sunday paper that features the latest fall coats at a chain of junior wear stores is an example of advertising. Other examples of advertising appear in magazines and on radio, TV, and billboards.

Visual merchandising consists of window and in-store displays, store layout, and store decor, such as carpet, paint, mirrors, lighting, and **logo** (the distinctive design of the store's name that appears on its signs, shopping bags, and advertisements). The way a store looks and its merchandise presentation can attract customers, create an interest in the merchandise, and move the customer to buy. Chapter 14 discusses visual merchandising in detail.

Personal selling is the one-on-one communication between a customer and salesperson. The salesperson talks with the customer to determine what

Figure 13-1 The Banana Republic's catalogs appealed to the customers' spirit of adventure as well as their sense of style. Reprinted courtesy of The Gap, Inc.

merchandise the customer wants, presents that merchandise, answers any customer questions, and closes the sale. This personal form of promotion allows a salesperson to receive feedback from the customer and tailor the selling message to that customer's specific situation. (See Chapter 15.)

Sales promotion is any activity that supplements advertising, personal selling, and visual merchandising. Examples of sales promotion include special sales, coupons, and premiums (buy one, get one free). Sales promotion activities give customers an added reason to buy, in addition to receiving the product. For example, the promotion that offers $25 off any purchase of $100 or more gives customers another reason to buy *that product* from *that store* on *that day*—to save money. (See Figure 13-2.)

Special events (sometimes called public relations) consists of any activity that builds goodwill toward a business, with the goal of attracting more people to shop and buy at the store. Examples of special events are seminars presented to working women on upcoming fashion trends for the office, visits by celebrities to promote their particular clothing line, fashion shows for local organizations, and teen boards.

Publicity is the free and voluntary mention of a company, product, or service by newspapers, magazines, and radio or television stations. For example, the fashion articles that appear in many newspapers mention the names of stores that provide the clothing for local fashion shows.

Stores combine different promotion methods, also called the **promotion mix,** to increase the likelihood that customers will want to stop, shop, and buy. For example, a department store may promote its annual spring sale by sending catalogs to customers' homes, running full-page ads in the local newspaper, and airing 30-second commercials on TV. It may also place large signs at the store's entrance and smaller signs on top of racks and fixtures where the merchandise is displayed. It may hold training classes for salespeople to make them aware of the spring sale fashions and their selling points. It may even send charge customers a coupon good for 20 percent off their first purchase on the first day of the sale. By coordinating the advertising, visual merchandising, selling, and sales promotion activities, the store presents a single, unified message to customers. A coordinated promotion mix that consistently presents the same message will get customers to the store, remind them of why they came, and help the salesperson finalize the sale.

■ ESTABLISH A PROMOTION PLAN

A successful promotion plan is the result of thorough planning and careful execution. The steps involved in establishing a promotion plan are:

1. Determine the promotional objectives
2. Develop the promotional budget
3. Establish the promotional schedule

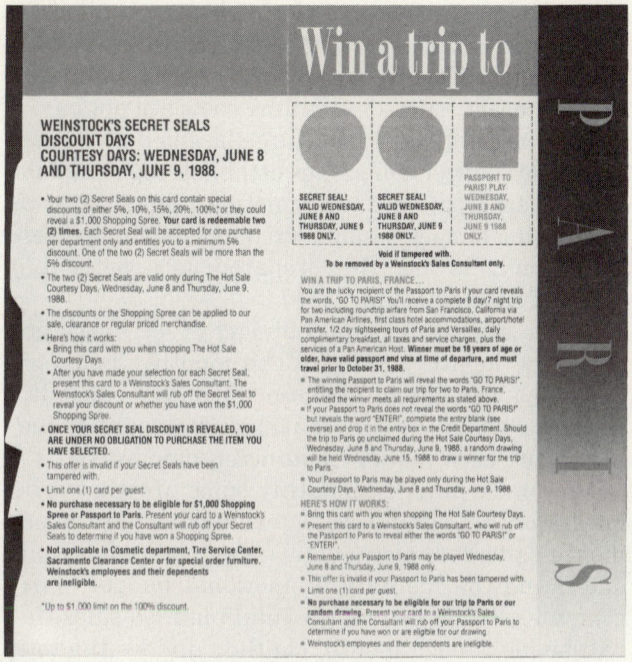

Figure 13-2 This direct mail ad gives customers an extra reason to shop at the store at a particular time. The ad is part of a coordinated sales promotion campaign. Courtesy of Weinstock's.

4. Select the merchandise to be promoted
5. Select the promotional method or methods
6. Assist in preparing the promotion (advertising, displays, and so on)
7. Coordinate the promotion mix
8. Evaluate the effectiveness of the promotion mix

■ DETERMINE THE PROMOTIONAL OBJECTIVES

The ultimate goal of every promotion is to increase the store's sales and provide a profit. Advertising, special events, and publicity encourage customers to come to the store. Visual merchandising, sales promotion, and the sales staff persuade customers to buy the merchandise. However, even the best planned and executed promotion mix may not result in improved sales and profit. Unless the other elements of the marketing mix, such as product, place, or price, are coordinated, customers may not purchase the merchandise featured in promotions. For example, the price of merchandise may be too high or the merchandise may not be on the selling floor because it was shipped late. Or the merchandise may be of poor quality, or there is an inadequate size or color assortment. Therefore, the promotion mix can only be held accountable for encouraging customers to stop and shop.

Before a company spends any of its promotional budget, it should identify its promotional objectives, or what it wants the promotion mix to accomplish. Some examples of promotional objectives are:

- Let customers know that the store has the latest fashions in stock
- Generate store traffic during slow selling times
- Encourage customers to come to the store during a major sale period
- Attract new customers to the store
- Encourage customers to buy regular-priced merchandise
- Encourage customers to buy prior season merchandise
- Educate customers on how to wear or accessorize new fashions
- Presell customers so salespeople have an easier time closing sales
- Build and maintain a particular image (such as fashion, quality, low prices, superior service, or any combination of these)
- Build customer goodwill
- Encourage customers to shop at a new location

In order to evaluate a promotion campaign, managers need to be able to measure results. Therefore numbers are often included in promotional objectives; for example, "To increase store traffic during the back-to-school promotion by 10 percent over last year."

Experience 13-1 ■ What's the Purpose?

Leaf through a local newspaper or fashion magazine. Locate and cut out or copy four clothing or accessory store ads. Each should support one of the following objectives:

1. Building or maintaining a fashion image
2. Generating store traffic during a major sale period
3. Presenting the latest fashion trends
4. Clearing out old merchandise

Attach the ads to a separate sheet of paper and write the promotional objective under each ad. On another sheet of paper, compare the ways the different ads look and are written. Do the ads make you want to buy the merchandise?

■ DEVELOP THE PROMOTIONAL BUDGET

The promotional budget is a plan of how much will be spent on promotions during the season. Like the six-month merchandising plan, the promotional budget is usually based on a season's planned sales. It is prepared twice a year. Store management budgets a percentage of anticipated sales to pay for promotional activities, usually 3 to 10 percent. The exact percentage depends on the season's promotional objectives, the amount spent in prior seasons, and competition. For example, opening a new store or adding a new line of merchandise requires a higher budget. Stores selling fashion goods also tend to budget a higher amount.

During the season, a manager should review the promotional plan to be sure it is meeting its objectives. The promotional budget can be adjusted up or down, depending on how sales are going. For example, when sales exceed plan, management may decide to increase the budget and feature a 16-page, full color, direct mail catalog, rather than a 4-page, black-and-white advertising spread in the newspaper. When sales don't meet plan, management may decide to reduce or reallocate a portion of the promotion budget.

Computing the Budget

Suppose an exclusive men's store called Jerome's had sales of $542,000 last fall season. The owner, Jerry Walsh, anticipates a 6 percent sales increase this fall, for sales of approximately $575,000. For the past three seasons, the store has budgeted 7 percent of its anticipated sales to pay for all promotions. This season's planned promotions are basically the same as last year's and there has not been a significant change in competition. Therefore, the store continues to budget 7 percent of planned sales. It allocates $40,250 ($575,000 × 0.07) for advertising, visual merchandising, sales promotion, and special events for the fall season. (Personal selling costs are not part of the promotion budget; they are classified as wages.)

(*Note:* Larger companies establish separate budgets for each part of the promotion mix, such as an advertising budget, visual merchandising budget, etc. These budgets are also prepared as a percent of planned sales.)

Once the season's total budget is set, Jerry prepares the monthly budget. The monthly budget is based on a percentage of planned monthly sales. It is then analyzed to be sure that there is enough money to support each month's planned promotions.

Jerome's planned monthly sales are shown below. (Monthly sales were determined when the six-month merchandise plan was prepared.) To determine the monthly promotional budget, Jerry computes the percent of the total planned sales that will be made each month. He then divides the total fall promotional budget of $40,250 into the same proportion as the percent of total sales. (See Table 1, Column B.) For example, because he anticipates that 13 percent of his total sales will occur in August, Jerry allocates 13 percent of his total promotional budget to August.

After computing this budget, Jerry analyzes each month's allocation to be sure there is sufficient money to support the planned promotions. For example, Jerry is planning a major suit sale in August. He knows that $5,233 is not enough money to pay for the newspaper advertising he wants to run. Jerry decides to reduce the December budget by 20 percent, to $9,789, and add the difference, $2,447, to the August budget. This additional money will allow him

■ **Table 1**
Monthly Promotional Budget

Month	A Monthly Sales	B % of Total Sales by Month	C Monthly Promotional Budget	D Revised Promotional Budget*
August	$ 75,000	13.0%	$ 5,233	$ 7,680
September	$ 58,000	10.1%	$ 4,065	$ 4,065
October	$ 79,000	13.7%	$ 5,514	$ 5,514
November	$140,000	24.4%	$ 9,821	$ 9,821
December	$175,000	30.4%	$12,236	$ 9,789
January	$ 48,000	8.4%	$ 3,381	$ 3,381
Total	$575,000	100.0%	$40,250	$40,210

*Due to rounding.

to purchase additional advertising and increase the chances of a successful suit event.

(Many clothing stores do not plan December's promotional budget on an exact percent of total sales basis. By reducing December's budget, a manager can spread out promotional money to the other months and provide extra support to other important selling events.)

The Advertising Budget

Most clothing stores use the majority of their promotional budget to pay for advertising. Also, the majority of their advertising budget is spent on preparing and placing newspaper ads.

Stores can increase their advertising budget through cooperative advertising. In "co-op" advertising, the manufacturer shares the cost of the ad with the store that promotes its product. The amount of cooperative advertising money offered to a retail store is usually based on a percentage of the *total merchandise orders placed* for that season.

Not all vendors offer co-op advertising. However, those who do must be willing to provide this advertising support to all of its distributors in equal portions. Often the advertising money is equal to 3 percent of the season's total orders.

The store's ads must meet the manufacturers' specifications before the co-op money is paid. Usually the manufacturer requires a copy of the ad before it sends the retailer its share of the cost. In larger stores, the advertising department keeps track of the co-op money; in smaller stores, the buyer or owner keeps the records and follows up with the vendors.

Cooperative advertising can be a real challenge to a clothing store. With "co-op," a store can advertise more often or place larger ads. Co-op advertising is also good for manufacturers because it costs less to advertise their merchandise through a local retailer than if they advertised themselves. (Local newspapers, radio and TV stations charge different rates for local and national advertisers. Local retailers pay less than national companies for advertising space or time.) However, a buyer must be careful that he or she is not lured by co-op advertising into buying the wrong merchandise. Also, if a store decides to use the advertisements provided by the manufacturer rather than prepare its own ad, the buyer must be confident that the image projected in the manufacturer's ad is compatible with the store's image.

Experience 13-2 ■ Prepare the Budget

As the owner of Jessica's, a small specialty shop, you are responsible for preparing the promotional budget. You have decided to budget 6 percent of the season's planned sales for promotion. Using the monthly planned sales shown to the right, compute the monthly promotional budget.

How could you increase the amount of advertising without increasing your promotional budget?

____

____

____

	Planned Sales	Promotion Budget
February	$ 4,800	________
March	$ 8,800	________
April	$ 7,200	________
May	$ 9,700	________
June	$ 8,400	________
July	$ 6,300	________
Total	$45,200	________

■ ESTABLISH THE PROMOTION CALENDAR

The promotion calendar is a written guide that outlines when a store intends to run the season's promotions. This calendar contains very confidential information. Competitors are most interested in when another store is planning to promote merchandise; they may then use the information to plan their own promotions.

When preparing the promotion calendar, the owner or manager considers the following:

- Holidays
- Merchandise classification promotions
- Manufacturer promotions
- Special store promotions

There are certain holidays during the year when a clothing and accessory store expects to have high sales. At these holiday times, customers buy clothing and accessory gifts for others or merchandise for themselves. For example, February is known for Valentine's Day; May for Mother's Day; and June for Father's Day, graduations, and weddings. Easter shopping may occur anytime between March and May, depending on the year. November and December are busy Christmas and Hanukkah shopping months. President's Day, Memorial Day, Fourth of July, Labor Day, Columbus Day, and Veteran's Day are other holidays when many clothing stores generate sales by featuring merchandise at reduced prices.

There are also traditional times of the year when merchandise classifications are promoted. Swimsuits are advertised in February, coats in late July. August is a peak selling time for children's and junior clothing because of the start of the school year. February promotions feature the latest spring fashions in men's, women's, and children's ready-to-wear. Fall fashions are promoted in August.

Some clothing and accessory manufacturers offer retailers markdown and advertising allowances to promote their merchandise. These promotions may occur once or twice a year, at approximately the same time. Customers can recognize these promotions from the ads of different local stores that feature identical merchandise at the same sale price. These promotions are common in women's lingerie and hosiery.

Clothing stores also create their own special promotions to entice customers into their stores and produce sales. There are one-day sales, day and night sales, weekend sales, end of month (EOM) sales, private sales for charge customers, spring and fall sales, anniversary sales, and sidewalk sales. These events are designed to stimulate customer excitement in new and reduced-price merchandise.

Clothing stores that want to promote their regular-price new fashion merchandise most often place ads in the Sunday newspaper. Friday is the most popular day to advertise weekend promotions that feature merchandise at regular and sale prices. Some stores establish one day each week when their catalog is distributed with the daily paper.

The promotional calendar for the women's, men's, juniors, young men's, children's, and accessories areas will be different, even in the same store. This is because the important selling times for each of these areas is different.

While the calendar is being planned, the manager must make sure there is enough money in the promotion budget to pay for the planned advertising, displays, and any special events. Sometimes, the amount of promotion must be reduced to stay within budget. Major promotions, including large ads or ads that appear more frequently, can be run during the months with the highest sales.

■ SELECT THE MERCHANDISE TO PROMOTE

There are a number of factors that affect the selection of merchandise featured in promotions. First, the promotional objectives influence the specific merchandise featured in promotions. If a store's objective is to encourage customers to buy multiple items, it may feature a collection of separates, showing a coordinate jacket, blouse, skirt, belt, shoes, and purse, not just a blouse or a pair of shoes. Stores that are trying to attract customers away from competitors may feature the same merchandise as a competitor and focus on its lower selling price.

Second, the season's current fashions also determine what merchandise will be promoted. At the beginning of a season, clothing stores will advertise the new fashion goods that their buyers purchased, not the merchandise left over from last season. A fashion store that sells basic stock will run ads that feature basic items in the latest colors and styles.

Third, the quantity of merchandise available for sale also determines which items are promoted. Any merchandise that is featured in ads, displays, or sales promotion vehicles should be available in a complete style, color, and size assortment.

Buyers are responsible for buying a sufficient supply of merchandise to back up an ad or other type of promotion. They use information about the quantity sold in past promotions when placing orders for new merchandise. Buyers are also responsible for coordinating the delivery of merchandise with promotional events. Sometimes delivery problems prevent merchandise from arriving in time for promotions. For example, merchandise may be shipped late from the manufacturer, or only a partial order may have been shipped, or the merchandise is still being processed at the company's distribution center. When these problems occur, the promotion should be canceled or different merchandise that is available for sale should be substituted. If the promotion isn't canceled or other merchandise is not substituted, the store will have many angry customers. Eventually, the store's promotions will lose credibility with customers.

■ SELECT THE PROMOTIONAL METHOD

As you know, a clothing store can use advertising, visual merchandising, personal selling, sales promotion, and special events to promote its merchandise. Most clothing and accessory stores rely on advertising and special events to attract customers to their stores. After they are in the store, retailers rely on visual merchandising, sales promotion, and the sales staff to encourage customers to buy.

Advertising Media

Clothing stores use print and broadcast media to advertise their merchandise. Print media consist of newspapers, magazines, direct mail, and outdoor advertising. Broadcast media is radio, television, and in-store video and announcements. Each of these media has advantages and disadvantages.

Newspapers Both large and small clothing stores advertise in newspapers. Newspapers reach a broad group of people who live in the area that a store serves. In fact, over 96 percent of consumers read the paper over a five-day period, and ads are the most popular part of the paper with women and second only to the sports section with men. Stores can request their ads to appear on certain pages, called preferred positions, or in certain sections so as to attract the attention of their specific target market. The time it takes to prepare a newspaper ad and have it appear in the paper is short. Also, newspapers can handle last-minute changes and cancellations, an important advantage to a store that is experiencing delivery problems. Merchandise in a newspaper ad can be photographed or illustrated and a detailed description of the clothing and accessory items can be written. Readers can refer to the ad later, after reading it for the first time, for additional details. A big advantage of newspaper advertising is it can produce immediate action. A reader can see an ad in the morning paper and rush to the store before it opens or phone in a charge card order.

Some disadvantages of newspaper advertising are that the quality of the illustration or picture is marginal and the color reproduction is poor. Newspaper readers represent a very broad group and may not fit the target market of a store as well as the audience of another advertising medium. Also newspapers have a very short life and are often read in a hurry. Readers may quit reading the paper before seeing a store's ad. Also, readers may not remember the ad because it is competing with so much other information in the paper. Nevertheless, newspapers are the most common advertising medium used by clothing and accessory stores.

Magazines Magazines offer great reproduction quality. A photograph can be clear, crisp, and in the color of the clothing featured. Magazines have a targeted readership, that is, people with similar interests subscribe and read a particular magazine. Many magazines offer regional advertising, allowing a store to place its ad in the magazines that are distributed in a particular geographic area, such as the Southwest, or New England states. Magazines also have a longer life than newspapers; they may sit on a reader's table for weeks and may even be passed around from person to person. National fashion magazines, like *Vogue, Glamour,* and *Gentlemen's Quarterly,* are viewed as fashion authorities, and stores that advertise in them enhance their fashion image. A disad-

Figure 13-3 Direct mail pieces include letters, billing inserts, invitations to special events, and catalogs. © Jules Allen Photography.

vantage of magazine advertising is the amount of time that elapses from submitting the ad to the date of publication. Clothing and accessory merchandise featured in magazine ads must be carefully selected because last-minute changes cannot be made. Also, magazine ads do not usually move a reader to immediate action.

Direct Mail Direct mail is any form of advertising that is sent through the mail to prospective customers. Examples of direct mail are charge statement enclosures, postcards and letters, personal notes from salespeople, and catalogs. (See Figure 13-3.) The use of direct mail pieces has increased dramatically over the last few years because of the advantages it offers. First, direct mail can be sent to a store's true target market rather than a broad market that includes people who are not potential customers. The mailing list comes from a store's charge account records or preferred customer files. Sometimes all the charge customers receive the direct mail piece, sometimes only charge customers who have made similar purchases to the items being advertised. Stores that do not offer their own charge account purchase mailing lists that include potential customers with similar demographics and life styles.

Another advantage of direct mail is that it can be timed to correspond with a major in-store promotion, thereby motivating customers to come into the store immediately. The direct mail pieces can be printed on excellent quality paper that show a clothing or accessory item's style and color to its best advantage. Direct mail catalogs are often placed at the store's entrance to remind customers of the merchandise featured. (Some stores use pages from their catalogs for print ads and save the production costs on preparing another ad.) And direct mail does not compete with other print material as newspapers and magazines do.

A disadvantage of direct mail is the increasingly high cost of mailing and printing. Maintaining a good mailing list is also very expensive. Because the quantity of direct mail cards, letters, and catalogs sent to a customer's home has increased so much, many customers consider direct mail to be "junk" mail. They throw away the direct mail pieces without opening them.

Outdoor Advertising Outdoor advertising, such as billboards and transit signs supplement the other advertising media used by clothing and accessory stores. (See Figure 13-4.) Billboards are an excellent reminder medium. They offer repetition of the message because the same people pass by on a regular basis. However, the people who see the message may not be the best fit of the store's target market. A billboard's message must be short and simple with an attractive illustration to catch the driver's or passenger's eye. Transit signs inside a subway or bus can also remind customers to shop at a particular store. Signs on the outside of a bus must be designed similar to billboards.

Figure 13-4 Outdoor advertising requires a short, simple message that the potential customer can read at a glance. Courtesy of Robinson's.

Radio Many clothing and accessory stores whose target market is teenagers use radio to promote their merchandise. Radio is also a good vehicle to reach a target market of working adults during "drive time," when they are driving to and from work. Advantages of radio are:

- It allows retailers to reach selected audiences (different radio stations appeal to a variety of interests).
- It is portable and can go anywhere the potential customer goes.
- It is flexible, and the retailer can choose the time of day when its message will air.
- No other message competes for the listener's attention.

Disadvantages are:

- The merchandise style, detail, and color cannot be seen, only described.
- After the message is over, there is nothing a listener can refer to.
- If the listener's mind is on other things, the entire message may be missed.

Television Major clothing and accessory stores use television to advertise their merchandise. Television can show the clothing and accessory merchandise in an exciting way, including how it should be worn and how it looks on the body. Because TV allows a viewer to see the merchandise and hear a description, it is similar to an expert salesperson presenting the merchandise. Cable television provides retailers with the ability to reach a more specific audience than the national networks. Stores can also select the time when the ad will be aired, thereby increasing the chances of reaching its target market. The cost of preparing television commercials and buying the air time is very high and prevents many companies from using this advertising medium. Most television ads are made and produced by the advertising agency of the manufacturer, and the local clothing firm inserts its name and logo into the commercial. This is one example of cooperation between the manufacturer and retailer. Also TV commercials are not available for reference; when a viewer misses something, he or she has to wait until the ad runs again. Sometimes the same ad is replayed so often that viewers are annoyed and no longer pay attention to the advertising message.

In-Store Video In-store video has become a popular advertising medium. Videos help promote the merchandise after the customer is in the store. Many manufacturers produce in-store videos and distribute them to retailers free of charge. Buyers are responsible for requesting these promotion vehicles. Some videos are educational, teaching customers how to wear a clothing or accessory item. Others present the latest fashion show from a famous designer.

In-Store Announcements In-store announcements about merchandise are the least expensive advertising medium. These announcements, which require a public address (PA) system, motivate customers who are in the store to stop and examine particular merchandise. An example of an in-store announcement is "Shoppers! Our entire selection of young men's T-shirts has been reduced 25 percent. These shirts are made of 100 percent combed cotton and are available in eight fashion colors. Be sure to visit our Young Men's Department on the second floor for this exceptional value." The stores that make in-store announcements often have a discount image.

Selecting the Advertising Media

Stores choose the advertising media that will accomplish their promotional objectives and reach the largest number of people in their target market at the lowest cost. A medium should cover their entire market area with a minimum of coverage outside their market. Stores match their target market's demographics and psychographics with the descriptions of circulation (readers), listeners, or viewers kept by newspapers, magazines, outdoor advertising companies, and radio and television stations. These descriptions usually include the sex, age, marital status, occupation, education, income level, interests, and lifestyle of the audience reached.

In addition to cost considerations, stores may decide against using a particular advertising medium because of its disadvantages. For example, some clothing and accessory stores avoid using radio because the merchandise cannot be shown. Some stores dislike billboards because they can't present an adequate selling message. Some avoid magazines because they aren't flexible to change. Other stores prefer particular magazines or newspapers because those publications enhance their fashion image. Some choose TV because of the excitement it can generate.

No matter what the combination of advertising media planned, its cost must be compatible with the promotional budget.

Experience 13-3 ■ In-Store Promotion

Managers may be responsible for preparing in-store announcements. Try your skill at preparing a message for one of the following clothing items:

- Women's scoop-neck tank top, made of body-hugging Lycra and polyester, available in red, yellow, pink, turquoise, and orange; sizes small, medium, and large; $14 in the Junior Department.
- Men's poplin slacks, belted with plain or pleated front in cotton/polyester blend, sizes 30 to 42 in assorted colors; regular price $24, sale price $18.99.

After writing the announcement, time it to be sure that is no more than 10 to 20 seconds. (Customers will not listen to long messages.) Read the advertising message into a tape recorder. Evaluate your message using the following chart. Circle the appropriate responses.

Description	No				Yes
1. Does ad effectively capture attention?	1	2	3	4	5
2. Does ad explain the merchandise benefits in detail, yet is short enough to keep the customer's attention?	1	2	3	4	5
3. Does ad flow smoothly?	1	2	3	4	5
4. When read aloud, is ad clear and easily understood?	1	2	3	4	5

Preparing and Proofing Ads

Some stores use an advertising agency to design and create ads. Other stores have a complete in-house advertising department that handles all the specific advertising tasks. Smaller stores work with the local newspaper advertising department to create their ads. In all these situations, the buyer or manager is the product specialist, providing the advertising professionals with information that will help them create customer interest in the merchandise.

Many clothing stores require buyers to complete a form that describes the merchandise to be advertised, the manufacturer, style, colors, fabric, size assortment, and price. At meetings with the advertising specialists who prepare the ads, they discuss the benefits that the merchandise offers customers and important fashion trends. (See Figure 13-5.)

The advertising specialists are responsible for designing and producing ads that feature the merchandise that store management has selected. For example, using information provided by the buyer, the creative director decides on the ad's theme or central idea. Copywriters create the headline, which attracts the readers' attention to the ad. They also write the copy, describing the merchandise style and assortment and emphasizing the benefits. Artists illustrate the merchandise. Layout specialists attractively arrange the illustrations, headline, copy, store logo, and store location and hours into the advertising space purchased and according to the image the store wants to project. Media buyers purchase space in the newspapers and follow up on any discounts they should receive or extra charges for color, ads placed in special sections of the paper, or preferred positions. (See Figure 13-6.)

After the ad is created, the buyer or assistant buyer is responsible for reviewing the ad to be sure it is accurate and all the important information has been included. Buyers read the proofs, or copies of the print ads, before they are sent to the newspaper, magazine, or catalog printers. Ads are proofed for the following elements:

Price Is the price correct? Smart buyers check the price ticket or their purchase order rather than rely on their memory.

Headline and Body Copy Has the copywriter omitted any important facts? Does the copy have a message that can be believed by the customer?

Size and Color Assortment Do the sizes and colors listed reflect the available stock assortment? The buyer should once again verify the availability of merchandise to back up the ad.

Illustration Is the picture of the merchandise realistic? If a person is pictured in the ad, does he or she resemble a typical user of the product?

ZCMI NEWSPAPER ADVERTISING INFORMATION

DEPT. Junior Sportswear DEPT. NO. 150 BUYER Margaret Clawson

	Day of Week	Date	Space
DESERET NEWS	Sunday	Aug. 21	half page
TRIBUNE	Sunday	Aug. 21	half page
STANDARD EXAMINER	Sunday	Aug. 21	half page
PROVO HERALD	Sunday	Aug. 21	half page
LOGAN HERALD JOURNAL			

IMPORTANT This form must be in the advertising office 18 days before running date (add one extra day for holidays)
Color ads and double trucks 21 days before running date

IF REPEAT ATTACH TEARSHEET

INCLUDED WITH AD COPY (check):
MDSE. ☒
GLOSSIES ☐
SLIX OR MATS ☐

IF OLD ART TO BE USED GIVE DATE OLD AD RAN ——

MAIL ORDER COUPON
YES ☐ NO ☒

CO-OP ☒Yes ☐No
☐Yes, BUYER YOU MUST:
(1) INCLUDE ALL CO-OP REQUIREMENTS
(2) ATTACH CO-OP FORM #280

ITEM: What is the item(s)? Jacket, shirt, pants

BRAND NAME(S): French Connection

IS THIS: (Check one) Regular Promotion X Sale ____

WHAT'S THE BIG IDEA? (What is the one thing you want the reader to remember about this product or products.)
The fashion look is menswear and plaids.

COPY INFORMATION Please include sizes, colors, fabrics (other vital information to meet FTC regs.), patterns. List unique selling points and how these are translated into customer benefits. List how to use the product to best advantage, or how to take care of it.

	SPECIFY "REG." "LIST" "IF PERFECT"	SELLING PRICE
Shirt-jewel neck, silk crepe de chine		$108
t-shirt with long sleeves in ivory,		
sizes 4-14		
Pants-100% wool - wide leg trousers		$156.00
with high waist, sizes 4-14, subtle		
plaid of olive/violet		
Jacket - shorter double-breasted jacket		$244.00
in large-scale plaid pattern, 100%		
wool, sizes 4-14 in olive/violet plaid		

Figure 13-5 This is an example of an advertising request. It should give all the information needed to prepare a desired ad. Courtesy ZCMI, Salt Lake City, UT.

Store Name, Location, and Hours Are the store name and logo, location, and hours the store is open included in the ad? Stores with multiple locations exclude those outlets that do not carry the merchandise.

Besides checking an ad for accuracy, buyers or their assistants should evaluate the ad's selling power also. Will the ad persuade customers in the target market to visit the store and examine the merchandise featured?

Special Events

Clothing and accessory stores use a variety of special events to attract customers to their stores, including fashion shows, customer seminars, contests, **trunk shows** (exclusive designer or vendor repre-

Figure 13-6 Can you identify the contributions of Fayva's advertising specialists to this ad? How does the ad reinforce the store's image? Courtesy of Fayva ®

sentatives show customers samples of the merchandise line and take orders), and celebrity appearances. Besides generating in-store traffic, these events also enhance the store's image in the community. (See Figure 13-7.)

When deciding to become involved in a particular special event, the store management should weigh the cost of the event, in terms of staff time and other expenses, against the return the event will bring. Every manager hopes a special event pays off with increased sales. Some events give the store good publicity that builds a loyal customer following.

NEWS

SPECIAL EVENTS SUMMER/FALL CALENDAR, 1988

MACY'S CALIFORNIA

Date	Day	Time/Location	Event
MAY			
31 - June 2	Tues. Wed. Thur.	Noon, San Francisco, East 1st floor	**GQ Magazine Seminars** A three-day program of fitness seminars, fashion segments and celebrity appearances highlights Macy's new Cycle Shop, featuring cycling accessories from the trendy Lycra riding shorts to helmets with changeable "skins."
JUNE			
3	Friday	12:30 P.M. San Francisco	**Robin Leach Personal Appearance** Robin Leach, the celebrated host of Lifestyles of the Rich and Famous, appears at Macy's to greet the public and sign photographs as a celebrity guest of Giorgio Beverly Hills men's VIP fragrance.
5	Sunday	7:30 A.M. Moscone Playground--Chestnut and Buchanan Streets San Francisco	**Macy's-AYH Great San Francisco Bike Adventure** Whether coasting, cruising or creeping, cyclists in the Great San Francisco Bike Adventure enjoy a city without cars along a 15-mile tour of San Francisco. Proceeds to benefit American Youth Hostels (AYH).
11	Tuesday	1:00 P.M. San Francisco	**Brazilian Dancers At Macy's** Macy's and Brazilian Fragrance tip their hats to the Mission Economic and Cultural Association's Carnaval '88 with a troop of Brazilian dancers.

Press Relations • 790 Market St., 5th Floor • San Francisco, CA 94102 • (415)393-3455

macy's

CALIFORNIA

Figure 13-7 This news release lists a variety of events that are planned to bring customers into the store and promote its image in the community. Courtesy of Macy's California.

Experience 13-4 ■ Give the Right Information

As the buyer of women's designer merchandise, you have purchased a complete assortment of a popular fashion designer's current line. You want to run a newspaper ad that features this merchandise.

1. Choose a current famous fashion designer.
2. Identify merchandise from that designer's current merchandise line. (Refer to a current *Women's Wear Daily,* the *New York Times* or *Los Angeles Times,* or local papers to locate pictures and descriptions of the merchandise.)
3. Explain the merchandise to another classmate, who will play the role of creative director. (A creative director is responsible for determining the theme of the ad.) Be sure to include a description of the fashion designer's philosophy as well as the price, color, style, fabric, and details of the design. Also provide some ideas of what you would like the headline and copy to say.

At the end of your presentation, ask your teacher to determine whether you provided sufficient information for the creative director to design the ad.

Experience 13-5 ■ Make It an Event

Plan a special event for a local clothing store or area within a local store, such as Women's Dresses. Possible ideas for the event are a fashion show for a local organization, a seminar that presents information on how to create a work wardrobe, or a Christmas coloring contest for kids. Look through newspapers or talk with clothing and accessory managers for other creative special event ideas that would persuade the target market to visit the store.

Use a separate sheet of paper to describe the details of the event. Include the following information:

1. A detailed description of the event, including what will happen first, second, third, etc. State where the event will be held, who will conduct it, when it will be conducted, any rules of the event, and how people will be notified about it.
2. The event's objective.
3. Description of the target market for the event (it should be the same as the store's target market).
4. Everything you need to create a successful event, such as refreshments, room, supplies, talent, etc.

■ COORDINATE THE PROMOTION MIX

A coordinated promotion mix increases the store's chances of achieving its promotional objectives. A coordinated mix includes (1) advertising, special events, and sales promotion that create sufficient interest and desire to bring customers into the store, (2) displays that remind customers why they came to the store, and (3) an informed sales staff that can answer questions and persuade customers to buy.

Coordinate Displays With Other Promotional Activities

After planning the promotional calendar, clothing stores often prepare a separate visual merchandising or display calendar that outlines when in-store and window displays will be built and taken down. By following the display calendar, a store makes sure the merchandise featured in ads or other promotion vehicles is prominently displayed and easy to find. Stores often place catalogs or large display signs at the store's entrance or other heavy traffic areas to remind customers about the featured merchandise. All displays and signing should be set up prior to opening on the day the promotion begins.

Coordination With the Sales Staff

Many sales are lost because the manager failed to tell salespeople about merchandise that was being promoted. Managers should hold regular meetings to discuss merchandise that will be featured in ads or other promotion vehicles. They should discuss questions that salespeople may be asked and possible answers to those questions, including where the merchandise is located and the availability of additional stock. Merchandise information, such as whether the item runs true to size, how the clothing should be cared for, and what other items are available to coordinate with the items featured, is also important. Often, managers role play with their salespeople a sales presentation for the featured merchandise, covering the price, brand name, available sizes, color and styles, and the merchandise benefits (what the merchandise will do for the customer). Managers also discuss special events and the salespeople's involvement in upcoming events and sales promotions. Many managers place the day's ads or catalogs near the cash wrap desk so that salespeople can refer to them throughout the day and provide accurate answers to customer questions.

If you work in a clothing and accessory store, make it a habit to look through newspapers for your store's ads. Also, walk through your department when you come to work each day. Make a mental picture of where featured merchandise is located and if items have been moved. This routine will help you answer customers' questions more thoroughly.

■ EVALUATE THE PROMOTION'S EFFECTIVENESS

Clothing stores use a variety of techniques to evaluate the effectiveness of their promotion mix. The most common evaluation technique is to compare the store or department sales generated during the promotion with sales prior to the promotion. An increase in sales can at least be partially attributed to the promotion mix. For example, suppose an in-store back-to-school event resulted in sales in the Junior and Young Men's Areas of $2,975. Sales for the same period prior to the promotion were $2,375. The promotion would be credited with increasing the sales.

In addition to the total department sales made during a promotion, most stores keep an advertising file to help them evaluate past promotions and plan future ones. It includes tear sheets (actual ads that have run in newspapers and magazines) as well as information about the weather, promotions of competitors, and any other facts that may have affected the sale of merchandise. The quantity of merchandise featured in the ad sold during the promotion is also recorded. (Quantity sold data may be taken from point-of-sale records or can be computed. To compute quantity sold, most stores count the merchandise on hand before the promotion began and the merchandise on hand 3 days after the promotion. The difference is the quantity sold.)

Many retailers rely on direct mail techniques to help them evaluate their promotions. For example, a store may send customers postcards or coupons that must be returned to receive a reduced price. After the promotion, the store computes how many customers who were sent the coupon actually used it (number of coupons redeemed divided by number of coupons distributed). This "redemption statistic" can be compared across promotions to determine which was more successful and should be repeated.

A store can also test the effectiveness of a particular ad by mailing an equal number of two different ads for the same merchandise to customers in its target market. Both ads would include a different color coupon that customers must return to the store to receive a special price. Based on the number of coupons received, a manager can determine which of the two ads produced the best response.

Even though there are many ways to evaluate the effectiveness of the promotion mix, determining how a particular promotion influenced sales is extremely difficult. Customers may not respond to an ad immediately. They may visit a store months after they see or hear about the promotion. Also, the promotion mix is not solely responsible for increased sales. The other elements of the marketing mix (product, place, price, and people) may be the reasons customers purchase clothing and accessories at a particular store on a particular day.

Experience 13-6 ■ Did It Work?

In the space provided, describe four ways you could evaluate the effectiveness of a direct mail catalog. Be specific and creative.

1. ______________________________

2. ______________________________

3. ______________________________

4. ______________________________

On a separate sheet of paper, describe the major problem with trying to evaluate a promotion's effectiveness.

Chapter Note

1. *Direct Marketing,* November 1985, p. 76; *Continental,* February 1987, pp. 21–24; *Savvy,* February 1986, pp. 42–46.

CHAPTER 14

Visual Merchandising

The shoppers on New York's Fifth Avenue take them for granted. But if you ask the out-of-towners, they'll tell you that New York has some of the most creative, attractive, thought-provoking store windows in the country.

According to Bernard Salms, business display manager at Bloomingdale's store on Lexington Avenue and 59th Street in Manhattan, "The windows are the first look a customer gets before she [or he] comes into a store, and it's that impression that makes that store different from any other."

Like other retailers, Bloomingdale's uses its windows to reinforce its advertising. "A combination of window display and advertising brings the customer in," Salms says. "Usually a customer sees the merchandise in the Sunday *New York Times*. We also put it in the window, and then when the customer gets into the department, there's usually another point-of-sale display there. So the whole thing ties in; it's like a whole story. The customers see it in the paper, they see it in the window, they see it on the floor, and hopefully, they'll buy it."[1]

Window displays are just one of the ways stores create their own special image, setting themselves apart from the competition. Like advertising and personal selling, displays are part of the store's promotion mix, and their success is measured by how the merchandise sells.

Display is defined as the visual presentation of merchandise. It belongs to a broader category referred to as visual merchandising. **Visual merchandising** includes all the parts of a store that customers see: the store's interior and exterior design, its fixtures, paint and wall coverings, carpet, lighting, mirrors, merchandise displays, and department layout.

Although all visual merchandising elements are important in helping a store create a unique image in the customers' mind, this chapter will focus on creating and maintaining displays.

■ DISPLAY

Clothing and accessory store displays can be classified into three types: window displays, interior displays, and self-selection displays. Window displays are designed to capture the attention of people passing on the street and persuade them to come into the store. Interior displays encourage customers to buy when they are in the store and are located close to the merchandise assortment and cash register. Self-selection displays include the shelves, tables, counters, and fixtures inside the store that hold the merchandise assortment. These displays are designed to allow customers to handle the merchandise.

Your responsibilities for creating and building displays will vary depending on the store where you work. In chain department stores, there is usually a separate visual merchandise staff that has been trained to arrange the merchandise and display props in an attractive way. The displays are planned by the visual merchandise director and are installed in the stores by visual merchandisers. Sales staff and managers are not permitted to build major displays. However, they may be able to create small counter top or ledge displays within their own departments. Although the displays in a department store company may all have the same theme, the actual displays are usually not identical. Each store's displays are adjusted to accommodate its unique layout and target market.

In chain speciality stores, the displays are designed at the company's headquarters by visual merchandise specialists for all the company's stores. Some chains hire display professionals who are responsible for installing the identical planned interior and window displays at stores in a particular geographic area. Other chains send a display kit to the store manager. The kit includes sketches of how the finished displays look and step-by-step instructions on how to build the displays. Merchandise to be included is specified. Special props or signs used in the display are also included. The store manager and the sales staff follow the instructions and build the displays exactly as outlined. The display kits are one of the ways chain stores are able to present a uniform image to the public.

In small, independently owned and operated clothing stores, the owner or manager is responsible for displays. This person may either do the planning and installation of the window and interior displays, assign the job to an employee, or hire a local display specialist. This type of store provides an entry-level worker the most opportunity to create and build displays.

At all clothing and accessory stores, the manager is responsible for the arrangement of merchandise and fixtures within the departments. With the help of the sales staff, managers move fixtures and place merchandise on different racks, tables, and counters almost daily. These changes make a department look like it just received new merchandise.

■ SELF-SELECTION DISPLAYS

Part of your job at a clothing and accessory store will be to build and maintain self-selection displays. Because customers choose merchandise from these displays, it is important that they be arranged in a logical, orderly manner. When merchandise included on tables, counters, shelves, or inside enclosed glass cases is neat and properly signed, customers can quickly find what they are looking for, without the help of a salesperson. In fact, one of the major benefits of self-selection displays is that salespeople can devote their time to customers who need help selecting merchandise.

Shelf, Glass Cube, and Table Arrangements

So that customers may quickly find the merchandise they are looking for, many stores establish stocking guidelines for their shelves, glass cubes, and tables. Usually stores organize their stock by classification, style, color, and size.

All the merchandise on a shelf, glass cube, or table should be the same classification or type of merchandise, such as all denim pants or all dress shirts. All the merchandise in one side of a cube or in a group of shelves should be the same style. For example, in the denim pant classification, all the pants on one side of the cube could be the button-fly, boot leg cut. The pants on the other side could be the zip front, regular cut. Merchandise should also be arranged by color from left to right according to the store's color guidelines. Many stores arrange their merchandise starting with the light colors on the left, then to the brights, and then to the dark colors. Solids should be kept separate from patterned merchandise. Finally, the merchandise should be sized vertically, meaning the sizes should progress from top to bottom and from left to right. Only one size should be in each bin or shelf. (See Figure 14-1.)

Sale merchandise is usually displayed on tables. In table arrangements, the merchandise is stacked in separate piles according to size. The stacks should not be so high that merchandise falls over, usually not more than 10 inches. Because table merchandise gets messy quickly, it has to be resized and refolded many times a day.

To keep their displays neat and attractive, many stores establish specific folding guidelines. Some stores provide their employees with folding boards, similar to clipboards without the clip, so that all the merchandise can be folded to a uniform size, have sharp creases, and be stacked neatly. Shirts and sweaters are usually folded so that the front is facing up and the sleeves are tucked inside. Pants are folded first by the crease and then just above the knee. Merchandise should be folded so that price and

Figure 14-1 When merchandise is displayed in groups of like styles, colors, and sizes, customers can easily find what they want. Courtesy of Benetton

size tickets are visible. This prevents customers from unfolding many items to find their correct size.

Counter Arrangements

Accessory items, such as inexpensive fashion jewelry, small leather goods, scarves, gloves, and hats, are often displayed on the top or inside of glass counters, also called caselines. Sometimes these items are displayed on special accessory fixtures provided by manufacturers. (See Figure 14-2.) These fixtures show off the merchandise to its best advantage because the fixtures are designed to showcase the merchandise and prevent items from falling off. Buyers are responsible for asking manufacturers for these fixtures and negotiating the price. Sometimes merchandise boxes are used to make counter displays, with the boxes stacked in piles and the top boxes open to show the merchandise.

Stores use locking caselines to display expensive accessory merchandise. When possible, most clothing and accessory stores include only one classification of merchandise per caseline, such as all designer handbags. In caselines merchandise looks most appealing when it is displayed in a symmetrical manner, meaning the arrangement of merchandise on the right is identical to the arrangement on the left. Both sides of the display are the same distance from the center and are in similar order and size. (This is also called formal balance.) Also, the merchandise should be organized by color from left to right when viewing the front of the caseline.

Locking cases help stores reduce shoplifting of expensive merchandise. Stores often establish written guidelines as to what merchandise must be kept under glass or behind counters. For example, a store may establish a rule that all gloves over $50 be locked in the case.

Rack Arrangements

Clothing is displayed on a variety of racks. (See Figure 14-3.) There are T-stands, half-round racks, three-way racks, quads, round racks, A-frame racks, faceouts, wall displays, and modular rack systems. Each fixture has a specific use.

T-stands are used to show new fashion trends, newly received merchandise and best sellers. They hold a limited quantity of merchandise. There are two kinds of T-stands: waterfall T-stands and straight T-stands.

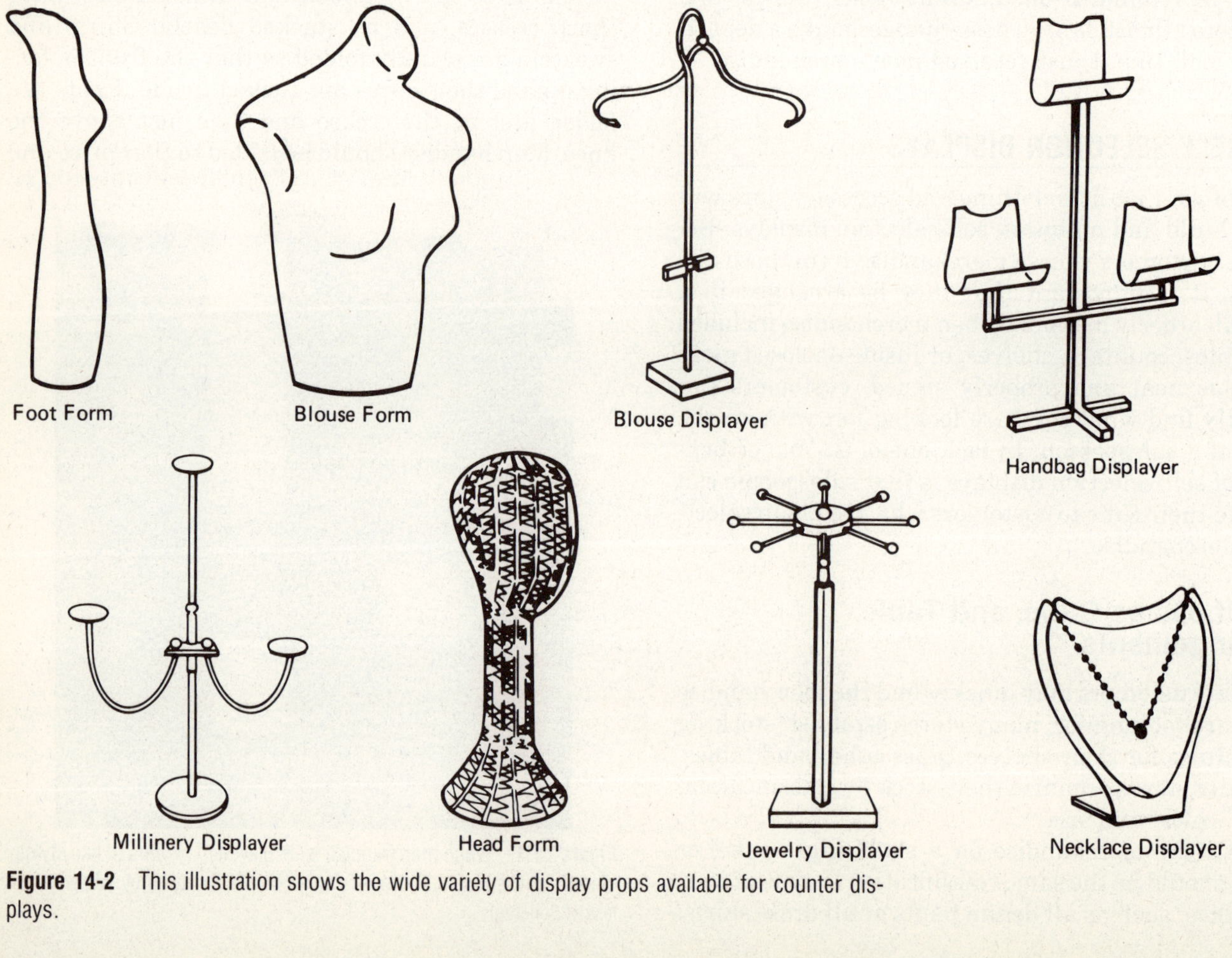

Figure 14-2 This illustration shows the wide variety of display props available for counter displays.

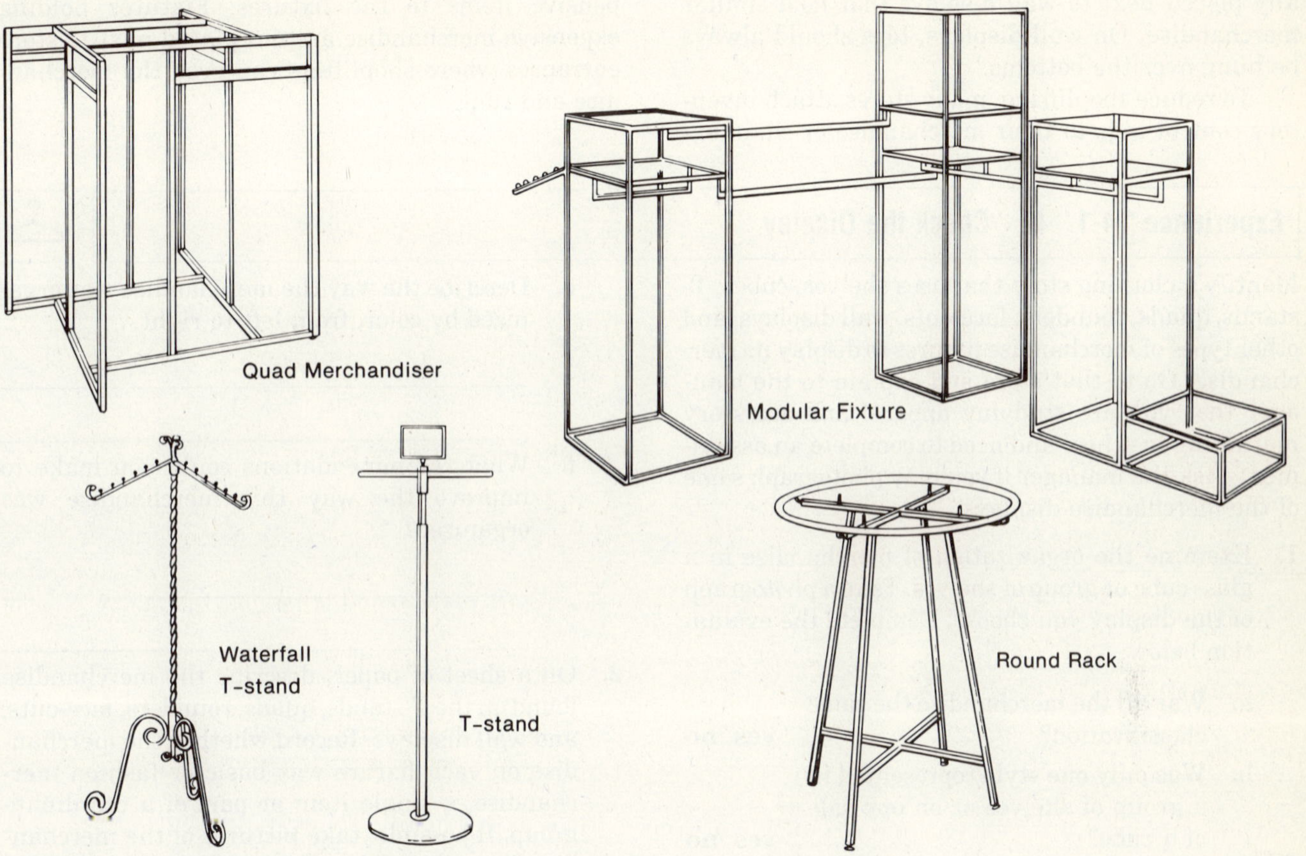

Figure 14-3 Display fixtures, such as those shown above, can be arranged in a variety of ways to give a new look to stores or departments.

Both can be adjusted in height and have adjustable arms. For the most effective presentation, follow these principles when displaying merchandise on T-stands:

1. The merchandise presented should be limited to one style or color per arm; for example, tops should be grouped on one arm, bottoms on the other arm.
2. Merchandise should be sized small to large, front to back, on each arm.
3. The fixtures should not be overloaded with merchandise.
4. The first top on a T-stand should be layered, such as a blouse that is layered with a vest and jacket.
5. Hangers should all be the same style and hanging in the same direction.
6. Arms on a T-stand should be adjusted so that the lower arm faces forward.
7. T-stands should be placed on the department's perimeter and along the department's interior aisles.

Half-round racks are used for men's suits or women's dresses. All merchandise should face in the same direction with the price ticket to the outside. The last item on the rack should face front-side out for maximum selling exposure.

Three-way racks are used to display coordinates. For example, slacks and skirts can be grouped on one arm, jackets and vests or sweaters on another, and blouses on the third. Merchandise should be sized small to large, left to right.

Quads are also used to display coordinates. For maximum selling, tops should be grouped on opposite arms, bottoms on opposite arms. Each arm should hold one color or style.

Round racks, also called **rounders,** work best when they contain only one classification, such as jackets. Usually merchandise is arranged by color in a counterclockwise direction and sized small to large. All merchandise should face in the same direction with the price ticket on the outside. Many stores place a glass shelf on top of this rack and, using a prop, display one of the items from the rounder.

Face-outs project from the wall and show the entire front of a clothing item. They work best when all the merchandise on an arm is the same style and color. On a waterfall face-out, there should be only one item of merchandise per stop. Face-outs are usually placed next to wall displays that hold similar merchandise. On wall displays, tops should always be hung over the bottoms.

To reduce shoplifting, many stores attach inventory control tags to their merchandise or chain expensive items to the fixtures. Fixtures holding expensive merchandise are not placed next to store entrances where shoplifters can grab the merchandise and run.

Experience 14-1 ■ Check the Display

Identify a clothing store that uses shelves, cubes, T-stands, quads, rounders, face-outs, wall displays, and other types of merchandise fixtures to display its merchandise. Go to that store and explain to the manager that you are studying apparel and accessory marketing in school and need to complete an assignment. Ask the manager if you may photograph some of the merchandise displays.

1. Examine the organization of merchandise in a glass cube or group of shelves. Take a photograph of the display you choose. Complete the evaluation below.

   a. Was all the merchandise the same classification? **yes no**

   b. Was only one style represented in a group of shelves or on one side of a cube? **yes no**

   c. Was the merchandise sized vertically? **yes no**

   d. Was only one size of merchandise contained on each section of the shelf or in each cube? **yes no**

   e. Describe the way the merchandise was organized by color, from left to right.

   ____

   ____

   f. What recommendations could you make to improve the way this merchandise was organized?

   ____

   ____

2. On a sheet of paper, describe the merchandise found on the T-stands, quads, rounders, face-outs, and wall displays. Record whether the merchandise on each fixture was basic or fashion merchandise, a single item or part of a coordinate group. If possible, take pictures of the merchandise fixtures you choose to analyze. Based on the information given in this chapter, was the merchandise in the store displayed on the correct fixture? What recommendations could you make to improve the way the merchandise was organized on the fixtures? Provide facts to support your answers.

■ SIGNS

Signs are a very important part of a store's promotional mix. They attract the customers' attention and help customers find the merchandise they want. To increase the chances that customers will purchase advertised merchandise, items that are featured in ads should always be signed. Signs must be set-up prior to the opening of business on a sale day and removed when the sale is over.

Depending on the store, signs may be requested by the buyer, visual merchandising staff, or manager. They are made by a professional printer. Placing the sign on the correct fixtures is the responsibility of the area or department manager.

Many stores establish uniform signing standards so that all their signs have a coordinated look and project the desired image. For example, a store may place poster-size signs (22 inches by 28 inches) in heavy traffic areas to remind customers of storewide events and major sales. They may use 5½ inch by 7 inch signs to describe the merchandise or indicate that merchandise on a particular fixture is on sale. These signs, also called "toppers" because they sit on top of counters and fixtures, can hold a one-half-inch heading and four lines of one-quarter-inch print. Upper and lower case letters are used to make the signs easy to read.

All signs should meet the following guidelines:

- All information on signs should be printed.
- The information on the sign, such as price and brand name, should be correct.
- All information should be correctly spelled.
- All signs should look new, that is they should be clean, clear, unmarked, and without wrinkles or bends.
- All signs should be placed in the correct-sized

sign holder; signs should never be taped or stapled to walls, racks, or fixtures.

- Sign holders that are visible to customers on both sides should contain two signs.
- All empty sign holders should be removed from merchandise racks and fixtures.

Signs that contain any errors should not be used. New signs should be ordered. Changes or additional information should never be handwritten on signs.

If you are responsible for deciding what should be written on signs, follow the guidelines in Figure 14-4. Always include information that customers want to know, such as what the merchandise will do for them. For example, the reason a customer would buy a scoop-neck sleeveless cotton sweater is that it's cool to wear. Customers are also interested in the price of the merchandise.

PREPARING MERCHANDISE SIGNS

A display sign should briefly state reasons that customers should buy the merchandise. When preparing signs, remember that just the name of the merchandise is not enough information. Avoid writing too much copy on the signs. Try to limit the words on a sign to two to three per line.

1. On the top line, write the reason the customer should buy the merchandise. This is the most important line of the sign. Write customer benefits. Ask yourself what the merchandise will do for the customer. If the merchandise item is the newest fashion accessory or never needs ironing, line 1 should say so.
2. Line 2 should identify the merchandise by name, brand name, or trade mark.
3. Lines 3 and 4 (or the additional lines on an 11-inch by 14-inch sign) should describe sales features that are not obvious to the customer. The construction, quality, ease of care, selection of colors, and size are ideas for these lines.
4. The last line of the sign may give the price and unit of quantity, such as $7.50 per package.

Figure 14-4 These guidelines will help you prepare copy for merchandise signs.

MAINTENANCE OF SELF-SELECTION DISPLAYS

One of the most common responsibilities of clothing and accessory employees is to keep the merchandise displayed on racks, counters, shelves, bins, and tables neat and orderly. The selling staff should check the merchandise many times a day to be sure it is properly organized according to the store's color and sizing standards. Signs should reflect the correct selling price. All merchandise on hangers should hang straight and be buttoned, zipped, and belted. All the hangers should face in the same direction and be spaced an even distance apart. Merchandise that is being returned to the floor from the fitting room should be made to look as though no one had tried on the clothing, and the merchandise should be returned to the appropriate rack. All folded merchandise should be folded according to the store's guidelines and look as if it was just placed on the shelf.

Salespeople also have housekeeping duties. All fixtures must be dusted and glass counters must be cleaned with glass cleaner every day.

Salespeople are usually responsible for replenishing basic stock fixtures with additional merchandise. The additional stock may be kept in **understock** (in the drawers under the counters or fixtures) or in **backstock** (in the stockroom). When refilling shelves, counters, or other fixtures, check the season code on the merchandise price tickets and bring the oldest merchandise to the selling floor. This FIFO stock rotation method will provide an organized flow of merchandise through the department and prevent old merchandise from accumulating in the stockroom.

When fashion goods are sold from displays, the salesperson may be asked to rearrange the remaining merchandise. For example, after a few weeks on the selling floor, the remaining pieces from a coordinate group will need to be color coordinated with other merchandise in stock. This will improve the sales of the existing merchandise and make room for newly received stock. Eventually, the clearance merchandise usually is placed on a rounder at the back of the department.

Experience 14-2 ■ Sign It

1. On a sheet of paper, write the information that you would include on a display sign that will appear on top of a rack of men's three-quarter-length suede lambskin coats. The coats have a zip-in acrylic pile lining as well as a rayon inner lining. The suede is specially treated to repel water and nonoily stains. Price: $250. Sizes: 36 to 44 regular and long. Color: tan and dark brown.

2. Prepare a 5½ inch by 7 inch sign for the suede coat display. If you have access to a sign-making machine, use the machine to make your sign. If not, use ¼-inch and ½-inch stencils and black felt-tipped pen to letter the sign on a sheet of unlined, heavy-weight paper or posterboard. Use upper- and lowercase letters. Be careful to make your signs neat, attractive, and accurate.

■ CREATING DISPLAYS

Mrs. Aron is the manager of Campus Corner, a speciality shop for junior-sized women. This week she ran an ad for three coordinate groups from a famous New York designer. To coordinate with the ad, Mrs. Aron decided to build various store displays. She also scheduled a staff meeting to inform the salespeople about the big promotion. (Mrs. Aron realizes that advertising, display, and personal selling all work together to produce sales. Advertising is the force that gets customers into the store. Displays catch customers' attention to come inside the store or department. And salespeople encourage customers to buy the merchandise.)

Mrs. Aron likes to give her salespeople a chance to plan and build displays. The salespeople like to prepare displays because it is a change from their regular job of selling and stock work. It also gives them a chance to express their creativity.

Before anyone builds window or interior displays, Mrs. Aron makes sure they carefully plan the displays, select the merchandise featured, choose the display location, and determine a display theme. She has them prepare sketches of what the display will look like and helps them evaluate the sketches to be sure that the display will accomplish its intended purpose. Then, with a clear idea of how the finished display will look, the salesperson assembles the merchandise, props, and other supplies and installs the display. Finally, the display is evaluated to see whether it achieved the desired result.

Display Planning

Clothing and accessory stores design their displays to convey a certain message to their customers. Window displays are often used to convey the season's fashion message, showing customers the latest styles or colors. Many windows show how new fashion merchandise should be worn. For example, when some customers were unsure how to coordinate shawls with their wardrobe, Mrs. Aron featured a shawl display in the window. This display helped customers feel confident about wearing a new fashion look. Because they have seen how to coordinate and accessorize a new look, customers are more likely to buy it.

The message or theme of some windows is the store's wide merchandise assortment. These stores include as many different looks in the window as possible. Sometimes stores want to communicate that they are having a sale. They place large signs that focus on price or sale items in their windows. (See Figure 14-5.)

Interior displays, which encourage customers to buy after they are in the store, may also have a variety of themes. Like window displays, interior displays may show how the merchandise should be worn, coordinated, and accessorized. For example, a display may feature mannequins dressed in the newest fashions and placed on a platform at the entrance to the department. An interior display can also be designed to show off the wide assortment of merchandise available, such as a wall display featuring identical shirts in eight different colors displayed on shirt forms. The purpose of another display may be to show how the merchandise looks in use. These displays are called **vignettes** and help customers imagine where and when the merchandise would be worn. For example, a vignette might display a mannequin dressed in a business suit and seated at an expensive executive-styled desk reading the *Wall Street Journal*. Customers who want to look like successful business executives may see themselves in the display and buy the same or similar clothing featured on the mannequin. A display may be designed to encourage impulse buying, such as buying a pair of earrings to match a new sweater, suggested by a display that features these items on a ledge behind the cash register. Another interior display may attract customers to the merchandise on a particular fixture, such as a bust form placed on top of a rounder to attract customers to a special purchase of blouses. Some interior displays are created to carry out a store's promotional theme. They feature merchandise from current ads and remind customers to purchase those items once they are in the store. Point-of-purchase displays are next to the merchandise assortment. They include stands, racks, or other types of display fixtures and signs that encourage customers to buy. (See Figure 14-6.)

Many stores tie in seasonal or storewide themes when they plan their displays. For example, at Christmas, a store may include identical Christmas trees with signs that say, "Make This a Christmas to Remember" in all of its displays. For its Mother's Day promotion, a store may include a sign saying "Thank You Mom…Mother's Day is May 8" in each of its women's clothing and accessory displays and ads. These promotional tie-ins present a uniform message to customers and give the store a coordinated look.

© Jules Allen Photography. Jules Allen Photography. © Views and Reviews.

Figure 14-5 Window displays give customers a preview of what the store has to offer.

The first step in creating effective, sales-generating displays is to identify the display's message or theme. The most effective displays have simple themes that are easy to understand.

Merchandise Selection

The exact merchandise selected for a display depends on the message the display is supposed to communicate. If a display is planned to show how a new fashion look should be coordinated and accessorized, the appropriate related items must be included. (Related items are items that naturally go together, such as a robe, nightgown, and slippers, or sweater, shirt, and tie.) If a display is designed to tell customers about the wide assortment of styles or colors, multiple styles or colors must be included in the display.

The number of items to feature in the display also depends on the store's image. Exclusive, high-fashion stores tend to feature single or relatively few items with a minimum number of props in their window displays. Moderate-priced stores often feature a wide assortment of clothing in their windows.

All clothing and accessories featured in displays should be seasonable and timely, representing the newest, freshest merchandise received. At high-fashion stores, the newest merchandise will be clothing and accessories at the beginning of the fashion cycle. At stores that sell moderately priced goods, the new arrivals will be in the rise or peak phase of the fashion cycle.

It is very important to select merchandise for display that is backed up by adequate stock on the selling floor. Customers who are attracted by a display expect to find additional clothing and accessories in a wide variety of similar styles and colors to make their merchandise selection. Many stores place T-stands and quads holding the full assortment next to a mannequin that is wearing the same clothing.

Another factor in selecting display merchandise is to choose items that are featured in upcoming ads. By coordinating their advertising and display calendars, stores give their merchandise maximum customer exposure. The display should be built no later than the day the ad or promotion begins.

Even when a store doesn't run ads, coordination between displays and other advertising is still possible. For instance, such magazines as *Seventeen, Vogue, Gentlemen's Quarterly,* and *Men's Wear* feature ads of the latest fashions. Customers who read these ads will remember them when they see the same merchandise displayed in the stores where they shop. Adding a sign, such as "As Seen in *Seventeen,*" to the merchandise display will also help remind customers about the fashions they just read about.

For your display at Campus Corner, you decide to feature merchandise from a coordinate group that you just checked in. The merchandise, which includes a

© Hambrecht Terrell, Int'l.

© Michael Lopez.

© The Gap.

© The Gap.

Figure 14-6 With a careful selection of fixtures and props, the whole selling floor can become part of an effective interior display.

cowl-neck pullover, matching cardigan, and lightweight wool slacks, is featured in an ad your store has placed in Sunday's newspaper. You have also seen similar sweater sets in many fashion magazines this month. The message you want to communicate is that your store carries this new, fashion look.

Mrs. Aron asks you what colors you will show. The coordinate group was purchased in three colors, dove gray, pale peach, and lavender. You realize that the color you choose depends on the accessory items that will be included in the display as well as the color of the display background. Dark or bright colors look best on a light-colored background. Light or neutral colors look best on a dark background. Because your background is dark, any of the merchandise in pale colors would make an attractive display.

If the background color were not appropriate, you would have to cover it with a material of another color before building the display.

Display Locations

In large clothing and accessory companies, display locations are determined by the store design department. This department decides the store's layouts, including where departments are located and where platforms, wall displays, vignettes, shadow boxes, and other display areas are built. (A **shadow box** is a small glass-enclosed case that is built into walls.) It chooses display locations in high traffic areas so that the merchandise featured receives maximum customer exposure.

What merchandise will be used for a display in a particular location is determined by the individual planing the display. Usually merchandise used in displays is contained within the area where identical or similar merchandise is stocked. For example, the logical location for a coat display is at the entrance to the coat area. This practice allows customers who are interested in the merchandise featured in interior displays to quickly locate and review the stock assortment.

When many different merchandise classifications are sold within the same area, the area's major display often features a group of related items. For example, the junior area of a large store may have a display that includes a casual top, slacks, vest, shoes, and fashion jewelry. Including related items in a display increases the chance that customers will want to purchase the "complete" look.

Some stores display merchandise outside the items' regular selling area, such as at the store entrance or in a heavily traveled aisle. These displays remind customers that the store sells the items and, hopefully, motivate customers to go to the department and make their merchandise selection. There are also "outposts" or major displays with an assortment of merchandise outside the items' normal selling department. Many large stores establish outposts at Christmas time. They feature commonly purchased items that are boxed, wrapped, and ready for gift-giving, such as knit hat and scarf sets.

The display location influences the type of merchandise featured. For example, only small items can be placed in shadow boxes and on narrow ledges. In order to protect expensive merchandise from shoplifters, it is displayed in glass counters or enclosed window displays. Expensive items are not usually displayed near department or store entrances. Because stores want to sell display merchandise, items that can be easily cleaned, such as vinyl coats or shoes, are best for a dirt-catching ledge. Because merchandise featured in windows that receive strong sun light fade and may not be salable, stores may limit the items included in these display locations.

Experience 14-3 ■ Plan a Related Merchandise Display

1. Assume you are responsible for planning an interior display that includes two female mannequins that stand together on a platform at the entrance of the store. The store specializes in selling the newest fashions.
   a. Based on this season's current fashion story, decide what clothing and accessory items you will use to dress the mannequins. Try to use as many related merchandise items as possible without making the display cluttered. Use the most popular colors and accessories. (Read current fashion magazines or visit similar clothing stores if you need ideas on what items to include.) Write your ideas on a separate sheet of paper.
   b. Sketch the mannequins and their clothing and accessories on the same sheet of paper. Color the sketches. (Or cut out the clothing and accessories from fashion magazines.)
2. Assume you are dressing two male mannequins that stand together on a platform at the entrance of a sporting goods store. It's the season to promote ski wear. Follow steps a and b for planning this display.

Display Sketches

Before building your display at Campus Corner, Mrs. Aron has you plan it on paper. Making a sketch of the display will save time and help you create a more effective, pleasing display.

You get a pencil and paper and begin to draw a scaled-down version of the wall display space where you will be building your display. The wall area is 8 feet high and 4 feet wide. To make your drawing in the exact proportion to the actual space, you decide to let one foot equal one inch. You then make rough sketches of the sweaters, slacks, and accessories in the available space. You make three sketches before you decide which merchandise arrangement looks best. You sketch your displays so that the clothing looks natural, the way it looks when it is worn. Displays achieve the natural look by showing the clothing with natural blends and contoured to the body. You try to give your display a feeling of movement, to make the display more exciting to look at.

Display sketches let you try out different display arrangements without taking the time to build a display and then changing it. They also let you evaluate the display's color, balance, proportion, and harmony.

Many display professionals first learned how to combine colors in a display by studying the color wheel. (See Figure 14-7.) The color wheel shows the relationship between colors. It contains all the colors in the solar spectrum, including the primary colors of red, blue, and yellow. The color wheel helps you imagine monochromatic color schemes, which are a combination of various shades and tints of only one color. An outfit of navy pants and pale blue shirt is an example of a **monochromatic color scheme.** Although monochromatic color displays are the easiest to create, they aren't the most eye-catching. **Analogous color schemes** are a combination of colors next to each other on the color wheel. An example is a red blazer with navy dress and red-violet scarf. Another color combination is the **complementary color scheme,** a combination of colors that are opposite each other on the color wheel. A bright yellow leotard with purple tights is an example of a complementary color scheme. Complementary and analogous color schemes attract customer attention. They are more exciting to look at than monochromatic color schemes.

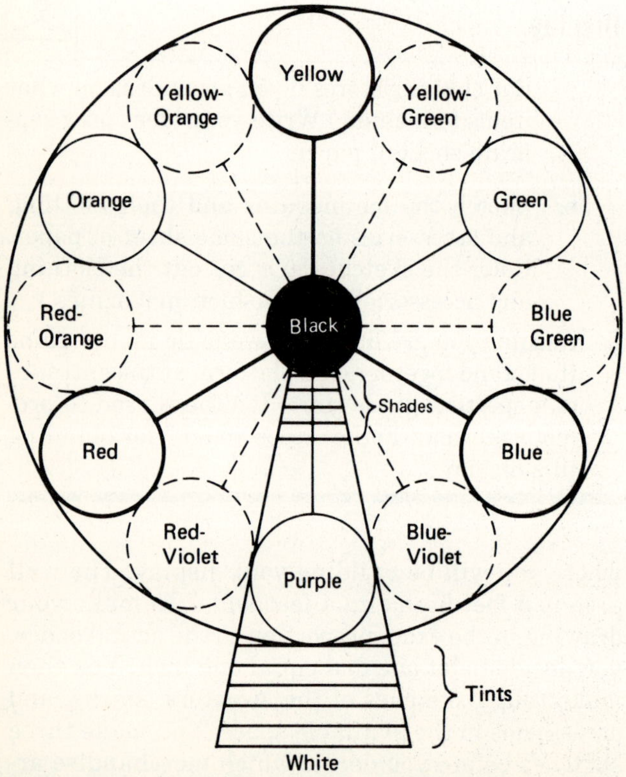

Figure 14-7 Color relationships are illustrated in this color wheel.

Balance refers to the arrangement of items in a display. As discussed earlier, displays with formal balance (symmetrical displays) are arranged so that both sides appear identical. Informal balance, where the size or number of items on both sides of the display are not identical, is usually a more interesting display arrangement. (See Figure 14-8.)

Sketches also help you evaluate the display's proportion. **Proportion** is the relationship each part of the display has to the total. Which item catches your attention first? This item, called the dominant item, should be the merchandise you want your customers to look at first. After looking at the dominant item, notice where your eye goes. The eye should move easily over the entire display, seeing all the merchandise featured. Your eye shouldn't stop at a particular prop or turn away because the display is too cluttered or crowded. Also, the display should have harmony, meaning all the items in the display should work together to convey the same theme.

If you plan your display with an understanding of these principles, you should be able to design it to attract attention, look attractive, tell the story you want, and hold the customer's interest. Remember that the design and color scheme are to be planned to promote the merchandise; the merchandise does not support the design.

Of course, once you have chosen the display sketch you want to build, you must decide whether the required props fit into the budget. The display budget is used to buy props, backdrop materials, lighting, mannequins, and other display supplies. Sometimes displays must be redesigned because the cost of the props is too expensive.

After you have selected your final sketch and have checked that it has all of the elements of good display design, you start working with your merchandise. You don't immediately start pinning up your display, however. First on a flat surface, you experiment with how you will pin the clothing to the wall. You bend the sweaters and pants in natural places, such as at the elbow and knee, to see how they will look. You experiment with folding the neck of the cowl-necked sweater to show it off to its best advantage. Mrs. Aron tells you to gather the sweaters and pants in the back to give them a "thin," tapered look. This display tip makes the merchandise appear more desirable to customers.

Figure 14-8 Informal balance usually results in a more interesting display than formal balance. © John Curtis/The Stock Market.

Experience 14-4 ■ Make a Sketch—Build a Display

As manager of the Menswear Department, you are responsible for the merchandise displays. Assume you are building a men's pant and shirt display for your Father's Day promotion. The display area is a wall, 8 feet high by 4 feet wide. On a separate sheet of paper, create four scaled-down sketches of the clothing display. Include related items and any appropriate signs, props, or background materials. Using color pencils or crayons, color your sketches. Choose the sketch you like most. Collect the clothing and accessories needed to build the display. Experiment with various ways of folding the clothing.

Using a wall space 8 feet by 4 feet, pin up the display you sketched and arranged. You will need T-pins, background paper, and staples to build the display. (Staple the background paper to the display space.) When you have completed your display, answer the following questions in the space provided.

1. Does your display show the merchandise as it would be worn?

____

2. Does your display tell a story? What is the story?

____

____

3. Does your display attract attention? What part of the display attracts attention?

____

____

4. Identify the factors that make the display interesting to look at.

____

____

____

5. Are the colors in the display coordinated? Does the display have a particular color theme?

____

6. Does the display have a "thin" appearance?

____

7. Are the accessories appropriate?

____

8. Is the background appropriate?

____

9. Is the clothing firmly attached to the wall?

____

10. Are the T-pins out of view?

____

Building Displays

Before you build your display, Mrs. Aron explains that you must clean the display area and assemble and prepare the merchandise. You also have to gather all the supplies and equipment needed to pin up the wall display.

Use masking tape to pick up lint off the fabric background of a wall display. Always vacuum the floor of a carpeted window or platform display. Counters and ledges should be cleaned before displays are built on them. The glass on window displays and shadow boxes must be cleaned with glass cleaner inside and outside.

The clothing used in any display must be steam-pressed. Even the smallest wrinkles can detract from the appearance of clothing featured in displays.

Check mannequins to be sure they are in good condition. They should not be used if they are chipped or have any broken parts. Their hair should be arranged in a current, attractive style.

Mrs. Aron made a short checklist to help you with your wall display. It's shown in Figure 14-9.

DISPLAY CHECKLIST

1. Clean the display area.
2. Press the display clothing.
3. Gather your needed equipment: ladder, pins, fish-line, and so on.
4. Pin the merchandise to the wall.
   a. Using T-pins, pin the pants to the wall according to the display sketch. Be sure to give the clothing a thin appearance by folding in the waist. Keep the pins invisible.
   b. Arrange the cowl-necked pullover inside the matching cardigan before climbing onto the ladder. Use T-pins to pin the shoulders and neckline to the wall. Give the sweaters a thin appearance by folding in the sides. Keep the pins invisible.
   c. Pin the accessory items to the display. Keep the pins invisible.
   d. Clean the display area.

Figure 14-9 This checklist for your wall display can be adjusted to suit the installation of other displays.

Dressing Mannequins Many stores display their clothing on mannequins because they give customers a better idea of how merchandise drapes and looks on the body. There are many different styles of mannequins sold by display supply companies. Some mannequins stand, others sit, and some recline. All are very expensive props, some costing more than a thousand dollars each. That's why many stores only allow professional display specialists to work with mannequins.

Before you try to dress a mannequin, be sure you know how the mannequin comes apart. Take off the head, hands, and arms before detaching the torso and legs. Be careful, the head and hands are easy to drop and chip. When dressing mannequins, be sure to include coordinating hosiery, shoes, and accessories.

Making Counter Displays Many stores build small counter displays on top of caselines to attract attention to the merchandise assortment inside the counter. These displays often involve a variety of display props that help customers visualize how the merchandise looks when worn. Head forms display hats and scarves. Usually a small mirror is placed on the counter next to the display to encourage customers to try on the merchandise and see how it looks on them.

To make counter displays more interesting and realistic, and to increase sales, use related items. For example, display a hat, scarf, and sunglasses on a head form. Be sure to tie the scarf in an interesting way rather than drape it across the counter.

Counter displays should also be sketched before they are built. Their color scheme, balance, proportion, and harmony should be examined to be sure the display attracts customers' attention and interest in the merchandise.

Evaluating Displays

After a display is built, it should be evaluated to determine if it accomplished its intended purpose. Many visual merchandising professionals watch the customers' reactions to their displays and listen to their comments. They may keep a tally of the number of customers who pass by the display and notice it.

Tracking sales of merchandise featured in displays is another way most stores evaluate a display's effectiveness. By counting the available merchandise featured in the display three days after the display has been installed and comparing it to the merchandise assortment prior to the display, management can determine the sales contributed by the display. A display can usually be given full credit for producing sales when there was no other promotional activity in the store, such as advertising or price reductions. However, like the other parts of the promotional mix, a display may not be the major reason why customers purchase clothing and accessory merchandise.

Experience 14-5 ■ Be Fast...But Be Careful

If you have access to a mannequin, practice dressing it in clothing and related accessories. Time yourself as you undress and dress the mannequin in the same clothing and accessories three times. Be careful not to damage the mannequin or the clothing.

Record the time you took in the spaces provided.

Time it took to undress and dress a mannequin:

First practice ____________

Second practice ____________

Third practice ____________

■ MERCHANDISE AND FIXTURE PLACEMENT

Salespeople, department managers, and buyers make decisions daily about how merchandise and fixtures should be displayed in a department. How can new items receive the best customer exposure? How should the department be arranged so that customers feel comfortable? How much merchandise can a display fixture hold and still allow customers to easily examine the merchandise? How can shoplifting of the smaller items be reduced? When making these decisions, customer convenience, security, and ease of restocking are considered.

Merchandise Rotation and Arrangement

Merchandise items that are visible from the entrance of a department have a much better chance of being seen and purchased by customers. However, because there is limited space at the front of a department, merchandise must be rotated or moved. Merchandise that is not selling on quads or A-frames may become the hot item when displayed on T-stands at the front of the department. Moving merchandise to another rack is a better way to stimulate sales than markdowns.

New merchandise should always be displayed in the front of the department. To motivate customers who are shopping for marked-down items to walk through the department, clearance racks should be in the back. Merchandise in the same classification should be grouped on adjacent fixtures. Related merchandise classifications should be placed nearby. For example, the complete selection of moderately priced casual tops should be contained on fixtures that are next to each other. Casual bottoms should be displayed next to the casual tops, rather than next to unrelated items such as formal dresses or raincoats.

Most stores "remerchandise" their selling areas often, moving merchandise from one fixture to another. These floor moves give the department a fresh look. Regular customers get the feeling that there is always a selection of new merchandise in the store. Floor moves also consolidate remaining merchandise to make room for new goods.

Fixture Placement

Fixture placement affects the way customers move around the merchandise in a store. When racks are too close together, customers have trouble walking

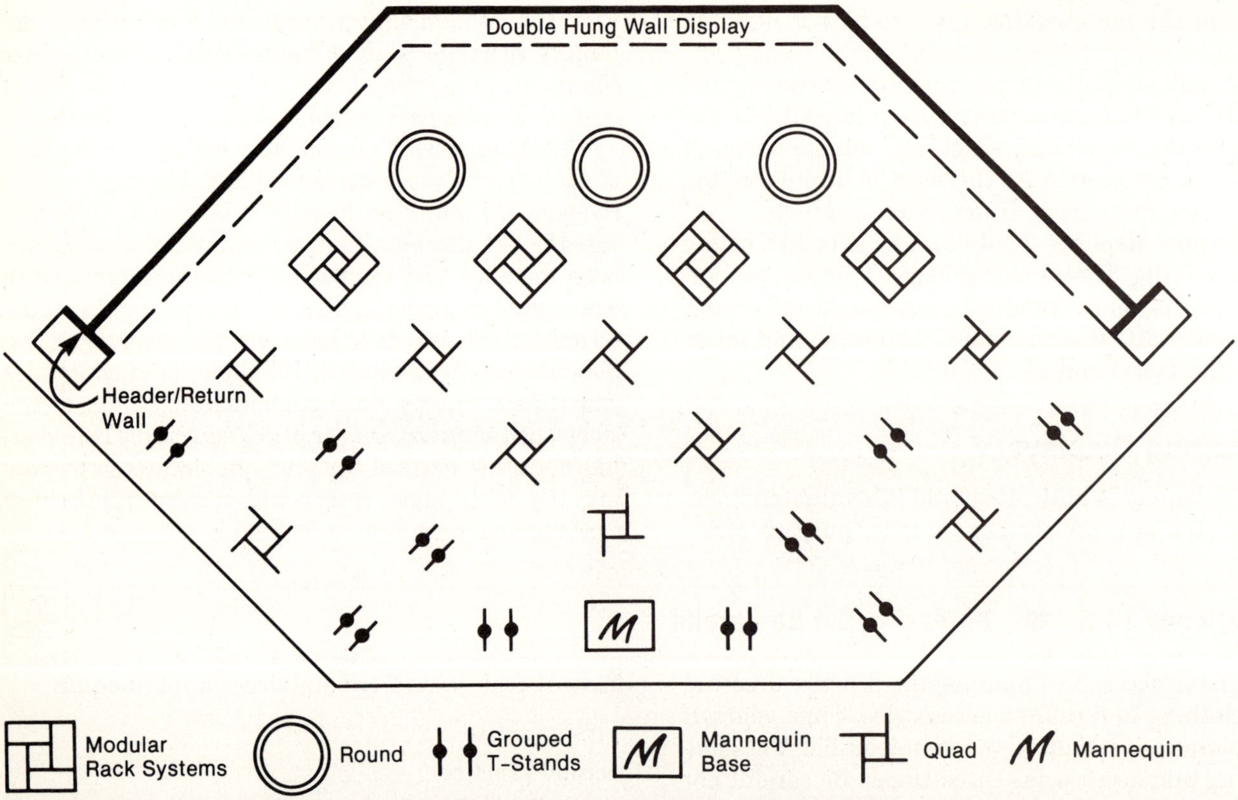

Figure 14-10 This floor plan is symmetrically arranged with enough room for customers to walk around the fixtures and see the merchandise on display.

down the aisles and moving among the fixtures. A crowded arrangement makes it difficult for them to find coordinating merchandise, walk to the fitting room, or locate the cash register.

Some guidelines for helping you plan fixture placements are:

1. Place the lighter fixtures, such as T-stands, in the front of the area.
2. Progress from the lighter fixtures to taller, heavier fixtures. For example, place quads behind T-stands. Place rounders after quads.
3. Arrange fixtures in a symmetrical configuration within the area. (See Figure 14-10.)
4. Keep the aisles wide enough for rolling racks and bins containing new stock to move onto the selling floor.

Experience 14-6 ■ Build a Department

Using the department floor plan on page 215, prepare a layout. Trace the fixtures in Figure 14-10 and cut out each fixture symbol. Make as many copies of each symbol as you need. Arrange the fixtures so that the department looks balanced, not lopsided. Stock the newest merchandise to the front of the department. Allow sufficient aisle space so that customers may move freely. Following are the merchandise items and fixtures needed. When you have developed a layout that you like, label each fixture symbol with the number that corresponds to the listed merchandise to be displayed on that fixture.

1. New arrival coordinate group in black and white, on a quad
2. New arrival dresses, on a half-round rack
3. Clearance dresses, on a round rack
4. Clearance skirts, sweaters, blouses, and pants, on a round rack
5. New arrival sweaters, bulky knits, on a round rack
6. Sweaters, on a round rack
7. New arrival coordinate group in red and gray, on a quad
8. Skirts, on a round rack
9. New arrival sweaters and blouses on a T-stand
10. New arrival coordinate group, black and white, on a T-stand

11. New arrival coordinate group, red and gray, on a T-stand
12. New arrival dresses on a T-stand
13. Jewelry rack, on a square rack
14. Jewelry rack, on a square rack (jewelry and scarf)
15. Hosiery rack, on a square rack

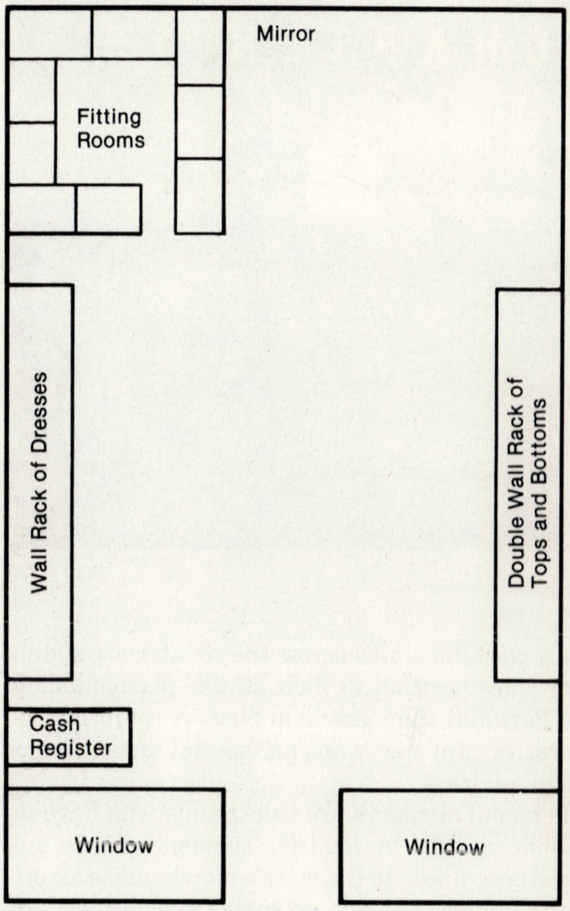

Chapter Note

1. *Chain Store Age Executive,* November 1986, pp. 108–109.

CHAPTER 15

Selling

Many clothing stores across the country are adding a new sales position to their staffs—personal shoppers. Personal shoppers help busy or confused customers, or anyone wanting special attention or fashion advice.

Personal shoppers are salespeople who have an excellent ability to identify customer needs and match those needs to the store's merchandise assortment. They also have an extensive knowledge of the merchandise sold throughout the entire store, including clothing, shoes, and accessories, and an up-to-the-minute awareness of fashion trends.

Usually personal shoppers telephone the customers listed in their client book, discuss some of the new merchandise the customers may be interested in, and make an appointment for the customers to come into the store and make their selections. Then they go throughout the store, gathering the items that would best meet the customers' needs. When the customers arrive in the store, the personal shoppers present the items, assist the customers in the fitting room, and demonstrate how an item should be worn and accessorized. The service is usually free. The stores that provide the personal shopping service, such as The Broadway Southwest, a department store chain headquartered in Phoenix, Arizona, benefit when customers purchase several hundred dollars' worth of merchandise at one time. The special attention makes customers feel important, which can result in regular store patronage.

The Broadway Southwest, like most clothing and accessory retailers, knows how important its sales staff is to its overall success. Broadway Southwest management realizes that although it has many employees who try to buy the most desirable merchandise, create the best displays, and write the most exciting ads, it is ultimately the salespeople who keep the customers happy and coming back. In many ways, salespeople have the most important job in a store—without satisfied customers, there would be no store.

If one of your goals is to be a professional salesperson, you have to be proud of what you do. You have to believe in selling. You also have to believe that customers need help in purchasing clothing and accessories. To provide that help, you have to know enough about the merchandise to explain why a particular item is the one a customer should buy. You have to believe the customer needs the merchandise you are trying to sell.

Some salespeople who have worked in retailing say that selling isn't so difficult. However, many people confuse selling with "order-taking" or "cashiering." Selling is helping the customer obtain maximum personal satisfaction for the money spent.

■ CUSTOMER TYPES

It takes many customers to make a store a success. Each customer who walks through the door is an individual, with different needs and feelings. Because all customers do not act or react the same way, some successful salespeople classify customers into certain types. This helps them determine the best way to help customers with their particular needs. Table 1, on page 217, lists some of these types.

Of course, this table doesn't include all of the types of customers you'll meet during a retail sales career. Each customer is an individual and must be treated as such.

■ **Table 1**
Typical Customer Types

Customer Type	Method for Handling
Ms. Iwanna Argue will give you all sorts of reasons why your merchandise is bad.	Keep cool, calm, and courteous. Don't try to win an argument, for you might lose a sale. Use Ms. Argue's own arguments to prove your points. "You're right, classic styling is important to consider when buying pants. This classic trouser style in 100% wool gabardine will give you long years of wear and always be fashionable."
Mrs. Hesi Tant is never sure if she should buy. Just when you think she's made a decision, she changes her mind.	Help this customer make a decision. Give her your opinion. Reassure her. "That's a very good choice, Mrs. Tant. I'm sure you'll be pleased with your purchase."
Mr. Izit A. Bargain is always looking for a sale or a good deal.	Know the quality of merchandise you sell. Sell him on the additional quality and service you can offer him.
Mr. Si Lent is the toughest customer of all. He just doesn't talk, even though he's quietly thinking about what you're saying.	Ask open-ended questions that give Mr. Si Lent a chance to talk. Don't hesitate in trying to close the sale. He only needs to say one word—yes.
Miss Ina Hurry is impatient and easily irritated if you're slow.	Give her fast service and the important facts. Don't bore her with details, or she will leave.

■ SELLING CLOTHING AND ACCESSORIES

Most stores realize that salespeople need sales training. They believe selling is a skill that can be learned and improved. They conduct sales seminars that demonstrate and discuss the different parts of a sale, including the approach, determining the customer's needs, the product presentation, handling objections, closing, and suggestion selling. Of course, without sufficient product knowledge, even the best salesperson will have problems making the sale.

A sample sales presentation that demonstrates each part of the sale and good product information follows:

Salesperson approaches the customer.

Salesperson: Good morning. How are you today?

Customer is examining a glass cube of boys' shirts.

Customer: Fine, thanks.

Salesperson: What size can I help you find?

Customer: Size 10. My nephew's birthday is next week. My sister mentioned that Robbie could use a dress shirt.

Salesperson: Do you think Robbie would prefer short sleeves or long sleeves?

Customer: With summer coming, probably short sleeves. They are so much cooler.

Without hesitation, the salesperson leads the customer to another shirt counter of solid blue and white dress shirts.

Salesperson: These cotton/poly blends are our most popular shirts. Because cotton is a cool, absorbent fabric, it is especially comfortable to wear in the summer. And the addition of polyester makes the shirt easy to care for. This shirt can be machine-washed in warm water and tumble-dried. Robbie's mother will appreciate the no-ironing.

Customer: I suppose, but this shirt is so plain. I prefer the brightly colored patterned shirts that I've seen in the ads. Robbie already has dress shirts like this, so I want something different.

The salesperson moves to the next fixture, where the patterned shirts are stocked. She quickly locates the size 10s and pulls out a blue, green, and yellow plaid.

Salesperson: Lots of young boys really like this shirt. It would look nice with solid blue, green, or yellow pants. Your nephew could also wear gray, black, or white pants with this shirt.

The salesperson hands the shirt to the customer. She handles the merchandise with care and respect to show that the shirt is valuable.

Customer: I like this...but maybe it's a bit too bright. What about a pinstripe? They are always a safe choice.

Salesperson: We have blue and white, red and white, and gray and white pinstripe shirts (taking the customer to another fixture). These shirts are also a cotton/poly blend, so they are very easy to care for.

Customer: The pinstripes are so...so conservative. I suppose I prefer the plaid shirt.

Salesperson: That's a good choice (walking the customer back to the plaid shirt fixture). It's so versatile, your nephew will be able to wear the shirt with many colors. I'd suggest a yellow tie as a finishing touch for a dress outfit. (The salesperson puts the tie next to the shirt.) Younger boys often prefer these clip-on ties. This tie gets its interesting design from a special weaving process. How do you like it?

Customer: Yes, it's nice. In fact, I like it very much.

The salesperson picks up the shirt and tie and walks the customer to the cash wrap desk.

Salesperson: I'll take the price tags off this merchandise for you and put the shirt and tie in a gift box. That should reduce your wait at the gift wrap desk. Will this be on your charge?

Approaching the Customer

The first step of a sale is called the **approach.** In an effective approach, the salesperson gets the customer's attention and makes him or her feel welcome in the store.

The best approaches are prompt. Many stores have established a performance standard for salespeople that specifies that they should approach customers within one minute of entering the department. However, when the salesperson is helping another customer, this may not be possible.

Even if you are busy with someone else, acknowledge all customers who enter your department. Look up and make eye contact with them. Say "I'll be with you in a minute." And then continue to keep looking up and making eye contact so customers don't think they've been forgotten.

There are three types of approaches: greeting, service, and merchandise. The greeting approach involves saying such things as "Hi," "Hello," "Good morning," and "How are you?" The approach used in the sample sales presentation, "Hi, how are you today?" is an example of the greeting approach. This friendly greeting makes customers feel welcome.

The service approach involves asking the customer a question or offering to be of service to the customer; for example, "What may I show you today?" A combination greeting and service approach is "Good evening. I'm Patti Ortega. Perhaps I can help you find what you are looking for today."

There are some service approaches that should be avoided, such as questions that can be answered with a simple yes or no, like "May I help you?" Customers have been programmed to respond with the usual "No thanks, just looking." Instead, ask open-ended questions that begin with the words *who, what, why,* and *how.* They make the sales presentation a two-way conversation by giving the customer a chance to talk and the salesperson a chance to find out what the customer wants.

In the merchandise approach, the salesperson says something specific about the merchandise the customer is examining. A typical merchandise approach is "We just received this spring suit. Isn't the color great?" The merchandise approach is not effective all the time, however. It is inappropriate for customers who are not looking at particular items or for times when customers approach the salesperson. In these cases, the service or greeting approach is best.

The approach won't make a sale, but it can lose a sale. Therefore, it is important to learn how to begin a conversation with a customer. Successful salespeople have learned to approach customers with a friendly, helpful attitude and a suitable remark.

Experience 15-1 ■ Say What's Appropriate

Assume you are a salesperson in a young men's sportswear store. Have classmates be the customers. Your store sells casual and dress clothes. Use a tape recorder and record how you would approach customers in each of the following situations. Also, record why you believe each approach was appropriate. (If you do not have a tape recorder, write your dialog on a sheet of paper. Explain why you think your remarks are appropriate.)

1. Customer, not particularly interested in any one item, wanders around tables and racks.
2. Customer stops to look at a table of sale pants. Customer remains at table.
3. Customer stops to look at a table of sale sweaters. After a minute or two, customer walks to the wall of sport coats and then to the shirt racks.
4. You are showing polo shirts to a customer when another customer enters your store. A third customer enters and walks directly to the cash register.

Determining Needs

Before you can successfully sell a particular item, you need to know if the merchandise would be appropriate in satisfying customer needs. Find out what customers want by questioning, observing, and listening. You need to know whom the item is for; where the clothing or accessories will be worn; what size is needed; what other items the customer already owns and wants to wear with the new merchandise; and what color, style, and fabric are preferred by the customer.

One of the quickest ways to determine customer needs is by asking open-ended questions. These questions encourage customers to talk because they cannot be answered with a simple yes or no. Some examples of open-ended questions are: "What color did you have in mind?" "Where will you wear the dress?" and "What other items do you want to coordinate with it?"

By asking questions, a salesperson finds out that all customers are not alike. Decided shoppers have a very specific idea of what they want to purchase. The customer who was looking for a dress shirt for her nephew in the situation described earlier was a decided shopper. Undecided shoppers have only a general idea of what they want. For example, an undecided shopper may be looking for a birthday present for his father but will consider buying anything a man can use. Shoppers who are browsers are not looking for anything in particular. They may be just bargain-hunting. Once the salesperson finds out whether a customer is a decided shopper, undecided shopper, or browser, he or she can better meet that customer's needs. (See Figure 15-1.)

After asking a few questions, a salesperson may discover that the customer wants something the store doesn't stock. When that happens, the salesperson should try to suggest some substitutes. For example, suppose the customer wants a jewel-neck teal blouse. If you don't have any jewel-neck blouses, but do have teal camp shirts, show her the camp shirt. Color may be more important to the customer than style, and she will decide to buy it. However, if the customer's mind is set on a jewel-neck teal blouse, and no substitutes are acceptable, write down the customer's request on a want slip. These forms, discussed in Chapter 11, help buyers plan their future purchases.

Figure 15-1 When helping a customer select clothing for an infant, it is especially important to determine the size that will be needed when the item is in season. © Richard Hackett/Courtesy of Kids Kids Kids.

Sometimes, even after asking open-ended questions, customers' needs are not clearly identified. In situations like this, a salesperson may learn more about what the customer wants during the sales presentation or when the customer makes an objection. The salesperson should begin showing merchandise once he or she has a general idea of what the customer wants and then watch the customer's reaction. A scowl means "Forget it"; a twinkle in the eye may mean "This is what I've been looking for."

Experience 15-2 ■ Which Size, Style, Color?

Assume you are a salesperson. While you are straightening the merchandise on a display table, a man enters the store. He spends some time looking at worsted wool dress slacks. What five questions could you ask him to determine his needs?

1. ______________________
2. ______________________
3. ______________________
4. ______________________
5. ______________________

■ BUYING MOTIVES

Buying motives are reasons why people purchase something. Very few people purchase clothing just to stay warm. Rather, people have a variety of reasons to justify their clothing purchases. These reasons can be classified as rational or emotional needs. Some of the **rational needs** are to save money (economy), to save time (convenience), for health reasons, for warmth, and for modesty. Some of the **emotional needs** are for pleasure, to attract attention, to be different, to be a follower, for variety, to be a leader, to attract the opposite sex, and to win recognition or social approval.

Everyone has rational and emotional needs in different degrees. That's what makes every customer unique and different. Discovering buying motives is not difficult. You can discover buying motives by listening carefully to what a customer says and observing how he or she reacts to what you say.

For example, suppose a woman is shopping for a new dress for work. One of her rational buying needs may be to find a dress that coordinates well with her existing wardrobe, so she won't have to spend additional time and money putting together an outfit. Another rational need is to find a dress with long sleeves so she won't be cold when sitting in her drafty office. An emotional need would be to choose a dress in a style and color that flatters her figure and personal coloring. Another emotional need would be to buy a dress that is made by a popular manufacturer, so she will be thought of as fashionable when she wears it. By describing the dress in a way that appeals to the customer's buying motives, you can make the buying decision easier for the customer.

Experience 15-3 ■ Buying Motivators

Determine which emotional or rational buying motives these customers seem to be expressing.

1. Have you received your first shipment of designer fashions for spring?

   ____

2. A friend of mine bought a crepe de chine dress here about a week ago. Do you still have any in stock?

   ____

3. I already have red, green, and black turtleneck sweaters. I want something different.

   ____

4. I heard you were having a sale on leather wallets. Do you have any of the tri-fold kind?

   ____

5. I'm looking for a pair of riding boots. I have plans to go riding with someone very special.

   ____

■ PRESENTING MERCHANDISE

After approaching customers and discovering what merchandise they need or want, you should follow up with the presentation. During the presentation, show customers how the merchandise will satisfy their needs. To make an effective presentation, you need to know your product and be familiar with special selling skills.

Merchandise Knowledge

Do you recall how the salesperson selling boys' dress shirts showed selling confidence? That's the kind of confidence you get when you know your merchandise. Merchandise knowledge will help you choose the product best suited for the customer. It will help you explain product features and benefits and overcome customer objections. Customers want to know certain information about the clothing and accessories they buy. They will ask such questions as the following:

About Fabric How will it wear? Will it hold its shape? Will it shrink? Can it be washed? Will it fade? Does it need to be ironed? What kind of material is this?

About Construction Is it true to size? Can it be altered?

About Fashion Is it fashionable?

About Coordination What do you have that goes with this?

About Fit Are the shoulders supposed to fit like this? Isn't it too long?

Some customers also want to know the history of the product or the many different ways to wear an item.

To learn about the assortment of clothing and accessory products you sell, prepare a product analysis chart. It's easy to prepare. List all the facts or features about a product on the left side of the chart. Across from each fact, write an advantage or benefit of it. (See Table 2.)

Product analysis charts help you prepare for a sale. They help you know exactly what you want to say about a product on the spur of the moment. When a customer asks an unexpected question, you won't

■ **Table 2**
Product Analysis Chart

Product: Two-Piece Print Dress

Sales Features: Facts	Selling Points: Benefits
Fabric: 100% polyester	It is easy to take care of; it won't cost you extra money for dry cleaning.
Construction: shoulder pads attached with velcro; no hanging threads or construction flaws; well-sewn buttons and seams	It requires no extra care during laundering; shoulder pads can be removed for laundering.
Style: wrap-blouson top with accordian-pleated skirt	It is fashionable.
Color: cream background with black and silver geometric design	It is attractive and has a popular color scheme; it is easy to coordinate with basic accessories, such as black shoes; it won't cost extra money to accessorize.
Manufacturer: Liz Claiborne	This manufacturer is very popular.
Price: $140	It is a good value for the price.

lose the sale because you couldn't answer. By analyzing the merchandise beforehand, you can coordinate other clothing or accessories that complement the merchandise. However, just because you know the merchandise, doesn't mean you'll be able to sell it. You need to have special selling skills too.

Merchandise Presentation

The following rules will help you be a more effective salesperson:

1. Limit the amount of merchandise shown.
2. Point out the obvious features first.
3. Point out the hidden features.
4. Be enthusiastic.
5. Show the medium-priced item first, unless the customer requests otherwise.
6. Talk about benefits.
7. Handle the merchandise with care and respect.
8. Let the customer participate in the sale.
9. Appeal to the customer's senses.
10. Help customers in the fitting room.

Point One: Limit the amount of merchandise shown
Customers who are considering ten items have a more difficult time making a selection than customers who are considering just two or three. Too many alternatives confuse customers. Some customers will leave the store rather than be forced to choose from among too many products. (See Figure 15-2.)

Figure 15-2 If the salesperson overwhelms the customer with choices, the customer may choose not to buy anything at all. © Richard Hackett/Courtesy of Bagatelle Shops.

In the selling situation in the beginning of this chapter, there may have been 15 shirt styles in the department. Yet the salesperson showed only three.

The customer purchased one of those three shirts because the salesperson correctly determined the customer's needs. The salesperson made it easy for the customer to buy.

Point Two: Point out the obvious features first An obvious feature is the first thing that catches the customer's eye. It may be a color or a special style detail. For a dress, it may be the cool, summery color. For a pair of jeans, it may be the label on the back pocket. By pointing out the obvious features first, you are drawing the customer's attention to the merchandise. By commenting on a garment's color or style, you are reinforcing what the customer can see and already knows. If you prepare a product analysis chart on a merchandise item, you can determine the features to be emphasized. This will also help you determine benefits, which is explained in Point 6.

Point Three: Point out the hidden features Most customers are unaware of the hidden features of a clothing item, such as its quality construction or fabric. And they won't know about the less-than-obvious features unless a salesperson tells them. Hidden features are things customers are not interested in at first, but they help customers reach a decision to buy. They help customers justify spending their money.

Point Four: Be enthusiastic This point, be enthusiastic, applies to everything salespeople do. If you're excited about the merchandise, the excitement will be carried over to the customer. Customers can sense when salespeople are genuinely interested in merchandise they sell. They increase their desire for the merchandise when they see the salesperson thinks the merchandise is valuable.

Point Five: Show the medium-priced item first, unless the customer requests otherwise Most clothing and accessory stores carry three price lines: high, medium, and low. If customers don't mention a price range, show them the medium-priced range. Then listen carefully to what customers say and how they react to the medium-priced product. If customers mention, "This is a bit more than I wanted to spend," take them to the next lower-priced line. If customers say, "Is this the best quality item in stock?" show them a higher-priced line. Avoid asking them, "How much do you want to spend?" That question limits a salesperson; it doesn't give you freedom to "trade-up" to a better-quality but more expensive item.

Point Six: Talk about benefits Benefit statements tell customers what they will receive when they buy that dress, or coat, or sweater. They answer the customer's question, "What's in it for me?" Customers want to hear what items will do for them. Some customers really don't care about the name of the fabric, which is a feature or fact, as long as the clothing is easy to care for, the benefit. By talking about the benefits that are important to the customer, you will be able to create desire and interest in the merchandise so that the customer wants to buy it. Words such as *attractive, flattering, comfortable, versatile, suitable, appropriate,* and *good value* are useful to describe benefits. Salespeople should be skillful in relating merchandise benefits to customers' needs. Customers often reveal their wants and needs in the course of the sales presentation.

By focusing on the benefits that meet the customer's needs, you can help the customer and make the sale. For example, suppose you have a male customer who has mentioned he is interested in a comfortable, long-wearing shoe. You are about to try on his foot a shoe that has the following features: calfskin upper, leather lining, half leather and half rubber heel, leather sole, and steel shank. You could make the customer aware of the comfort benefits of the shoes by saying:

"The shoe you will be trying is one of our finest. It is made of calfskin, a most comfortable leather. Calfskin is a durable leather that does not scuff easily, making the shoe easy to care for and a good value. The lining is also made of leather. Because leather can breathe, your foot will stay cool and dry. The leather lining adds to your comfort. The shoe has a steel shank. This is a metal support that runs from the center of the heel to the ball. It adds support to your foot. It also has a leather sole that flexes and adds to your walking comfort. The heel, which is half leather and half rubber, is designed to resist wear, making your shoes a longer lasting and a very wise investment. This shoe will keep its shape and fit longer than many others. It will continue to look attractive for a long time."

Customers are very selective in their spending. By stating the benefits, you give customers a reason to make a purchase.

Point Seven: Handle the merchandise with care and respect No one wants to spend money for something that someone else doesn't think is valuable. If a salesperson handles merchandise carefully, as though it is truly valuable, customers will believe the merchandise has real value too. They will be more likely to make the purchase.

Point Eight: Let the customer participate in the sale Customers are more likely to buy when they participate in the sale. A professional salesperson will involve the customer with the merchandise. For example, have the customers try on scarves and jewelry and look at themselves in a mirror. When show ing coats, help the customer off with his or her coat and on with the merchandise. Then have the customer look in a mirror to see how the coat looks. Get the customer involved with a handbag by having the customer open the fastener to see how it works. By offering the bag to the customer for inspection, the salesperson is letting the customer experience a feeling of ownership. When the customer holds the bag next to his or her body and looks in a mirror, the customer can actually see how attractive the bag is. Many salespeople discover that letting the customer participate in the sale often convinces them to buy.

Point Nine: Appeal to the customer's senses By letting customers participate in the sale, you will be appealing to their senses of sight, touch, and hearing. The more senses you can appeal to, the better you will be able to hold a customer's attention. Customers will get a feeling of ownership of the item that can be hard to resist.

For example, when showing a customer a 100 percent silk dress, you can appeal to her sense of touch and sight. Silk feels good against the skin, drapes well over the body, and moves nicely. Silk also takes dye beautifully, which accounts for the bright, deep hues of silk clothing. Silk makes a person feel special. By describing the merchandise in a way that appeals to the customer's senses, you help the customer make the decision to buy.

Point Ten: Help the customer in the fitting room Full-service stores want their salespeople to carry the clothing the customer is interested in and lead the customer to a clean, empty fitting room. After hanging the garments on hooks, the salesperson should close the fitting room door or curtain and allow the customer enough time to try on the clothing. The salesperson should check back with the customer periodically. By seeing how the new clothing items look on the customer, the salesperson can tell if the customer needs a larger or smaller size. The salesperson can also use the time the customer is trying on clothes to locate other items on the selling floor that will coordinate with the items the customer has chosen. While the customer is wearing the clothing in the fitting room, a salesperson should point out the benefits the merchandise offers. An example of appropriate comments are as follows:

> *Salesperson*: Mrs. Parker, those pants fit perfectly! Look how well they fit at the waist. And they're the perfect length. How do they feel?
>
> *Customer*: They are comfortable. Do you think taupe pants would go with a mauve jacket. I want to wear them to work.
>
> *Salesperson*: Without seeing the jacket, it's hard to tell. But taupe is a neutral, versatile color. Not only is taupe attractive on you, it's also a big color this fall. These pants are dressy enough for work, yet they can look casual too.

Customers appreciate the extra attention they receive in the fitting room. The salesperson's presence also acts as a deterrent to shoplifting because the salesperson can monitor the merchandise that goes into and comes out of the fitting room.

Experience 15-4 ■ Present the Merchandise

Role play how you, as a salesperson, would present merchandise to a customer. Ask another student to pose as a customer.

1. Choose a clothing or accessory item your customer may be interested in purchasing. Bring that item to class for your sales presentation.
2. On a sheet of paper, prepare a product analysis chart for your product. Be sure to include features and benefits about its fabric, construction, style, color, and price. Also identify other clothing or accessories that would coordinate with the product.
3. Using a tape recorder, record how you would present the merchandise to your customer. Assume you have already greeted the customer and determined the customer wants to see the product you are about to present. Be sure to follow the ten points of a presentation on pages 221–223.
4. After the merchandise presentation, ask your customer to evaluate your performance by completing the following evaluation form.
   a. Did the salesperson talk about the merchandise benefits? Give three examples.

   ____

   ____

   ____

   b. Give two examples that show the salesperson handled the merchandise with care and respect.

   ____

   ____

   ____

c. How did the salesperson let you participate in the sale?

___

___

___

d. How did the salesperson appeal to your senses of sight, touch, smell, and hearing?

___

___

___

e. How would you rate this salesperson's enthusiasm? (Circle one)

Very enthusiastic   Not so enthusiastic   Somewhat enthusiastic

f. Any suggestions for improving the presentation?

___

___

___

___

■ OVERCOMING OBJECTIONS

The salesperson's job would be very easy if, after every presentation, customers said, "I'll take it." Usually, however, customers have questions or comments about the merchandise. When these questions or comments indicate reasons why customers may honestly refuse or hesitate to purchase merchandise, they are called **objections** and you will need to know how to handle them. Remember that objections are a natural part of the sale. They indicate that the customer is sincerely interested in the merchandise. In some situations, customers may raise an objection because they want more information before deciding to buy.

Kinds of Customer Objections

Customer objections to merchandise can be classified in three categories. These are (1) need, (2) price, and (3) objections to particular features. A salesperson must successfully overcome these objections to make the sale.

Need Objections When customers say, "I don't need that," they cannot see how the merchandise will benefit them. When customers try to put off buying merchandise, they don't think they need it now. Some common objections salespeople hear are, "I don't think I'll get enough wear out of it," or "I'd like to think it over."

Anticipate need objections. Find out why customers are uncertain about making the purchase. Say, "You seem to be uncertain about making this purchase. Is there something else I can answer for you?"

Remember that when a salesperson and customer meet, there is always a sale. Either the salesperson sells the customer the merchandise or the customer sells the salesperson on the idea that he or she doesn't need the merchandise.

Price Objections Salespeople often hear price objections, such as "It costs too much." If a customer really can't afford the merchandise, there's not much a salesperson can do. If you have not pointed out the value of the merchandise, show customers that the merchandise offers full value for the money spent. People do not buy price; they buy value and benefits. Price is related to style, fit, construction, quality, and store services.

Don't apologize for the price. Explain why the price is justified. Consider the following example.

Customer: This blazer is too expensive.

Salesperson: I admit that this is not an inexpensive jacket. Let me tell you why. This is 100 percent worsted wool, a warm, comfortable, popular fabric. Because this jacket has classic, tailored lines, it will be fashionable for years to come. All the seams are hand-stitched. This quality construction feature adds to the durability and appearance of the jacket. The jacket also has a full lining, which definitely adds to the jacket's appearance and fit.

Objections to Particular Features Throughout a sale, customers may object to various features of the merchandise shown. Some samples of these objections are "I don't like the color," or "That style makes me look terrible." Sometimes customer objections about style, color, and fit help you better determine customer needs. When you show an item the customer objects to, you have a better idea of what the customer really does want to purchase. This helps you choose more appropriate merchandise.

Some objections about the fabric or construction may be based on wrong information. The salesperson should try to overcome such objections. Consider the following example:

Customer: This really is a beautiful coat, and it

is so comfortable. But it's too lightweight to keep me very warm.

Salesperson: I felt the same as you did until I bought a coat like this. It's made of lightweight wool. You may not know that wool absorbs dampness from the air in the cold weather. You'll feel warmer when wearing wool than when you wear other fabrics.

Customer: This suit is all right, I suppose. But I always wear [brand name] originals.

Salesperson: You have good taste, sir. That brand is excellent, but you may want to consider this brand. As you can see, this suit has an exceptional style and cut. This designer is new to the fashion world, but already has an excellent reputation.

General Tips on Handling Objections

Every customer objection must be handled with respect. You are a salesperson; you are not supposed to win an argument with a customer but lose the sale. To overcome customer objections, follow these simple rules:

- Let customers state their objections.
- Be sure you understand the objection. It may be helpful to restate the objection in your own words.
- Respect the customer's objection. By doing this, you will show the customer you think the objection is important. Also, you show the customer that he or she is not wrong in making the statement. Showing respect for the customer's opinions avoids a possible argument.
- Mention additional information that the customer may not have considered.

Some examples of how salespeople can effectively overcome objections by using these rules are:

Customer: I don't like the length of these new dresses. They make me feel dumpy.

Salesperson: I can understand why you say that. New styles are hard to get used to. However, in a short while you'll be seeing this skirt length all over town. In fact, you may feel uncomfortable wearing the length you wear now.

Anticipate Objections

If you can anticipate possible customer objections, you will be better able to handle them. To help you overcome objections, prepare an objection analysis chart. List the type of objection, what the customer's objection sounds like, and what you can say to overcome the objection successfully. See Table 3 for an example of an objection analysis chart.

Objections Versus Excuses

Salespeople must be able to distinguish between customer objections and customer excuses. Objections are honest, real reasons why the customer hesitates to purchase the clothing or accessories you are presenting. Excuses are reasons customers have made up for not purchasing the product. The best way to find out if the customer has a real objection is to ask questions.

For example, suppose the customer says, "These shoes aren't really the style I had in mind." The salesperson could say, "Can you describe the style you wanted?" If the customer answers with a definite answer, the customer has a real objection. For example, if the customer says, "I want a more casual looking shoe—perhaps a loafer," the customer has a real objection. The reply, "I'm not sure what I want" may be an excuse.

Experience 15-5 ■ Customer Objections

Using a tape recorder, record how you, as a clothing salesperson, would overcome the following customer objections. Assume you are showing the customer a $95 silk blouse.

1. "It's a nice blouse. I'm just not sure I need it right now. I want to think about this. Perhaps I'll be back to get it later."
2. "Ninety-five dollars is a lot of money. In fact, I think this blouse costs a lot more than it's worth."
3. "Silk is so hard to care for, and dry cleaning is awfully expensive."

■ CLOSING A SALE

If a salesperson has successfully determined the customer's needs, presented the merchandise, and overcome any objections, the customer should be ready to buy. Most customers, however, don't say, "I'll take it." They need a salesperson's help to decide to buy.

Usually, a salesperson who is paying close at-

■ **Table 3**
Objection Analysis Chart

Product: Man's two-piece gray worsted wool suit

Type of Objection	Customer's Possible Objection	Salesperson's Suggested Answer
Need	"I like the suit. I just don't think I'll get enough wear out of it."	"You mentioned that you will soon be graduating from college and will be interviewing with employers for a job. This suit will serve your interviewing needs well. It is also versatile enough to wear to work or for an evening out. This jacket which is a blazer style, can also be worn with coordinating pants."
Price	"This suit costs a bit more than I wanted to pay."	"I agree that this suit is not inexpensive. However, it has many quality features that add to its value. For example, this is a quality wool that will retain its crisp, neat look and wear well. The manufacturer is also a leading designer. Much time and care have been taken to size this suit to give you a good fit."
Particular feature: Fabric	"Won't this suit be too hot during the summer? It's wool, isn't it?"	"Yes, but it's a very lightweight wool. Many summer suits are made of this fabric because it absorbs moisture to cool the body."
Particular feature: Color	"The color is so drab."	"Gray is a very versatile color in a wardrobe. You can wear it with a variety of colored shirts and ties to create a number of different looks."

tention to customers can tell when they are ready to buy. Customers make positive statements about the merchandise or their actions indicate they want to buy. For example, handling the merchandise again, after it's been set aside, is a buying clue. When the buying signals indicate that customers are ready to buy, the salesperson should try to close the sale.

There are many ways to close a sale. If the customer has been examining two different items, a salesperson may say, "Which do you like better, the striped or the plaid?" If the customer is sending buying signals that indicate he or she wants to buy, a salesperson could ask, "Should I charge this to your account?" or "Would you like me to take off the price so you can have it gift wrapped?"

Be ready to close the sale as soon as the customer seems ready to buy. Some customers are ready to buy much sooner than others, so don't drag out the sale longer than needed. Salespersons who don't know when to stop talking and start closing talk customers out of buying.

Experience 15-6 ■ It's Time to Close

You are a salesperson in the men's shoe department. You have successfully approached a customer and found out the customer wants a loafer. You have sized the customer's foot, located the shoe in the stockroom, and explained the shoe's comfort and wear features. The customer keeps nodding his head in approval, agreeing with everything you say. List three ways you could close this sale.

1. ______________________________

2. ______________________________

3. ______________________________

■ SUGGESTION SELLING

In suggestion selling, the salesperson may suggest related items that complement or accessorize the customer's first purchase, such as a scarf to go with a new dress. Or the salesperson may suggest an additional number of the item the customer is purchasing, such as multiple pairs of socks or packages of underwear. Whatever the suggestion, it should be made after the customer has decided to buy the first item and definitely before the customer goes to the cash wrap-stand to pay for the merchandise.

For example, suppose a customer has decided to buy a pair of taupe gabardine pants. The salesperson can suggestion sell by saying, "We just received a cotton sweater that matches these pants perfectly. Let me show you how great they look together."

You will get the best results from your suggestions if you remember these points:

- Make your suggestion specific, don't say, "Is there anything else?" Say, "Let me show you."
- Make your suggestions customer-oriented. Tell customers how they will benefit if additional merchandise is purchased.
- Suggest with enthusiasm. If you're not excited about the additional merchandise, why should the customer want to see it?

When a customer is in the mood to buy, the salesperson should suggest additional items. These suggestions benefit the customer, who is able to avoid additional trips to the store to buy the related items he or she needs. And the suggestions benefit the store with increased sales.

Figure 15-3 By selling a customer related or additional items, the customer benefits from having a more complete wardrobe. And, the store gains from the larger sale. © Richard Hackett/Courtesy of Urban Outfitters.

Experience 15-7 ■ Suggestion Selling: It Helps the Customer and You

On a separate sheet of paper, list three merchandise items from your department that you could suggestion sell to the following customers.

1. The customer has purchased a pair of school shoes for his daughter. (You work in the Children's Shoe Department.)
2. The customer has purchased a scarf. (You work in the Women's Accessory Department.)
3. The customer has purchased a man's blue terry cloth bathrobe. (You work in the Men's Furnishings Department.)

■ BUILDING GOODWILL

Every store wants its customers to leave with a good feeling about the salespeople and merchandise. As a salesperson, you can build this good feeling by treating all customers courteously. Tell customers you appreciate their business. Say, "Thank you, and please come in again." Some stores encourage salespeople to write their customers thank you notes. Other stores provide their salespeople with business cards that the salesperson can give to the customer. Try to identify your regular customers and call them by name. Ask how they like an item they purchased in the recent past. They will be impressed by your ability to remember them.

At Burdine's, a chain of department stores in Florida, salespeople end the sale by asking customers if they would like to receive advance information about new merchandise or special sales. Most customers are flattered by the special attention and agree to fill out a client card. When the salesperson locates a clothing or accessory item he or she feels the customer would be interested in, or when there is a special event happening in the store, the salesperson telephones the customer or sends a postcard. Salespeople who build a client file and then follow up with those customers regarding information the customer wants to know increase their personal sales and the store's profit. In addition, customers who receive this extra service tell their friends, and the salesperson's clientele grows.

■ MERCHANDISE RETURNS

Customers return merchandise to a clothing store for many reasons. Sometimes, the merchandise is defective. For example, a shirt may be constructed poorly or may be made from a poor-quality fabric that fell apart when it was laundered. Sometimes customers make a mistake when choosing merchandise. Whey they arrive home with their purchase, they become dissatisfied with the color, size, or style of the item. Some people return gifts that don't fit or don't suit their taste.

Most stores establish their own return policy that guides employees in deciding which returned merchandise items to accept from customers. Usually, all stores accept the return of defective merchandise, except items that are sold at final closeout, "all-sales-final" sales or are marked "as is." Often stores allow merchandise to be returned if the return is made within a reasonable time period, such as 1 or 2 weeks after it was purchased. Many stores require the sales slip to accompany the returned merchandise as a proof of purchase before giving the customer cash back. Unless the returned item is defective, most stores require that the merchandise not have been worn.

Because returns are a drain on profits, store management tries to reduce the number of returns as much as possible. One way of doing this is by training salespeople to help customers purchase merchandise that best fits their needs and wants.

During their sales training, salespeople should be warned against misrepresenting merchandise, that is, making false claims about a merchandise item. Salespeople should also realize that every customer return is another potential sale. Because the item the customer is returning is not exactly right, the customer is still in the market for another merchandise item—one that better fits his or her needs.

Whenever you deal with a customer who is returning merchandise, be as cheerful and pleasant as you were when the customer originally made the purchase. Listen to why the customer wants to return the merchandise. If the merchandise meets your store's return policy, process the return by completing your store's return forms. Try to turn each return into an exchange. At the time of the return, always suggest other merchandise that will better meet your customer's needs.

Experience 15-8 ■ Put It All Together

Choose a particular clothing or accessory item. Bring that item to class for your sales presentation. On a sheet of paper, prepare a product analysis chart for the item. On another sheet of paper, prepare an objection analysis chart.

Bring to class five related items that you could suggestion sell with the item. Ask a classmate to pose as your customer. Set the stage for the customer. Explain the type of store, price range of your merchandise, and type of merchandise the customer is going to buy. Ask the classmate to act like a customer who seriously wants to buy the merchandise. Ask the customer to come up with some real objections about the merchandise.

Role play a sales presentation. Use a tape recorder to record your sales presentation. Do as follows:

1. Use an appropriate approach.
2. Determine the customer's needs.
3. Point out the benefits of owning the merchandise.
4. Demonstrate the merchandise.
5. Overcome objections.
6. Close the sale.
7. Suggestion sell.
8. Build goodwill.

CHAPTER 16

Management and Administration

Managers have tough jobs. They have to choose the right people to work in their stores and train them to be productive. And it's the manager who has to counsel employees when they aren't doing a good job and fire employees who can't perform the job duties. The manager is also responsible for watching expenses, not going over budget, and making sure the store makes a profit. No wonder the manager in the Cathy comic strip hides in a "secret little room!" Clothing and accessory store managers have lots of responsibility, and in order to operate a well-run, profitable store, they must be well-trained.

■ WHAT IS A MANAGER?

There are many different tasks involved in keeping a clothing and accessory store operating smoothly and efficiently. Some of these tasks are buying, receiving, stocking, and selling merchandise. Others are providing security and credit and controlling finances. In a small store, all these tasks are done by the owner and probably two or three employees. In large stores, however, there may be hundreds of people doing these tasks. In all stores, someone must coordinate the work of the other employees in order to achieve the store's goals. This "someone" is the manager. (See Figures 16-2 and 16-3.)

Although there are many different types of managers, with many different titles, they all basically perform the same management tasks. They are all involved in planning, organizing, directing, and controlling the work of other employees and store operations. Some of the management tasks to be discussed in this chapter are:

- Recruiting and hiring for various jobs
- Supervising and evaluating employees
- Planning and controlling store expenses

Figure 16-1 Cathy's view of management. Cathy © 1986 Universal Press Syndicate.

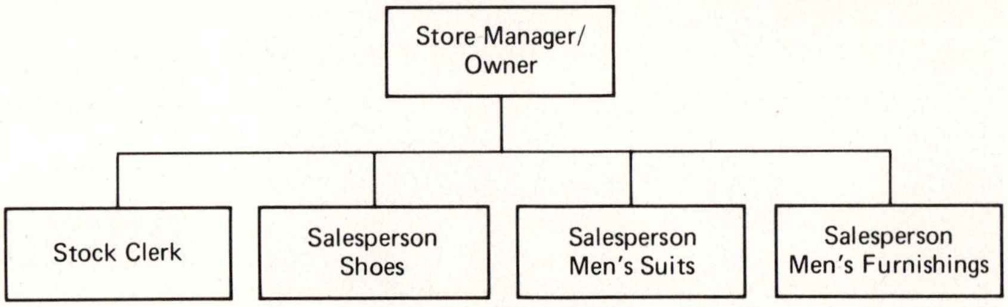

Figure 16-2 Here's a typical example of an organization chart of a manager-owned store.

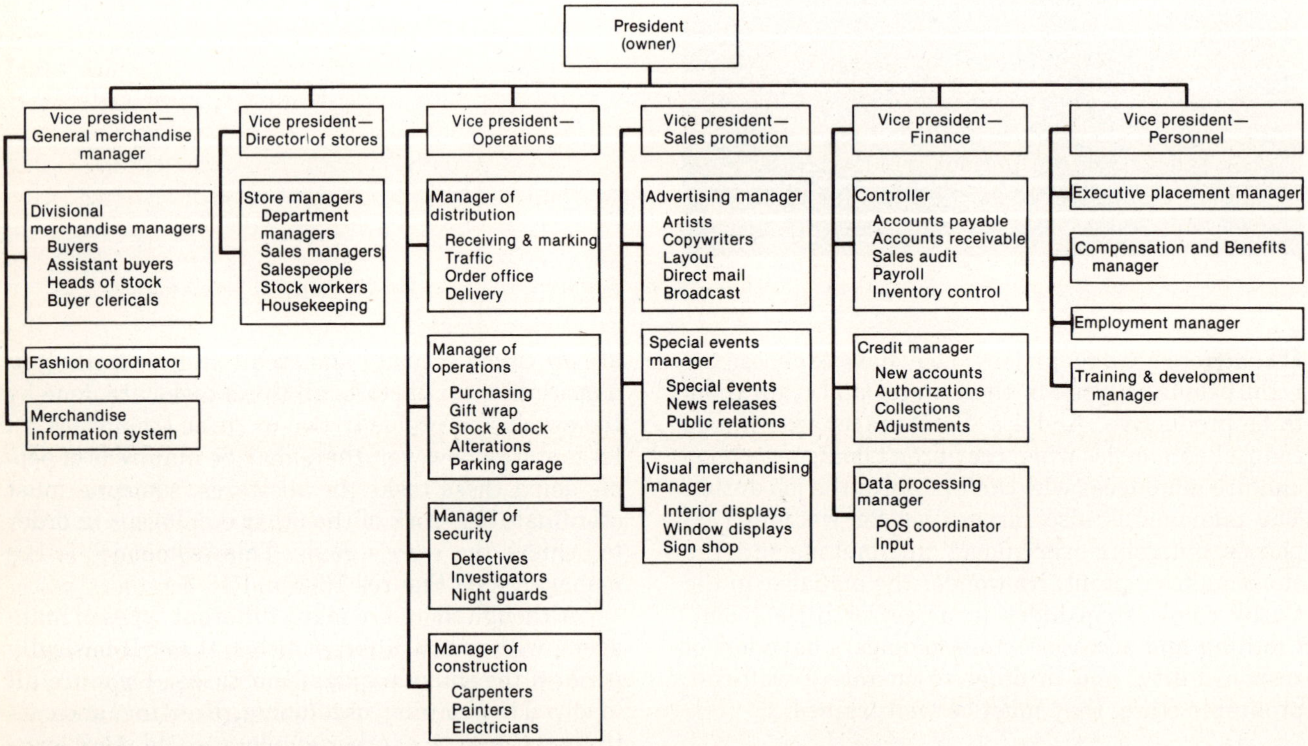

Figure 16-3 In a large store, supervisors, managers, directors, vice presidents, and a president manage the work.

Other management tasks, including training and motivating employees to do a good job, were reviewed in Chapter 3.

■ RECRUITING AND HIRING

In small stores, the owner or store manager is usually responsible for recruiting and hiring new employees. In larger stores, this task is often handled by the personnel manager. The personnel manager often works with the store manager and department manager in deciding which applicant is best suited for the job.

The basic procedures for finding and hiring the best person for the available job are the same, no matter how large or small the clothing store is. The manager in charge of hiring is looking for a person who is physically, mentally, and emotionally able to perform the duties of the job. By choosing the right person to fill a job, the store reduces employee turnover and training costs. The store also gains an employee who is satisfied with his or her job and who has a good effect on the other employees' morale.

Although each clothing store has different recruiting and employment guidelines, most stores use employment application forms and job interviews when deciding which applicants to hire. Before starting the recruiting process, many stores prepare job descriptions.

Preparing Job Descriptions

A job description lists in detail the responsibilities of the job and the skills required by the worker to perform the job. For example, some of the responsibilities listed on the job description for a salesperson in the accessory department may include the following:

- Meet the department's sales productivity standard (that is, sales per hour)
- Explain merchandise features and benefits to customers
- Suggest related merchandise to customers
- Enter transactions into the register
- Make change accurately
- Handle customer complaints tactfully
- Prepare effective displays

Experience 16-1 ■ What Is the Job?

1. If you are employed by a clothing and accessory store, prepare a job description for your job. If you are not employed, visit a local clothing store and prepare a job description for one of the sales positions. It will be necessary for you to talk to the salespeople to find out what types of work they are expected to do. Be sure the job descrip tion you prepare includes all the responsibilities you or the salesperson are expected to perform.
2. After you have prepared the job description, show it to your boss. If you don't work, ask the manager of the clothing store you visited to read and evaluate your job description. Find out if you were able to identify the basic responsibilities of the job. Add any that were omitted. Hand in your completed job description to your instructor for evaluation.

Identifying Recruiting Sources

Employees at a clothing store can be classified into three groups: management, sales, and sales-support. (This last group includes stock workers, housekeepers, credit employees, secretaries, display assistants, tailors, and so on.) When there is a job opening in any of these groups, the person in charge of recruiting and hiring looks for qualified people inside and outside the store.

Often, managers fill an opening by promoting a person who already works for the store. The promotion usually requires the employee to accept greater responsibility and do more difficult work. When there are no qualified current employees, the manager looks to other recruiting sources to find qualified applicants.

Most stores accept applications for future openings at specific times during the week. When there is an opening, the manager reviews these applications for qualified candidates. Many stores advertise their job openings in the local newspaper. Sometimes stores recruit new entry-level management employees through personnel recruiters who visit schools or colleges. Other stores recruit new employees from marketing education, fashion merchandising, and vocational school programs. Some stores find qualified applicants through employment agencies.

Reading Employment Applications

Most stores ask all job applicants to complete an employment application (called an "ap" for short) before conducting an interview. The ap gives the manager information about the job applicant's work history. Managers read the completed application carefully before meeting the job applicant. (The applicant may also have submitted a résumé sheet with a cover letter.) They take special notice of the applicant's handwriting and the accuracy and neatness with which the form was completed. They believe that a hard-to-read or incomplete application is an indication that they will have a hard time reading the forms that the applicant must complete when working at the store.

The completed application forms help managers decide what questions they should ask job applicants during the job interview. For example, if an applicant writes "worked as salesperson at Bank's Clothing Store," it is natural for the manager to ask about the applicant's duties at the store.

Conducting Job Interviews

As a manager, you will interview job applicants. Your interviews will be more meaningful and easier to conduct if you can make the applicant feel comfortable. Smile, shake hands with the applicant, and ask him or her to sit down.

Begin the interview meeting with general conversation. For example, you may want to start off by talking about the weather or a local event the applicant would probably know about (such as an important football game or some other impersonal topic). Or, you may offer the applicant a soft drink or cup of coffee. Any subject that does not refer to the applicant's eligibility for the job will ease his or her nervousness.

As the interviewer, you should guide the job interview. It is your responsibility to find out if the applicant has the ability and desire to do the job as well as to tell the applicant about the job and company.

Finding Out About the Applicant After you have put the applicant at ease, find out about the applicant's qualifications. Even though much information about the applicant is written on the employment application form, you will want more details about the applicant. Some of the questions you may ask will relate to the applicant's past jobs, courses in school, ability to work the required schedule, communication and math skills, ability to sell, and motivation for working at your store. Some usual interview questions are:

- What kind of work did you do at your last job?
- Why did you leave that job?
- What kind of work would you like to be doing in 5 years?
- In school or in any of your past jobs, were you ever in charge of planning and organizing a program or project? Tell me about it. How did it turn out?
- Have you ever supervised people? When?
- Would you be able to work every Thursday evening and either Saturday or Sunday each week?
- How soon can you start?

When asking questions, try to use words the applicant can understand. Ask the questions one at a time so that the applicant will not be confused.

Explaining the Job Some of the information you should provide the applicant are:

Duties the applicant will be expected to complete, such as explaining product information, answering customer questions, making sales presentations, operating cash registers, handling money, and making change.

Working conditions, such as the surroundings the applicant will work in, the hours the applicant is expected to work, and the types of people with whom the applicant will work.

Fringe benefits, such as vacation time, insurance programs, and employee discounts.

Wages, such as how much and how often the applicant will be paid and when paydays are.

Opportunities for advancement, such as other positions into which the applicant could be promoted.

The interviewer should describe the job to the applicant as realistically as possible and be careful not to oversell the job by painting a rosy picture that just isn't true. After explaining the job, the interviewer should find out if the applicant has any questions about the job. Those questions should be answered at this time.

Closing the Interview Close the interview in a polite, cordial way. If, during the interview, you are convinced that the applicant is the best person for the job, you may want to end the interview by offering the job to the applicant. And you may arrange the time when the applicant should report to work.

Sometimes an interviewer wants more time to decide whether or not to hire the job applicant. In this case, you may end the interview by saying, "I appreciate the time you spent with me this afternoon discussing the opening in our shoe department. We have had several applicants for the job. I want to review everyone's qualifications before making my final decision. I should know by the end of this week."

It's important to tell the applicant when the decision will be made. It is not fair or ethical to keep an applicant waiting any longer than necessary to find out your decision. While the applicant is waiting to hear from you, he or she may pass up other job opportunities.

Sometimes you will interview a job applicant who is obviously not qualified for the job. At the end of the interview, it will be your responsibility to tell the applicant that she or he does not meet the job requirements. This need not be an unpleasant task. Be tactful and honest. The applicant probably realizes that there are several people applying for an opening and not everyone can be hired. In situations like this, you may say, "I appreciate your coming in to talk with me about this opening. I do not believe, however, that you would be happy working as a salesperson. The job involves standing and operating the electronic cash register. During our conversation, you said you prefer to have a job where you can sit down and that you were not comfortable operating a cash register at your last job. I will keep your application on file and call you in the future if there are any other job openings that match your skills and interests. Again, thanks for coming in."

Choosing the Best Applicant

The manager in charge of hiring should carefully consider the needs of the job and hire the applicant who best fits those needs. Sometimes the decision is easy because one applicant is obviously the best qualified for the job. Other times, one applicant may not have as much experience as another, but is more enthusiastic and eager to do a good job.

Following the Employment Laws

There are various federal and state laws that apply to recruiting and hiring new employees. All businesses must abide by these laws. For example, the Equal Employment Opportunity Act prohibits discrimination in any matter of employment because of race, color, sex, religion, or national origin. Other federal laws prohibit discrimination against people who are over 40 or who are pregnant. Stores that discriminate against any applicant because of these reasons can be prosecuted.

Under the Federal Labor Standards Act, the government has established a minimum hourly wage for nonsupervisory employees. Because of the labor shortage that is affecting retailers, most clothing and accessory stores pay more than minimum wage. These employees, including salespeople or stock workers, must receive one and one-half times their regular pay rate for hours worked beyond 40 during a work week. The law requires employers to give equal pay for equal work to both men and women.

Retailers may employ minors who are at least 16 years old in any job, except in those jobs declared hazardous, such as stock jobs that require work around a trash compactor. Because each state has its own employment laws, including those that determine when meal and rest breaks must be given, contact your state government offices for additional information.

New-Hire Paperwork

Usually, the first day on the job, the manager has the employee complete many new-hire forms, including a form called W-4, Employee's Withholding Allowance Certificate, and I-9. The W-4 form, which asks for the employee's social security number, is used to determine how much to deduct from the employee's paycheck for taxes. The manager asks to see proof that the new employee is eligible to work in the United States, such as a social security card and driver's license, and writes this information on the Form I-9. (Any company that employs illegal aliens will be fined.) If the company offers health insurance, life insurance, or other fringe-benefit programs, the manager explains these benefits to the employee. The employee also receives a discount card, entitling him or her to reductions in the price of merchandise sold at the store.

■ SUPERVISING EMPLOYEES

As a manager, one of your major responsibilities is to supervise people. Your employees will look to you for guidance, direction, and encouragement. You are also expected to plan and organize your own work as well as that of your employees. And you schedule your employees' working hours, delegate responsibility, and evaluate job performance.

Planning Work

Planning means deciding in advance what needs to be done, who will do it, and when it will be done. To be successful, you must be a good planner. You must learn how to set your goals and those of your employees. And you must identify how to achieve those goals.

Principles of Planning You will be involved in planning what you and your employees should accomplish every day as well as what should be accomplished in the future. The principles of short-range (within the next 6 months) and long-range (in the future) planning are basically the same. They are as follows:

Identify exactly what you want to accomplish. Make your goals as specific as possible. For example, instead of setting the goal "increase sales in the Women's Sportswear Department," set the specific goal "increase the sale of blouses in the Women's Sportswear Department by 10 percent by the end of the season."

If you want to accomplish more than one goal, prioritize your work. That is, decide what goal is most urgent and write a *1* next to that goal. Write a *2* by the next most urgent goal, and so on.

Identify what you must do to accomplish each goal. It's easier to reach a goal that is divided into smaller tasks. Set realistic deadlines for completing each subgoal. By setting a time limit for accomplishing each subgoal, you will be able to achieve your major goal.

When planning, realize that even the best plans sometimes do not work out. You may have overlooked an important fact that affects your plans or you may have set an unrealistic deadline. Circumstances beyond your control may require you to change your plan. Remember to remain flexible. Evaluate your plans occasionally to see if they are being met or need to be revised.

Scheduling Employees

As a manager, you prepare your employees' work schedule. This is a written schedule showing the days and hours when each employee is scheduled to work. You want to have enough employees to handle customer sales and store security. You don't want so many employees that some have nothing to do. Clothing and accessory retail stores follow several different methods of scheduling employees. Many stores use last year's sales, this year's sales trend, and a salary-sales ratio when preparing employee work schedules.

Computing Planned Sales Suppose, for example, you are the manager of the Junior Sportswear Department. Last year, on the third Monday of June, your department sold $900 of merchandise. This year, your store is averaging a 20 percent increase in business over last year. (This is your sales trend.) Thus, your planned sales figure for the day is $1,080.

$$\$900 \times 0.20 = \$180$$
$$\$900 + \$180 = \$1,080$$

Computing Salary-Sales Ratios Let's also suppose that your store operates on a salary-sales ratio of 10 percent. That means that the total salaries of your salespeople should equal 10 percent of your day's sales. Using a 10 percent salary-sales ratio, your total salaries should not exceed $108.

$$\$1,080 \times 0.10 = \$108$$

If all employees in your department earn $5.50 per hour, you have enough money to pay for 19.6 hours of sales work for this particular third Monday in June.

Of course, you do not have to use all the time available.

Computing the Hours to Be Worked Once you know how much time you can schedule for employees (19.6 hours in this situation), you need to determine when the employees will work. Suppose that your store is open from 10 a.m. until 9 p.m. on the third Monday of June. You need at least one employee in the department at 9:45 a.m. to open the cash register. You also need one employee to stay from 9 p.m. to 9:15 p.m. to close out the cash register. Therefore, let's assume you need at least one person in the department from 9:45 a.m. to 9:15 p.m., a total of 11½ hours. To avoid paying overtime, the person who opens the store will be different from the person who closes the store.

During the day, you will need more than one salesperson in the department. Usually, the busy sales times of the day, or peaks, are from noon to 2 p.m. and 6 p.m. to 8 p.m. More salespeople are needed to help the customers. There must also be enough salespeople working so that all workers can take meal and rest breaks.

Completing the Staffing Sheet Most stores have some type of staffing sheet on which the manager writes each employee's name and the hours when that employee will work. This sheet is prepared in advance and should be posted where all employees can read it.

One possible schedule for the Junior Sportswear Department for the third Monday of June is listed on the staffing sheet in Figure 16-4. Notice that the time each employee comes to work, goes home, and has a meal break is different. Like the store in Figure 16-4, many clothing stores schedule their employees to work 3-, 4-, 5-, 7½-, or 8-hour shifts. Stores schedule their employees so as to have maximum coverage of the sales floor during the busiest times of the day.

	Sales	$1,080	
	Hours	19.5	
	Productivity	$55	
	Monday		Tues

Salesperson	Open	Meal	Close	Open	M
Jamie Parsons	noon	4:00	9:00		
Lynn Culver	——		v a c	a t i	o
Marsha Ortega	9:45		12:45		
Jon Greene	1:00		5:30		
Billy Washington	5:15		9:15		

Figure 16-4 The department manager schedules the sales staff so that there are enough people to serve the customers adequately, but each salesperson is productive.

Scheduling Meal and Rest Breaks Most clothing stores allow employees who work more than 6 hours either a 30-, 45-, or 60-minute meal break. Employees who work more than 4 hours receive one 15-minute rest break for every 4 hours worked. The rest breaks are not usually written on the schedule sheet; they are taken during slow times.

Before your employees take their break, be sure there are enough salespeople remaining on the sales floor to handle sales and store security. If there are not enough salespeople to handle the work load, call in other salespeople from slow departments to help. Or reschedule the employee's break to a later, slower time.

Another Scheduling Method

Stores that do not use projected sales or salary-sales ratios usually rely on the common sense of their managers when they prepare work schedules. The managers realize that there must be at least one person in the department at all times, no matter what the sales are, for security reasons. When this person goes to lunch, dinner, or on a rest break, a store floater may be assigned to enter the department and take over. A **floater** goes from department to department when needed to fill in for the regular salespeople. A group of floaters are sometimes called the "flying squad."

A Final Word About Scheduling

Scheduling employees is a tricky business. Even though there are elaborate formulas to help managers determine how many people should work and when they should work, these formulas are not foolproof. Sometimes a department will have more workers than needed, even when the manager has followed the formula. This problem shows that it's very hard to guess how many customers will be shopping in your store at any particular time of the day.

Experience 16-2 ■ How Many Workers?

As a manager, you must prepare a written schedule for your ten salespeople. Your store uses last year's sales, this year's sales trend, and a salary-sales ratio to determine how many salespeople can work on any particular day. You are responsible for determining when these workers will come to work, go on a meal break, and leave.

Last year's sales by day of the week and store hours are listed below. (You are closed on Sunday.) Two employees must be in the department at 9:30 a.m., and two must stay until 9:30 p.m. to close out the department. The busiest times of the day are noon to 2 p.m. and 6 p.m. to 8 p.m.

1. Compute the planned sales for the fourth week in July. Your department is averaging a 15 percent increase in sales over last year. Write the planned sales in the appropriate column.
2. Using the planned sales figures from no. 1 above, compute the maximum number of hours you will be allowed to schedule employees each day. Assume your employees earn an average of $5.00 per hour. Your store management asks you not to go above an 8 percent salary-sales ratio. Write the maximum number of hours to be scheduled in the space provided. (Round all numbers to the closest hour.)
3. Using the following information, as well as the maximum number of hours to be scheduled from number 2 above, complete the work schedule for the week. On a separate sheet of paper, make a weekly schedule using Figure 16-4 as a guide.

Assume your store allows each employee who works 6 hours a day a 1-hour lunch break. Of the 10 employees in your department, Carol, Lindsey, Michael, and Jason are full-time, averaging 37½ hours per week. Three employees, Amanda, Jennifer, and Mitchell, usually work 20 to 24 hours each week. The other three employees, Jill, Ian, and Bob, all average about 10 to 12 hours per week.

Jill, Ian, and Bob go to school; they can work evenings and weekends only. Carol prefers Monday off; Lindsey wants Wednesday off. Amanda can only work when her children are in school, between 9:30 and 2:30.

Day of Week	Store Hours	Last Year's Sales	This Year's Planned Sales	Maximum Number of Hours to Be Scheduled
Sunday	Closed			
Monday	10 a.m.–9 p.m.	$1,540	______	______
Tuesday	10 a.m.–9 p.m.	$1,080	______	______
Wednesday	10 a.m.–9 p.m.	$1,940	______	______
Thursday	10 a.m.–9 p.m.	$2,660	______	______
Friday	10 a.m.–9 p.m.	$3,420	______	______
Saturday	10 a.m.–9 p.m.	$3,980	______	______

Delegating Work to Others

Clothing and accessory managers are responsible for a variety of job duties, such as analyzing sales records, checking in new stock and interviewing job applicants. Because there is usually more work than time in which to do it, effective managers delegate some of their job duties to subordinates.

Delegation is giving to a subordinate the manager's authority or power to accomplish a particular task. For example, on the manager's day off, he or she may delegate to a salesperson receiving, checking-in, and display of new merchandise. Or the buyer may delegate to the assistant buyer the preparation of an advertising request.

When you delegate some of your work, you must:

1. Decide which tasks can be delegated.

2. Choose a capable person to do the job.
3. Assign the job.
4. Control the delegated task to be sure the desired outcome occurs.

The best tasks to delegate are routine duties that occur over and over. For example, basic stock counts that are required every week are easy tasks to delegate. Taking markdowns can also be delegated to careful, experienced employees. Some managers delegate minor decisions to long-term employees, such as allowing them to approve customer checks or okay customer returns. Some managers delegate jobs that take a lot of time to complete, such as building displays. There are some tasks that should not be delegated, including those involving confidential information and personnel matters. It usually doesn't pay to delegate infrequent tasks that require a great amount of skill and experience to complete. It takes more time to train a subordinate to complete an infrequent task than it does for you to do it yourself.

When deciding which employee will be given the additional authority and responsibility, consider the employee's ability and willingness to do the job. Some employees may already know how to complete the task and may be eager to do a good job. Other employees may be resentful about being asked to do something extra, even though they know how. Some employees may be willing to do what you ask but need training to accomplish the task. To ensure a good outcome, delegate your tasks to employees who are willing *and* able to do the job. If necessary, train a willing employee in a new task. This is an excellent way to build confidence and morale.

When assigning the work to an employee, ask the employee if he or she will help you complete a particular task. Don't order the employee to do it. When someone is asked to do something, he or she will usually do it and feel good about it. But when someone is ordered to do something, the person may become uncooperative or purposefully hinder the outcome. Explain in detail what you expect the person to do, and especially describe the outcome you want. Tell the employee how much time he or she has to complete the work, when the task should be finished, and how much freedom he or she has to complete the task. Be sure to tell other people whose cooperation is needed that you have assigned a particular task to a member of your staff.

After a task is delegated, the manager should follow up to be sure the desired outcome will occur. A good delegator checks back occasionally to find out how things are going but doesn't hover over the employee. Although the manager may offer to help by giving suggestions, a good delegator allows the employee the freedom to complete the task within the guidelines described. When the task is finished, the manager should evaluate the work, comparing it to the desired outcome and giving appropriate praise and suggestions for improvement.

For example, suppose you choose Stephanie to orient a new employee to your department during the week you are on vacation. Stephanie has worked in the store over 3 years and has always received outstanding ratings on her performance reviews. She is known for her willingness to help new hires and for her friendly attitude.

The week before you go on vacation, you ask Stephanie if on the next Monday she will orient Doug, a new employee, to the area. You ask her to introduce Doug to his co-workers, show him his work schedule, explain the arrangement of merchandise in the area, and review features and benefits of the best sellers and basic stock items. You tell her to allow 3 hours on Monday for the orientation. You also let the other employees know that Stephanie will be conducting this orientation for you, so they will cooperate with her and know what she is doing. When you return from vacation, you ask Stephanie how the orientation went. You also ask Doug some questions to be sure he understood and retained what Stephanie explained to him.

Managers who delegate effectively receive many rewards. They have more time to concentrate their efforts on essential, important matters. By delegating different tasks, they make their employees' jobs more interesting. They also help their staff master new skills and become more valuable employees. In fact, effective delegation is a good way to improve morale!

Experience 16-3 ■ Delegate!

1. Assume you are a manager in a clothing store. Identify a particular recurring, routine task that could be delegated to a salesperson. With one of your classmates playing the role of salesperson, demonstrate how you would delegate the task to this employee. Be sure to explain the outcome you expect and what guidelines the employee should follow to complete the job. Explain how much time the employee has to complete the task and when it should be completed. Encourage the employee to ask questions. Continue the discussion until you are confident the employee under-

stands what is to be accomplished. Use a tape recorder to record your conversation.

2. Identify other store employees who would need to be notified of the task you have delegated. Why is it important for you to tell others that you have delegated a task? Write your answer on a sheet of paper.
3. Identify three ways you could follow up to be sure the task was completed as you wanted. Use a sheet of paper to record your answer.

EVALUATING EMPLOYEE PERFORMANCE

Most employees want to know what their manager thinks about the work they are doing. They want to know whether they are doing a good job. They want their managers to give them feedback so that they know where they stand.

Today, many stores have established a formal procedure for evaluating employees. At least once each year, the manager rates each individual employee and conducts a formal discussion about the employee's performance. This meeting is called a **performance appraisal.** It is designed to provide a report of how well employees are performing in their jobs.

Criteria for Evaluating Employees

Many companies prepare a specific set of standards on which employees are evaluated. These standards are written on an employee performance rating form. (See Figure 16-5.) Before each performance appraisal meeting, the manager completes the employee's rating form. The manager should use objective data when completing the employee performance rating form. For example, the manager can use the employee's sales records to evaluate the quantity of the employee's work. The manager can refer to time cards to evaluate the employee's promptness.

Many employee performance rating forms ask the manager to evaluate certain traits of the employee, such as a cooperative attitude. When making this type of evaluation, a manager should back up his or her comments with specific examples.

When you complete an employee performance rating form, try to be fair and objective. Look at all the facts and don't let your emotions sway your judgment. Because someone's future may depend on the rating you give, take your job of completing employee performance rating forms seriously.

The Performance Appraisal Meeting

During a performance appraisal meeting, a manager should give the employee a firsthand report of the employee's progress. Areas in which improvement is needed should be discussed. The performance appraisal meeting should also create a deeper, better understanding and communication between the employee and manager.

A manager should encourage employees to participate in their performance appraisal meetings. To encourage participation, the manager may ask an employee how the company can make the employee's job more satisfying. Or the manager may ask about any on-the-job problems the employee has.

Both the manager and the employee should leave the appraisal meeting with a good feeling about what has happened. The meeting should build employee morale and give the employee a sense of direction and participation in the development of his or her career. It should make the employee realize that the quality and quantity of work done and his or her job performance really do matter and are appreciated.

FINANCIAL RESPONSIBILITIES

Besides hiring, supervising, and evaluating employees, many managers are responsible for preparing and administering a budget. A **budget** is a plan that shows how money should be spent. You prepared a promotional budget in Chapter 13, "Promotion." When budgets are followed, they help companies achieve their profit goal.

Managers may also be responsible for preparing financial reports, such as income statements. This report, also called a profit and loss statement, P&L, or operating statement, shows the amount of money a business earned during a specific time period. Managers analyze the financial reports to identify ways that profit can be increased.

Another financial responsibility of managers is depositing money in the bank, making sure the checks are endorsed properly and that the deposit slips are completed correctly. Stores that receive large amounts of cash often hire companies like Purolator or Loomis to pick up their deposits. This eliminates the risk that an employee will steal the money or be robbed.

Profit = Sales − Expenses

A clothing and accessory store earns income by selling merchandise. From this income, the store pays for its merchandise and all other expenses. Typical store expenses are salaries, taxes, rent, utilities, depreciation, advertising, display, supplies, fixtures, and equipment.

If there is money left over after a store has paid its bills, the store has earned a profit. If there isn't

RATING SCALE

When evaluating the associate, choose the statement that represents the frequency with which the associate has performed each responsibility, at the standard defined in the Sales Associate Job Description, and enter the corresponding letter in the appropriate space.

(E) EXCEPTIONAL - A rating of 'E' indicates the associate has made an outstanding contribution to the company, that is, the associate performed the responsibility at the standard about 95% of the time.

(V) VERY GOOD - A rating of 'V' indicates the associate has performed at a level above the company's standards, that is, the associate performed the responsibility at the standard about 80% of the time.

(G) GOOD - A rating of 'G' indicates the associate has performed at a level which meets the company's standards, that is, the associate performed the responsibility at the standard about 65% of the time.

(I) REQUIRES IMPROVEMENT - A rating of 'I' indicates the associate has performed at a level which is below the company's standards, that is, the associate performed the responsibility at the standard about 50% of the time.

(U) UNSATISFACTORY - A rating of 'U' indicates the associate has performed at a level which is not acceptable to the company, that is, the associate performed the responsibility at the standard less than 50% of the time.

(N) NOT APPLICABLE - This responsibility does not apply to the associate's job.

SALES ASSOCIATE RESPONSIBILITIES — RATING

SALES PER HOUR RESPONSIBILITIES _____

Selling productivity as measured by the Productivity By Pay Zone Report_____.
Rating scale: (U) More than 20% less than zone average; (I) 11 to 20% less than zone average; (G) Within 10% of zone average; (V) 11 to 20% higher than zone average; (E) Over 20% higher than zone average.
Pay Zone SPH __________.
Associate SPH __________.

Completes his/her Personal Sales Record every week_____.

SELLING RESPONSIBILITIES _____

Approaches_____. Determines needs_____. Relates product knowledge_____. Escorts the customer to the dressing room_____. Suggestion sells_____. Closes_____.

CASH WRAP RESPONSIBILITIES _____

Removes Sensormatic tags_____. Enters all transactions into the POS terminal correctly_____. Follows company policies and procedures for payment_____. Completes and distributes sales tickets_____. Creates good will_____.

CUSTOMER SERVICE RESPONSIBILITIES _____

Answers customer questions_____. Completes return and exchange transactions_____. Answers phone calls_____. Responds to the Sensormatic signal_____. Follows hold policy_____. Handles customer complaints_____.

Figure 16-5 Here is an example of the first page of an employee appraisal form. This form also provides space to evaluate other responsibilities, list the employee's strengths and areas that need improvement, and plan for the coming year's performance.

enough money to pay the bills, the store has suffered a loss. Most clothing stores hope to make a profit of 6 to 12 percent of total sales.

Expenses are very important in determining how much profit or loss a store earns. In the following examples, you can see what happens when sales remain the same but expenses increase.

Sales		Expenses		Profit or Loss
$100,000	–	$ 95,000	=	Profit of $5,000
$100,000	–	$ 99,500	=	Profit of $500
$100,000	–	$104,000	=	Loss of $4,000

As expenses go up, profit goes down. To obtain a profit, every employee must do his or her part to control expenses.

Preparing a Budget

When preparing a budget, managers try to estimate what expenses will be in the upcoming season. Some of these expenses are easy to estimate. For example, if rent has increased 5 percent every year for the past three years, a manager may assume that the rent will increase 5 percent this year also. If the sales people are paid a straight 8 percent commission on their sales, the manager multiplies the planned sales by 8 percent to compute the future salary expense.

Another large expense that is easy to budget is insurance. Store owners purchase insurance on their building and merchandise to protect it from fire, flood, and other types of risk. Although insurance is expensive and reduces the store's profit, most owners believe the cost is worth it. A major fire can destroy the building and all the merchandise, display props, and fixtures inside. Even merchandise that isn't burned will be damaged by water and smoke.

Stores also purchase liability insurance to protect themselves from lawsuits brought by customers who are injured in their store. Most stores have accident report forms that must be completed whenever a customer or employee has an accident on store property. Worker's compensation insurance guarantees the payment of medical expenses and salary to workers who are injured on the job. Stores pay unemployment insurance that provides weekly benefits to employees who have been laid off or have lost their jobs due to no fault of their own.

Experienced managers also budget for unexpected expenses that may come up during the season. An example of an unexpected cost is purchasing a new mannequin because a salesperson broke the arm of another mannequin while changing the window display. Although the manager didn't plan on buying a new display prop, the expense was unavoidable because the mannequin was needed immediately.

Administering the Budget

The manager who prepares the budget is usually the one who approves payment of a bill and sends it to the accounting department to be paid (the procedure in large companies) or actually writes the check and pays the bill (the procedure in small companies). This practice allows the manager to compare the difference in what was planned to the actual expense. If the manager notices some bills are higher than expected, he or she may have to reduce other expenses. For example, if expenses for advertising are over plan, the manager may decide to delay buying additional display props that were budgeted.

The manager who is responsible for paying payroll taxes and corporate income taxes must be sure these payments are made accurately and on time. Every business must make quarterly payments for the federal, state, and city income taxes that are withheld from employees' earnings. Stores must send in Social Security (also called FICA) payments each quarter. Every business is also required to make quarterly estimated corporate income tax payments to the Internal Revenue Service and the business's state and city tax divisions.

Preparing Income Statements

Most stores prepare monthly income statements to determine how much profit was earned. The first figure on the income statement is net sales. It includes all the merchandise purchased by customers during the month minus any returns. The **cost of goods sold,** which is the cost of the merchandise sold during the month, is the next figure. It is computed as in the following example:

	Opening inventory at cost	$ 43,228
+	Purchases	+ 36,351
		79,579
	Transportation charges for getting the merchandise to the store	4,870
	Total Purchase	84,449
+	Workroom or alteration charges (any costs of getting the merchandise ready for sale)	375
	Total	84,824
−	Closing inventory at cost	43,018
		41,806
−	Cash discount	727
=	Cost of Goods Sold	41,079

The next figure, the gross margin, is the difference between the net sales and cost of goods sold. This is the amount of money a business earned by selling merchandise. All the expenses of the store and the taxes must be paid from this figure. Expenses and a provision for income taxes are recorded on the income statement. Whatever is left over is the net profit earned. Gross margin is one of the factors most buyers are evaluated on.

Managers analyze each income and expense item to determine how the profit can be increased. For example, if prices were increased, net sales may go up. However, a price increase could also hurt sales because customers may decide to purchase their merchandise at another store. The gross margin can be increased by paying every bill in time to earn the cash discount. It can also be increased by finding different freight companies with lower rates.

Each entry on the income statement is shown as a percent of net sales. The net sales percentage is always 100 percent. Over time, a store establishes an acceptable percent of net sales for each item on the income statement. Management analyzes the percentages to identify any that are higher or lower than usual. The manager also finds out the reason for the variance and ways to bring the percentage back to the usual range. Every manager's goal is to increase the gross margin percentage and keep the expense percentages as low as possible.

■ CONTROLLING EXPENSES

Even though it's the manager who prepares and administers the budget, every employee has an effect on a company's expenses. To control expenses, stores establish specific guidelines for accepting checks and charges, reducing shortage, controlling supplies, and protecting fixtures and equipment.

Accepting Checks

Many customers use checks to pay for their purchases. Most customers' checks are good; some, however, are bad. A customer's check is bad when the amount of money the customer has in the bank is less than the amount written on the check. It is hard to collect bad checks. Clothing and accessory employees who carefully examine checks and customer identification before accepting this form of payment can help protect their store from bad checks. This practice reduces the store's expenses and protects its profits.

Credit Control

Today, over half the sales in many clothing stores involve credit. Credit services provide many advantages to customers. Credit makes it easier for customers to purchase merchandise they want and need when they want and need it. It eliminates the customers' worry about carrying large amounts of cash.

Credit also provides stores with many advantages. It produces sales that might never have been made because customers can make purchases even though they may not be able to pay for the purchase at that moment. Credit also makes a store's customer more loyal. The "charge," or credit, customer will continue to return to the store to make additional purchases.

Credit is an expense to the store. It costs the store money to set up a credit department and have employees administer the credit policies of the store. The cost of providing credit is further increased when some customers refuse to pay for their credit or charge purchases. That's why stores set up credit policies for their managers and salespeople to follow when accepting customers' charges. It is the clothing and accessory manager's responsibility to see that salespeople know the credit policies and handle the credit transactions according to these policies.

Store Credit Procedures Some clothing and accessory stores issue their own credit cards to customers. These stores have credit departments that approve credit applications, authorize store charges, and handle customer collection and billing. Once a customer's application is approved, the store issues the approved customer a plastic charge card that can be used to buy merchandise at any of the store's outlets.

The credit department at these stores establishes certain credit guidelines to keep the number of "bad" charges under control. All managers and salespeople must follow these guidelines when accepting a store credit card as payment for merchandise.

Each salesperson and manager has the authority to accept customer charges up to a certain limit. The largest amount for which you are permitted to accept charges without authorization is called the **floor limit.** When charges are more than the floor limit, the salesperson or manager must call the store's credit office for charge approval.

Bank Credit Card Procedures Many clothing and accessory stores also accept bank credit cards, such as Visa and MasterCard, and national credit cards, such as American Express. The guidelines for accepting these cards are similar to those for accepting retail store charge cards, but there are some differences.

On purchases over the store's floor limit, the salesperson must telephone the credit card company to get an authorization number before ringing up the sale. In many stores, computerized phone systems allow salespeople to access the authorization number. Computerized systems are programmed to decline charges to cards that have expired or have been reported as lost or stolen.

Shoplifting

When customers take merchandise from a store without paying for it, they are **shoplifting.** This criminal act decreases the amount of profit a store earns. In some stores, the value of merchandise that is stolen is as high as 5 percent of sales. Store employees can help reduce the cost of shoplifting by contacting their security department or manager when they see someone shoplift. The very best deterrent to shoplifting is an attentive salesperson. Managers are responsible for training employees to follow the store's policies regarding shoplifting.

Spotting Possible Shoplifters Most customers who enter a store are honest. They are more interested in examining the merchandise than in looking around to see if someone is watching them. Shoplifters are different, often more nervous, than the honest customers. They keep looking around to see who is watching them. They may look behind themselves as they leave the department. They either don't want the salespeople to see them at all or they do something to distract the salespeople's attention.

Anybody may be a shoplifter—the best-dressed one in the store, the wealthy person, or a regular customer. Managers should instruct salespeople to watch for any suspicious behavior.

Store Conditions That Make Shoplifting Easy There are certain conditions that make it easier for customers to shoplift. Shoplifters like working in a crowded department where people can't keep too close a watch on them. They also like stores with inattentive salespeople who pay little attention to customers. Store employees should make a special effort to acknowledge all customers who enter their department. Salespeople who are busy helping another customer should tell new customers, "I'll be with you in a moment." They should then keep glancing toward the customers so that the customers know they haven't been forgotten.

Ways to Prevent Shoplifting Most stores try hard to prevent shoplifting. They attach inventory control devices to their merchandise that set off an alarm when they pass through an exit. They display expensive items on locking hangers or in locked display cases, out of the customers' reach. They install cameras that record customers' actions and hire security agents to observe customers.

It is difficult for salespeople to keep track of all the merchandise, especially the merchandise that customers take to the dressing rooms. On busy days, many stores station a security person at the entrance to the dressing room. This person counts each of the garments and gives the customers a numbered tag corresponding to the number of items being tried on. When the customers come out of the dressing room, they must give the person the printed tag and the merchandise items. When there is no security person stationed at the dressing rooms, salespeople must be especially on the lookout for shoplifters.

Some ways you as a manager can help control shoplifting in your department are:

- Telling your salespeople to remove any garments left in the dressing rooms.
- Encouraging your salespeople to notice the clothing their customers take into the dressing rooms. Ask them to make a mental note of the style and quantity.
- Instructing your salespeople to check with dressing room customers often. Neglected customers may leave the store with merchandise for which they haven't paid.

How to Apprehend Shoplifters Almost every store has established policies and procedures that explain what employees should do if they see a shoplifter in action. Some of these rules are:

- Never accuse the person of stealing. If the employee is mistaken, the store may become involved in a lawsuit for false arrest.
- Call the security department or store manager immediately. Many stores employ people who are trained in handling shoplifters.
- Write down a description of the shoplifter. Include the height, general build, color of hair and eyes, and color and type of clothing. This information will help the security department, store manager, or police to locate the shoplifter.

Employee Theft

Another cause of higher expenses and lower profits is employee theft. As manager, you know that stores must trust their employees to work honestly with the store's money and merchandise. You will find some dishonest employees who betray their employer's trust by stealing from the store. Employee theft is a big problem. As manager, you should be aware of some methods and preventions of employee theft. Table 1 outlines some of these methods and preventions.

Clever, dishonest employees can think of many ways to steal from their employers. Prevention programs cannot stop all employee theft. Most stores realize that they must rely on honest employees and managers who report internal theft.

Controlling Supplies

Clothing and accessory stores use a wide variety of supplies. When added together, the cost of these supplies is expensive and can quickly drain profits. A careful use of supplies is necessary to control expenses.

For example, every time a salesperson gives a customer a large bag when a small bag will do, the salesperson increases the store's expenses by 1½ cents. If this happens twice a day, 5 days a week, 52 times a year, the store loses $7.80 a year. If 10 salespeople in the store do this, the store loses $78 a year. If 100 salespeople do this, the store loses $780 a year—a substantial drain on profits. Even a small savings in some expense areas can mean a substantial gain in profits.

Handling Merchandise, Fixtures, and Equipment

Another way employees can decrease store profits is by carelessly handling merchandise, fixtures, and equipment. For example, it's not unusual for an employee to rip a garment when putting it on a hanger, especially when in a hurry. It's very easy to break a glass shelf when changing displays. Even cash registers require repairs when they are abused.

■ **Table 1**
Employee Theft Methods and Preventions

Methods	Preventions
Outright theft of merchandise from the sales floor	Provide employees with a locker so that no purses, coats, or packages are allowed on the sales floor.
Outright theft of money from cash drawer	Carefully check the cash register audit tape to determine the flow of money.
Giving employee discounts to unauthorized persons	Require management approval on all employee discount sales.
Stealing merchandise from receiving and marking area	Keep receiving room area locked when receiving and marking employees are off duty; do not allow unauthorized persons in this area.

The money lost on damaged merchandise and broken fixtures or equipment can easily increase the store's expenses and thus dip into profits. The worst part about these expenses is that they all can be avoided if store employees are careful. As manager, show your employees the proper way to handle merchandise, fixtures, and equipment. Be sure employees know that it is better to take their time to do a good job rather than a fast job.

■ ARE YOU READY TO BE A MANAGER?

By now you should realize that a successful manager requires many special talents. The person must have exceptionally good "people skills" in order to recruit, hire, supervise, and evaluate his or her employees. In addition, the manager must also understand how to prepare and administer a budget to ensure the store makes a profit. Considering all this, are you ready for the job?

UNIT 4 PROJECT

Developing a Marketing Plan for a New Product

In Unit 4 you have become acquainted with the marketing skills needed to operate a successful clothing and accessories business. Developing a marketing plan for a new product is a critical application of these skills.

As in the Unit 1 Project, you will assume the role of the manager of Neighbour's Shoes. You have considered the needs of your current and potential customers, and you have decided to add a new product to your inventory: a popular model of running shoes.

In this project you will develop a marketing plan to launch your new offering. The assignments throughout this project and the forms on pages 251—252 will direct you as you draft your plan. You will review your business philosophy, your marketing research, and your target market; set objectives; plan your marketing mix; and assess your plan. Financial projections will guide your plan.

Computer Users: Computations can be done using the software "Career Competencies in Marketing Series, Second Edition: Marketing Plan."

Your Business Philosophy

A business's philosophy is the rationale it uses in decision making. For example, the philosophy will guide the business in applying the marketing concept and offering an attractive product or service to its customers. The philosophy is also what differentiates a business from its competitors. While Uncle Joe and Aunt Sarah never wrote down a formal philosophy, you have summarized Neighbour's guiding principle as providing "the quality and service that families deserve, at prices they can afford." By doing so, Neighbor's should earn a profit. You have been seeking growth opportunities that are consistent with these values.

Marketing Research and Target Marketing

Your marketing research in the Unit 1 Project showed that your current customers have lower incomes than the county median. (See pages 31-32.) Furthermore, the county is growing in places where richer people live. You expect most of the population gain in the next five years to be in Sumpter Heights and the other sections of the county where suburban developments are under construction.

By contrast, most of your customers come from Central Beach. You currently serve about 10 percent of the county's households, and your "loyalists" generate over half of your sales. Loyalists are families throughout the county who buy most of their shoes at your store. They make up 40 percent of your customers and tend to be wealthier than your other customers.

The loyalists have similar feelings about shoe buying to those of your customers in general. They value comfort and durability in a shoe, and friendly and personal service in a shoe store. However, they are less concerned about price than your lower-income customers.

Your analysis of customer attitudes showed that middle- and high-income customers resemble your loyalists, but that they also value brand names. Since Sumpter Heights is distinct geographically, and resembles your loyalists and high-income customers, you have decided to target that segment for a new product introduction. Naturally, many other people in the county, including your loyalists, will be interested in your new offering. You may wish to include other segments in your target and provide ways of reaching those customers as well in your marketing plan.

Setting Objectives

The success of a new product introduction can be measured only if you establish the standards of success. These standards or "objectives" generally include sales levels, market penetration, and profitability. Once established, these objectives guide the development of each element of the marketing mix.

Sales levels refer to the units or dollar volume you expect to sell. For instance, one sales objective might be to "sell 500 pairs of the new running shoes from March 1 through May 30."

Market penetration is the portion of the total market that will buy your product. For instance, an objective for market penetration might be to "sell the new running shoes to 7 percent of the households in Sumpter County."

Profitability relates to the percent of each sales dollar that is retained as profit. For instance, an objective for profitability might be to "achieve a 50 percent gross profit on sales of the new running shoes." (Profits will be discussed in detail under "Estimating Sales and Profits.")

As discussed in Chapter 13 (page 187), objectives for promotional plans should include nonfinancial objectives, as well. The success of an advertisement can be judged in terms of increased store traffic, as well as in terms of increased sales of a given product. An objective, therefore, might be to "increase store traffic from the mall by 100 potential customers during Easter week."

As you plan your strategies, set measurable objectives, and try to provide a means of judging the success of each promotional element.

Product Selection

The Mercury Shoe Company makes a full line of sports shoes, and you have selected their Wings model to introduce this spring. Wings are available in a full range of men's and women's sizes and boast a list of attractive features, including:

Nylon and leather uppers
Mesh ventilation
Padded ankle collar and tongue
Flared wedge heel
Double cushioned mid-sole
Resilient outer sole
Lightweight
Toe guard
Choice of white, gray, or blue
Mercury brand name

You expect the shoe to appeal to your loyalists because of its practical good value. The Sumpter Heights crowd will appreciate the Mercury name, as well.

Assignment 1 Review the features of the Mercury Wings and the qualities valued by your target customers. Complete the feature/benefit chart in your marketing plan (page 251). In the right column list the benefits of each feature. Indicate the three main benefits that will appeal to your customers by placing an asterisk (*) next to each.

Pricing

The price you charge your customers for Wings must be high enough to cover your cost of goods, overhead, and planned profits. Furthermore, customers must consider it to be the right price for the product.

You must also keep an eye on the competition. Wings are not offered locally, although they are advertised on Metro City television and radio. Comparable quality shoes in Sumpter County sell for $36.99 to $45.99. Better running shoes at Good Sports can sell in the $60 to $120 range.

Mercury Shoes offers a volume discount to encourage sales. If you order fewer than 151 pairs of Wings in a 13-week season, then your unit cost will be $16.99. This unit cost drops to $14.99 for orders of 151 to 300 pairs. For orders of 301 or more pairs Mercury charges $13.99 each.

Your rule of thumb has been to establish a price by doubling your cost and adding $2. You plan markdowns to encourage sales as the season progresses. You expect your average markup to be 48 percent to 52 percent, based on selling price.

Assignment 2 On the New Product Marketing Plan form (page 251) summarize the product cost (at different order quantities), your competitor's prices for comparable shoes, and your expected price range based on your "2 times cost + $2" formula.

On a separate sheet of paper describe your price strategy as high, middle, or low and explain why you have chosen this strategy.

Estimating Sales and Profits

A realistic estimate of the number of shoes you can sell in a season will let you predict your profitability. Every retailer would like each new product to become a best seller, but that is not always possible. Many factors should influence your sales estimate. Consider the past history of similarly priced items. Look at the number of people in your target market, their interest in similar products, and your market share in general. Also, consider secondary target markets; since people outside of Sumpter Heights listen to the radio, read newspapers, and shop at the mall. When you promote the product, some of them are bound to show interest!

Estimating Sales

Your best-selling women's shoe last spring sold 300 units. Your most popular men's shoe sold 200 units in the same period. Both styles sold in the $25 to $35 range. You will carry Wings in both men's and women's styles. With the right promotion, you figure that 500 units (300 women's+200 men's) is the most you might expect to sell this spring. But who will buy these shoes?

To build a sales estimate, first indicate the size of your target markets. Consider other geographic areas, income groups, and age groups. Also consider your current customers; after all, you know each other and you have many of them on your mailing list.

Of course, you must also avoid "double counting" as your target markets may overlap. For instance, you may be interested in both Sumpter Heights and in people between the ages of 35 and 64. Don't count the 35 to 64 year-olds in Sumpter Heights twice. As Unit 1 Project, Table 2 (page 29) shows, 4,570 of 7,592 households in Sumpter Heights are headed by people in this age group. If you count all the 35 to 64 year olds, then reduce your estimate of Sumpter Heights households by 4,570.

You need to adjust your market estimates for children (14 and under) in the population because your new product will be available only in adult sizes.

Finally, you must adjust for population growth. Project 1 showed figures from 19X0 and projections for 19X9. You can estimate the size of a targeted segment in 19X5, when you will introduce Wings in the spring season.

After you have determined the size of a target market, estimate the percent that might be interested in a new product like yours. You might guess that 25 percent of the people in Sumpter Heights will buy new running shoes in the year, and that 33 percent of those people will buy running shoes in the spring. To use round numbers, if you estimated that there are about 20,000 adults in Sumpter Heights in 19X5, you would estimate their springtime demand for running shoes to be 1,665 units (20,000 × 0.25 × 0.333 = 1,665).

What part of this demand can you expect to fulfill? Estimate your percent share of the market for your new product for each target. If you expect your share of running shoes sold to Sumpter Heights residents this spring to be 15 percent, you would estimate your sales to this target at 250 units (1,665 demand×0.15 share=250).

Repeat the process of estimating total demand and your share of each target market. Then total the sales estimates for all your targets. Finally, judge whether your overall estimate is realistic. Table 1 shows a sample sales estimate completed with rough numbers for market size.

Assignment 3 Build a sales estimate for your new product.

1. Complete the "Target Markets and Sales Estimate" section of the New Product Marketing Plan. Include estimates for Sumpter Heights, a current customer segment (for instance: loyalists, high spenders, or all customers), and one other segment. Use the data in the Unit 1 Project to estimate market sizes and adjust data to avoid double counting.

   In estimating your share of the demand from each segment, remember that your promotion will be aimed at Sumpter Heights residents, and will appeal to others with similar characteristics.

2. On a separate sheet indicate how you estimated each segment's size and how you avoided double counting.

Table 1
Sample Sales Estimate

		Estimated Demand			Estimated Sales	
Target Market(s)	Size	Year (%)	Spring (%)	Spring (Number)	Share (%)	Number
Sumpter Heights	20,000	25.0	33.3	1,665	15.0	250
Central Beach	50,000	10.0	30.0	1,500	12.0	180
Other	30,000	10.0	30.0	900	8.0	72
Total	100,000			4,065		502

Computer Users: Input data only in the shaded boxes on the Sales Estimate section of the New Product Marketing Plan. The spreadsheet will calculate the rest.

Profit Study

A profit study is a simplified income statement for a single product. Each product in your inventory can be seen as a miniature business, with its own sales and expenses. Once you have established a sales estimate in units, a selling price, and a unit cost, you can find the net sales in dollars, the cost of goods sold, and the gross profit.

As Table 2, *Sample Profit Study* shows, for the purpose of analysis, net sales is 100 percent. To find net sales, multiply the sales estimate by the selling price. Thus, if you expect to sell 500 units at $29.98 each, net sales equals $14,995.

Units × price= net sales

500 × $29.98 = $14,990

The cost of goods sold equals the units sold times their cost to the store. If you pay $13.99 per unit, cost of goods sold for 500 units is $6,995.

Units sold × cost per unit = cost of goods sold

500 × $13.99 = $6,995

The gross margin (or gross profit) is the difference between net sales and the cost of goods sold.

Net sales − cost of goods sold = gross margin

$14,990 − $6,995 = $7,995

Operating expenses (or general overhead) reduce gross margin to "net profit." You have decided to subtract 35 percent of net sales for general overhead.

Net sales × overhead charge rate
= overhead expense

$14,990 × .35 = $5,247

In addition, you must subtract the cost of any promotion of the product. If the promotion budget for the new product is $1,000, then its total expenses are $6,247 (overhead expense+$1,000).

Net profit equals gross profit less expenses.

Gross margin − total expenses = net profit

$7,995 − $6,247 = $1,748

Any figure in the profit study can be expressed as a percent of net sales. (See Table 1.) Thus, cost of goods sold is 46.7 percent (6,995÷14,990). Gross margin is 53.4 percent (7,995÷14,990). Net profit is 11.7 percent (1,748÷14,990).

Assignment 4 On the New Product Marketing Plan (page 251) complete a preliminary profit study. Assume the following:

Sales = 300 units Price = 29.98 Cost = 14.99

Overhead = 35 percent of net sales

Promotion = $1,000

Computer Users: Input data only in the shaded boxes on the Profit Study section of the New Product Marketing Plan. The spreadsheet will calculate the rest. Print out or transcribe the results.

You may wish to change some of the input to see the results at different sales levels or at different prices and costs, for instance.

Table 2
Sample Profit Study

Units to be sold 500
Average price per unit 29.98
Cost per unit 13.99

	Dollars		Percent*
Net sales (500 × 29.99)	14,990		100.0
Cost of goods sold	− 6,995		46.7
Gross profit		7,995	53.3
Overhead (35%)	5,247		35.0
Promotion	+1,000		6.7
Total expenses		−6,247	41.7
Net profit		1,748	11.7

* *Rounded to the nearest tenth of a percent.*

Place Marketing

For purposes of this plan, the place decision is limited to the location of the new product on the selling floor. Will you place the men's and women's styles in separate sections, or will you create a new "sports shoe" display. How will you call attention to the goods? Since these decisions involve display strategies, they will be discussed with other promotional techniques.

Promotion

Promotion includes all the activities that communicate to customers about your products and your business. Promotional elements include advertising, visual merchandising, sales promotion, personal selling, publicity, and public relations. Your choice of activities defines your promotional mix. Retailers coordinate activities to get the most out of their promotional budget.

Promotional Budgets

As you read in Chapter 13, clothing and accessories stores typically devote 3 percent to 10 percent of their projected sales to promotion. Since you are launching a new product, you will commit 10 percent of estimated net sales or $750, whichever is more.

To evaluate the effectiveness of each activity, you will set unit sales objectives as you allocate promotional dollars. For instance, if you spend $100 on advertising, you might expect to sell 20 to 50 additional units.

Fortunately, you will be able to stretch your advertising dollars. Mercury offers you a cooperative advertising budget up to 20 percent of the cost of the goods you ordered. They will pay for half the advertising of their product by your store up to that 20 percent maximum.

For instance, if you order 400 pairs of Wings, Mercury will match your advertising for Wings up to $1,119.20 ($13.99 cost per unit×400 units×20 percent).

Promotion and Profitability

Each unit that you sell should have a profit margin of several dollars. You have to decide how much of this profit you are willing to spend to pay for promotion. Of course, without promotion you will sell few, if any, shoes.

If you price your product at $30 and it costs you $14, your gross profit is $16/unit. Subtracting $10.50 (35 percent of $30) for overhead leaves $5.50 potential profit.

As you plan your promotional campaign, be *sure* that your costs do not exceed your potential profit.

Advertising

An advertising message may be conveyed through various media. You have chosen radio, newspapers, direct mail, and fliers for consideration. In order to communicate clearly with the buying public, an advertising message should be short, memorable, and consistent across media.

Assignment 5 On a separate sheet of paper write an advertising message that is powerful and can be used in newspaper or radio advertising to promote Wings at your store. Include the product name, selected features and benefits, price, store name, location, phone number, and business hours. You may also include a slogan.

Lay out a print ad, and record a 30-second radio spot using your basic advertising message.

Radio

Radio offers the ability to target your messages, since stations cater to distinct market segments. Most stations reach beyond the county, and their advertising rates reflect the size of their total audience.

The rate chart in Table 3 shows the cost for one minute of air time for four radio stations. To calculate the cost of shorter spots, simply divide. (For a 15-second spot, divide the minute rate by 4.)

Most effective advertising campaigns repeat their message, so you should consider the number of ads to run during each week.

Consult Table 3 to review the radio stations you are considering. At the bottom of the Radio Listenership portion of Table 3 is the estimated "base" audience, the number of people listening to any radio station in "drive" time, when most people are going to and from work. For each radio station "size" shows the estimated number of listeners, rounded to the nearest hundred.

Thus, 37,000 people listen to the radio in Sumpter County during drive time, and 1,800 of those people are tuned to SCC radio.

To calculate the cost of reaching 1,000 listeners, divide cost by the number of listeners and multiply by 1,000. To judge the value of advertising on one of these stations, estimate the number of additional sales that will result. Then, divide cost by the expected result to find the promotional cost of generating one additional sale.

For instance, ten 15-second spots during drive time on SCC radio will cost $50 ($20/minute/4=$5 per 15-second spot×10 ads). This message will reach 1,800 listeners for a cost per 1,000 of $27.78 ($50÷1,800×1000). Your experience has led you to expect additional sales of 1 percent to 3 percent of your listeners (or 18 to 54 sales). If you get 18 sales, the SCC ads cost $2.78 per sale ($50/18).

■ **Table 3**
Radio Stations

SCC (Campus Radio). Popular throughout the county. Transmission is weak, even to parts of Sumpter Heights. Varied audience because of diverse programming: jazz, rock, classical, as well as sports events. Popular among the "in" high school crowd.
SMT (Sumpter's Mellow Tunes). Mellow country and rock music aimed at the 25 to 45 year-olds. Broadcast from Sumpter Heights.
TLK (Talk Radio). Includes news, advice to parents, religion, sports talk, and health talk.
HPY (Happy Radio). Targeted to the 40 to 65 year-old audience; music, news, and some talk.

Advertising Rates Per Minute

	Drive Time	Off Peak
SCC	$20	$10
SMT	35	20
TLK	50	25
HPY	65	35

Radio Listenership Drive Time

	Central Beach	Sumpter Heights	"Other"	Sumpter County	Total Suburb*
SCC	900	400	500	1,800	NA
SMT	1,300	700	800	2,800	8,300
TLK	1,600	900	1,100	3,600	11,000
HPY	1,900	800	1,100	3,800	11,400
Base	18,000	8,500	10,500	37,000	111,000

* *Total three-county area.*

Assignment 6 On the New Product Marketing Plan, complete a weekly promotion plan for radio for the first week of the season. (See page 252.) Assume your radio budget is $250 for that week.

Newspapers Except for the *SCC Bugle*, the newspapers distributed in the county reache more households than any of the radio stations that you are considering. Table 4 describes Sumpter County's four newspapers. Select the publication(s) that will reach your target market(s). Compare the costs of reaching 1,000 readers. Project the response you expect in terms of additional sales. Then decide whether to advertise in one or more newpapers.

For instance, you might consider the *Herald*, the weekly with 10,600 readers in the county (3,500 in Sumpter Heights). The cost of a half-page ad is $220. The cost of reaching 1,000 readers with this ad is $20.75 ($220 ÷ 10, 600 × 1000). You have come to expect that 1 percent of the readers will buy an advertised product. This amounts to 106 additional sales of Wings. Each sale costs you $2.08 ($220 ÷ 106 sales).

Assignment 7 On the New Product Marketing Plan complete a weekly promotion plan for newspaper advertising for the first week of the season. (See page 252.) Assume your newspaper budget is $250 for that week.

Plan that you will repeat your first-week advertising in the second week of the season. Calculate your two-week budgets for both radio and for newspaper.

You expect to buy $1,800 of advertising for Wings in total ($900 of which will be paid for by Mercury). How much can you spend advertising Wings during the rest of the season?

Fliers and Direct Mail Fliers and direct mail can include coupons to promote sales. Furthermore, these ads can be targeted to specific customers. Their cost, however, can be relatively high, so the response rate must be large in order to pay off. Designing, printing, and distributing 5,000 fliers costs $400, or $80 per 1,000. To keep advertising costs to $2 per unit, you would need a response of 40 sales (4 percent) for every 1,000 fliers distributed.

Direct mail can be inexpensive for businesses that regularly mail billing statements to their customers. Neighbour's shoes, however, would have to add postage and envelopes to the cost of direct mail, putting the cost at over $200 per 1,000.

■ **Table 4**
Newspapers

SCC Bugle (Campus weekly). Tabloid; college news, features
Sumpter Herald (Weekly). Tabloid; local features, editorials; large readership in Sumpter Heights.
Central Post (Daily p.m.). Tabloid, sensational headlines; biggest readership in county.
Metro Times (Daily a.m. suburban edition). Standard paper; serious journalism; distributed in three suburban counties, including Sumpter. Big readership among commuters from Sumpter County to Metro City. Higher rates than other papers because of larger circulation. Do you want to reach the other two suburban counties served by this paper?

Newspaper Advertising Rates

	Size			
	⅛ page	¼ page	½ page	Full page
Bugle	$ 50	$ 75	$125	$300
Herald	75	125	220	350
Post	100	180	300	500
Times	150	270	450	750

Newspaper Circulation

	Central Beach		Sumpter Heights		"Other"		Total Sumpter County	
	Sales	Percent*	Sales	Percent	Sales	Percent	Sales	Percent
Bugle	—				—		8,000	22
Herald	3,800	20	3,500	46	3,300	46	10,600	28
Post	8,500	45	2,900	38	4,200	38	15,600	42
Times	4,800	25	3,000	40	3,900	35	11,700	31

* *Percents indicate proportions of households receiving each paper.*

Display

Many elements of display cost little or no money. Simply putting a new product on display on the selling floor will have a significant promotional effect. Furthermore, to make their product stand out, Mercury offers an attractive point-of-purchase display at no cost to you. You will use this tool and display Wings on the sports shoe shelves for men and women.

Sales Promotion

Sales promotion can include any number of activities and special offers to help move merchandise. Coupons can give your store an edge over competitors and also help you track the effectiveness of print advertising. Coupon redemption costs money, however.

Special events built around the holiday calendar can feature your products in a festive atmosphere. Special offers, like discounts on multiple purchases, can move several products at once.

Sales promotion works best when supported by advertising. The public has to know about your creative marketing techniques, if they are to work.

Assignment 8 You plan to offer a $2-off coupon for Wings to a select group of 200 people. You expect to generate 50 sales. The cost of creating and printing the coupon is $35. Calculate the total cost and cost per unit of this special promotion. Complete the sales promotion line under "Summary of Promotional Costs" of the New Product Marketing Plan form.

On a separate piece of paper describe at least one sales promotion idea. Consider when in the season you will use this sales promotion, and how to coordinate it with your advertising plan. Estimate the cost and increased sales expected for each sales promotion idea.

Personal Selling

Your sales staff must be aware of your promotional plans. The information that your staff gives customers should be consistent with your advertising and other promotional themes. Moreover, a well-informed sales staff can better serve customers and meet sales goals. Sales training in general will be discussed later under Personnel Strategies.

Publicity and Public Relations

Being associated with community activities and getting positive press coverage build good will and contribute to sales. A new running shoe gives you the opportunity to sponsor sporting events, encourage good health, and build profits.

Public relations can be related to other promotional activities, too. For instance, you can offer a discount to senior citizens who join the "mall walkers club," or advertise your store's participation in a community "fun run."

Assignment 9 You will become a sponsor of the Sumpter Runners Club's Spring Fun Run by contributing $100 to defray the cost of prizes for new members. You will be listed in the club's mailing to 1,000 households. You will also be pictured with the other sponsors and the club's president in the *Sumpter Herald* and in the *Central Post.* You expect to generate 50 sales through this publicity.

Calculate the costs and cost per unit, and complete the public relations line of the "Summary of Promotional Costs" on the New Product Marketing Plan.

On a separate sheet of paper, describe how you will coordinate other promotional elements to support your activity.

Personnel Strategies

Personnel strategies for promoting a new product will focus on sales training. You may consider spending some time training your staff in the features of a new product. What objections might some customers raise to the new product? How should they be answered? Carefully considering customer-staff interactions can improve sales.

The cost of sales training for a new product can be estimated as the value of the staff time taken. You might rely on a manufacturer's representative or a training video to convey the needed information.

Assignment 10 You pay your four part-time sales assistants an 8 percent commission, based on sales. This commission is part of your overhead charge. As an incentive for your staff to learn the new product and to increase sales, you are offering a 2 percent bonus on sales of Wings. You will also award two days off with pay to the assistant who sells the most Wings this spring.

You expect to generate 50 additional sales with this technique. Calculate its cost and the cost per unit, and complete the sales training line of the "Summary of Promotional Costs" on the New Product Marketing Plan.

Estimate that you will pay the commission bonus on two-thirds of all Wings sales. (You will sell one-third of the Wings yourself.)

Assessment

A marketing plan for a new product is a complex effort. To judge the effectiveness of a plan, you must build in meaningful objectives.

Typically, many elements of a marketing plan operate at the same time, and a retailer will not always be able to point at a single cause of sales performance. The effects of some promotional programs can be made measurable, for instance, by counting coupons redeemed.

As discussed in "Setting Objectives," goals should not be limited to unit sales increases.

Assignment 11 Review your assignments in this project and revise your ideas to create a *coordinated* marketing plan, which balances all parts of the marketing mix. Include a unit sales objective and one other objective for each element in your plan.

Create a sales estimate and set your average price.

Review and revise your Weekly Promotional Plan for the first two weeks of the season, and indicate how you will advertise and promote thereafter. You may rule a form similar to the Weekly Promotion Plan on page 252.

Finally, revise the profit study from Assignment 4. Base your unit costs on your sales estimate. Allocate 10 percent of net sales for promotion, or $750, whichever is greater. Remember to add Mercury's cooperative advertising budget to your media budget, but do not include this amount in your profit study. Rule a form similar to the profit study on page 251 for use in your revised profit study.

Computer Users: As in Assignment 4, complete only the shaded areas of the profit study.

Name ____________________ Date ____________________

NEW PRODUCT MARKETING PLAN: Mercury Wings Running Shoe

Features	Benefits
Nylon and Leather uppers	
Mesh ventilation	
Padded ankle collar and tongue	
Flared wedge heel	
Double cushioned mid-sole	
Resilient outer sole	
Light weight	
Toe guard	
Choice of white, gray, or blue	
The Mercury brand name	

Cost, Price, and Competition

Cost			
(Quantities)	0–150 pairs	151–300 pairs	301 or more pairs
Competition's Price Range	______ – ______	Your Price Range	______ – ______

Target Market(s) and Sales Estimate

		Estimated Demand			Estimated Sales	
Market	**Size**	**Year (%)**	**Spring (%)**	**Number**	**Share (%)**	**Number**
Sumpter Heights						
Total						

Profit Study

Units	
Price	
Cost	

	Dollars		Percent
Net sales			100.0
Cost of goods sold			
Gross profit			
Overhead (35%)			35.0
Promotion			
Total expenses			
Net profit			

(continued)

NEW PRODUCT MARKETING PLAN *(Continued)*

Weekly Promotion Plan

Radio Station	Number	Length	Cost	Cost/1,000	Objective	Cost/Unit
Example: SCC drive-time	10	15 sec	$50	$27.78	18 units	2.78
Total						

Newspaper	Number	Size	Cost	Cost/1,000	Objective	Cost/Unit
Example: *Herald*	1	½ pg	$220	$20.75	105 units	$2.08
Total						

Direct Mail and Fliers	Numbers	Cost	Cost/1,000	Objective	Cost/Unit

	Week 1	Week 2	Balance of Season	Total
Total advertising				

Summary of Promotional Costs

	Cost	Benefit	Cost/Unit
Advertising			
Display	0		
Sales promotion			
Sales training			
Public relations			
Total			